ISBN 978-0-331-33404-3
PIBN 11107654

ITHSONIAN MISCELLANEOUS COLLECTIONS.

——————— 658 ———————

INDEX

TO THE

LITERATURE

OF THE

SPECTROSCOPE.

ALFRED TUCKERMAN, Ph. D.

WASHINGTON:
PUBLISHED BY THE SMITHSONIAN INSTITUTION.
1888.

ITHSONIAN MISCELLANEOUS COLLECTIONS.

———— 658 ————

INDEX

TO THE

LITERATURE

OF THE

SPECTROSCOPE.

ALFRED TUCKERMAN, Ph. D.

WASHINGTON:

PUBLISHED BY THE SMITHSONIAN INSTITUTION.

1888.

PRINTED AND STEREOTYPED BY

JUDD & DETWEILER,

AT WASHINGTON, D. C.

ADVERTISEMENT.

With the rapid accumulation of scientific memoirs and discussions, published from year to year in numerous journals and society proceedings, a constantly larger expenditure of time and labor is required by both the investigator and the student, to learn the sources of information and the condition of discovery in any given field. Hence is felt the growing need of classified indexes to the work done in the various fields of research, and hence the corresponding tendency of the age to supply such demand.

The present work aims at a general survey of Spectroscopic Literature, with references to authorities in its more special subdivisions, and it has been prepared for the Institution by Mr. Tuckerman, without other remuneration than the expectation of serving the interests of scientific inquirers.

It has been brought down to the middle of the year 1887.

S. P. LANGLEY,
Secretary Smithsonian Institution.

WASHINGTON, *February*, 1888.

PREFACE.

This work is intended to be a list of all the books and smaller treatises, especially contributions to scientific periodicals, on the spectroscope and spectrum analysis from the beginning of our knowledge upon the subject until July, 1887; an Index or Bibliography of the Spectroscope and Spectrum Analysis.

It was begun at the suggestion of Dr. Wolcott Gibbs, whose work in connection with the subject is well known.

The object is to enable a chemist to find out at a glance all that has been published in any branch of his subject where the spectroscope is used, and what every writer has published.

The method pursued has been as follows: 1, to examine the bibliographies, booksellers' catalogues, and books on spectrum analysis for books; 2, to examine the scientific periodicals for the shorter treatises, the first and original contributions to the subject, and this was done volume by volume wherever there was no index to a series of years—as in the Comptes Rendus and the later volumes of the Annales de Chemie et de Physique and of (Poggendorff's, now Wiedemann's) Annalen der Physik und Chemie, as well as others. Use was made of the bibliography at the end of Roscoe's Spectrum Analysis, and in the reports of the British Association for 1881 and 1884, for such books and articles as the author could not find elsewhere. Credit is also due to the Astor Library and its managers for the means it afforded the author of making this Index.

After the greater part of the material was collected it was divided into such subjects as the titles indicated, in alphabetical order, easy finding being constantly kept in view. Titles have often been repeated more than once so as to make sure of their being found. Finally, at the suggestion of the Smithsonian Institution, the List of Authors was added.

The author hopes that his two objects, fullness and ready access of all the titles, will prove to have been gained.

New York, 1887.

TABLE OF CONTENTS.

viii TABLE OF CONTENTS.

TABLE OF CONTENTS.

(With the pages of the preceding Index
on which the titles of their works are
given.)

Number of titles 3,829
Number of authors 799

LITERATURE OF THE SPECTROSCOPE.

HISTORY.

Arago (Domenique François Jean), 1786–1853. Œuvres complètes, avec Tables, publiées d'après son ordre sous la direction de J. A. Barral. Paris et Leipzig, 1854–'62, 17 vols., ill., 8°.
> (Interesting here only in connection with polarized light.)

Barlocci.
> (Wrote on the influence of white light.)

Beccaria, 1716-81.
> (Wrote on the refraction of rock crystal, about 1750; see Ency. Brit., eighth edition I, 758.)

Becker (G. F.). Contribution to the History of Spectrum Analysis.
> Amer. Jour. Sci., (8) **16**, 892.

Bérard. Mem. de la Soc. d'Arcueil, 3 (1817); and Biot's Traité de Physique, **4**, 600–18, 673–4.
> (A full account of Bérard's experiments on the calorific rays of the spectrum.)

Berthold (G.). Zur Geschichte der Fluorescenz.
> Ann. Phys., u. Chem., **158**, 628.

Biot (J. B.). Traité de Physique expérimentale et mathématique. Paris, 1816, 4 vols., 8°.

——— — —. Mémoire sur les Lois générales de la double Réfraction et de la Polarization dans les Corps cristallisés. Paris, 1819, 4°.

——— — —. Mémoire sur la Polarization circulaire. Paris, 1832, 4°

——— — —. Mémoire sur la Polarization lamellaire. Paris, 1842, 4°.

Blair (Dr. Robert), 1787–1829. Edinburgh Transactions, III, 3.
> (He discovered the uses of muriatic acid mixed with antimony in correcting secondary spectra in telescopes.)

(1 T)

Boscovich (Roger Joseph).   Opuscula.   Bassano, 1784, 5 vols., 4°.   Opera pertinentia ad Opticam et Astronomiam (Astor Library).

Ency. Brit., eighth edition, I, 721-2, 753.

(He made a delicate micrometer with double refraction, about 1777, and observed the so-called Secondary Spectrum, consisting of purple and green light.)

Bouguer (Pierre), 1698-1758.   Essai d'Optique, sur la Gradation de la Lumière.   Paris, 1729, 8°; ed. La Caille, Paris, 1760, 4°.

Ency. Brit., eighth edition, I, 753-4.

(He published a number of treatises on the gradation of light.)

Brewster (Sir David), 1781-1868.   Treatise on Optics.   Edinburgh, 1831. New Analysis of Solar Light, indicating three primary colours, forming coincident spectra of equal length.   Edinburgh, 1834.

(See Life of B. by Mrs. Gordon.)

Buffon.

In his "Epoques de la Nature" he describes light and heat as known in his times.)

Delaunay.   Notice sur la Constitution de l'Universe.   Première Partie: Analyse Spectrale, Annuaire du Bureau des Longitudes, 1869, Paris, 8°.

(A masterly treatise on the subject at that time.)

Desains (P.), Recherches expérimentales sur les anneaux colorés de Newton.   Comptes Rendus, **78**, 219-21; Phil. Mag. (4) **47**, 236-7.

Dolland (John), 1706-61.   See Proc. Royal Soc., **50** (1757) 733, and Ency. Brit., eighth edition, I, 749-51.

(He discovered that dispersion depends not on the mean refraction but on the constitution of the diaphanous medium.)

Draper (Henry).   Obituary by G. F. Barker in Amer. Jour. Sci. (3) **25**, 89.

Draper (J. W.).   Early Contributions to Spectrum Photography.   Nature, **10**, 243-4.

Dutirou (l'abbé).   Memoire sur la détermination des indices de réfraction des sept raies de Fraunhofer dans une série nombreuse de verres.

Annales de Chimie et de Physique, (3) **28** (1850) 176.

Exner (K.).   Die Fraunhofer'schen Ringe, die Quetelet'schen Streifen und verwandte Erscheinungen.

Sitzungsber. de. Wiener Akad. **76**, II, 522.

Faye. Note sur l'Association nouvellement fondée en Italie sous le titre de "Societa dei Spettroscopisti Italiani." Comptes Rendus, **74**, 913–18, 1240–3.

    (See Tacchini, Comptes Rendus, **74**, 1287.)

Forbes (James D.). On the Refraction and Polarization of Heat. Edinburgh Trans., **13** (1836), 131–68, 446–72.

————— — —. Note relative to the supposed Origin of the Deficient Rays in the Solar Spectrum. Phil. Mag. (1836) 453.

————— — —. Researches on Heat. Edinburgh Trans., **14** (1840), 176-208, **15** (1844), 1–27.

————— — —. Article in Ency. Brit., eighth edition, on Sir David Brewster.

Fraunhofer (Joseph von), 1787–1826. "Bestimmung des Brechungs- und Farbenzerstreuungs-Vermögens verschiedener Glasarten in Bezug auf die Vervollkommung achromatischer Fernröhre. Von Jos. Fraunhofer in Benedictbaiern." Denkschriften der k. Akad. der Wissenshaften zu München für die Jahre 1814 and 1815. Band V, 193–226, mit drey Kupfertafeln, München, 1817, 4°. (Fraunhofer's annouhcement of his discovery of the dark lines of the spectrum of sunlight.)

    J. von Utschneider, Kurtzer Umriss der Lebensgeschichte des Herrn Dr. J. von Fraunhofer, Munich, 1826.

    Merz, Das Leben und Wirken Fraunhofer, Landshut, 1865.

    See Works of Sir David Brewster.

————— — —. Neue Modificationen des Lichtes durch gegenseitige Einwirkung und Beugung der Strahlen, und Gesetze derselben, München (no date).

    Edinburgh Jour. Science, No. **13**, 109, **15**, 7, new series No. **13**, 101.

Gerding (Th.). Geschichte der Chemie. Leipzig, 1867, 8°.

Herschel (A. S.). Progress of Spectrum Analysis. Chem. News, **19**, 157; Jour. Franklin Inst., **88**, 49, 136.

————— — —. Progress of Meteor Spectroscopy. Nature, **24**, 507–8.

Herschel (Sir John Frederick William), 1792–1871. On the Absorption of Light by coloured Media, and on the Colours of the prismatic Spectrum exhibited by certain Flames; with an Account of a ready Mode of determining the absolute dispersive Power of any Medium, by direct experiment. Edinburgh Trans., **9** (1823), 445.

Herschel (Sir John Frederick William). Homogeneous yellow and orange Spaces in the Spectrum. Phil. Trans., **90** (1800), 255.

—— —. Investigation of the Powers of the prismatic Colours to heat and illuminate Objects; with Remarks that prove the different Refrangibility of radiant Heat. To which is added, an Inquiry into the Method of viewing the Sun advantageously with Telescopes of large Apertures and high magnifying Powers. Phil. Trans., **90** (1800), 255–283.

—— —. Experiments on the Refrangibility of the invisible Rays of the Sun. Phil. Trans., **90** (1800), 284–292.

—— —. Experiments on the solar and on the terrestrial Rays that occasion Heat; with a comparative View of the Laws to which Light and Heat, or rather the Rays which occasion them are subject, in order to determine whether they are the same or different. Phil. Trans., **90** (1800), 293–326, 437–538.

Hoppe-Seyler (F.). Die Spectralanalyse. Ein Vortrag. Berlin, 1869, 8°.

Hunt (T. Sterry). Chemistry of the heavenly Bodies since the Time of Newton. Proc. Cambridge Philosoph. Soc., **4**, 129–139; Amer. Jour. Sci., (3) **23**, 123–138; Ann. Chim. et Phys., (5) **28**, 105.

Huyghens (Christian), 1629–95. Opera Varia, Leyden, 1724, 2 vols., 4°. Opera reliqua, Amsterdam, 1728, 2 vols., 4°.

Jahresbericht der Chemie (Liebig's), Jahre 1863, 113; 1866, 78.

Johnson (A.). On Newton, Wollaston, and Fraunhofer's Lines. Nature, **26**, 572; Beiblätter, **7**, 65 (Abs.).

Kirchhoff (G.). Geschichtliches über Spectralanalyse. Ann. Physik u. Chemie, **118**, 94, 102; Phil. Mag., (4) **25**, 250.

Kopp (H.). Entwickelung der Chemie in der neueren Zeit. München, 1871–3, 8°.

Ladd (William). On the Results of Spectrum Analysis as applied to the heavenly bodies. A Lecture delivered before the British Association at the Nottingham Meeting, August 24, 1866. London, 1866, 8°, with photographs of the stellar spectra.
Chem. News, **14**, 173, 199, 209, 235.

Lamansky (S.). Geschichtliches über das Wärmespectrum der Sonne.
Ann. Phys. u. Chem., **146**, 200, 207, 209.

Lambert (Johann Heinrich),1728–77. Photometria. Augsburg,1760,8°.

Liveing (G. D.) and Dewar (J.). Note on the History of the Carbon Spectrum. Proc. Royal Soc., **30**, 490–4; Beiblätter, **5**, 118–22; Nature, **23**, 265–6, 338.

Lloyd (Prof.). Report on Physical Optics. Fourth Rept. British Assoc., 1834, pp. 295–414.

Malus (E. L.), Paris, 1775–1812. Théorie de la double Réfraction de la Lumière dans les Substances cristallisés, Paris, 1810, 4°.
(See Ency. Brit., 8th ed., I, 754, for an account of him.)

Marie (L'abbé). Nouvelle découverte sur la lumière, pour en mesurer et compter les degrés. Paris, 1700, 8°.
(Gave the first ideas about photometry.)

Maskelyne. Account of a new Instrument for measuring small Angles, called the Prismatic Micrometer. Phil. Trans., **47** (1777), 799.

Mayer (A. M.). The History of Young's Discovery of his Theory of Colour. Phil. Mag., (5) **1**, 111–127.

Meldola (R.). Contributions to the chemical History of the aromatic Derivatives of Methane. Jour. Chem. Soc., **41**, 187–201.

Melloni (Macédoine). See Annales de Chimie et de Physique, **53** (1833), 5–72; do., **48**, 198, Recherches sur plusieurs phénomènes entreprises au moyen du thermomultiplicateur; do., **48**, 385; do., **55**, 337; do., **60**, 402, 410–18; do., **61**, 411; do., **65**, 5; do., **68**, 107; do., **70**, 435; do., **72**, 40, 334; do., **74**, 18, 331; do., **75**, 337.
(Melloni was famous chiefly for his thermomultiplier.)

Miller (William Allen). Recent Spectrum Discoveries, 1863. Jour. Franklin Inst., **76**, 29; Chem. News, 1863.

Morichini (Domenico Pino), 1773–1830. Sopra la forza magnetizzatrice del lembo estremo del colore violetto. Milano, 1802.
(A collection of his works was published by Pirotta of Milan in 1836.)

Mousson (A.). Resumé de nos connaissances actuelles sur le spectre. Archives de Genève (1861).

Newton (Sir Isaac). Collected Works. Optics, Chap. II, sections 1–3; vol. 3 of Latin edition, London, 1779–85, 5 vols., 4°.

Nobili, worked with Melloni, above.

Poggendorff (J. C.). Handwörterbuch der exacten Wissenschaften. Leipzig, 1858–63, 2 vols., lex. 8°.

Powell (Rev. Baden). Report on Radiant Heat. British Association Repts., 1, 295.

—— ——. Researches towards establishing a Theory of the Dispersion of Light. (1835) 549, (1837) 288, (1839) 1.

Priestley (Dr. Joseph). An Account of all the prismatic Colours, made by electrical Explosions on the Surface of Pieces of Metal. Phil. Trans., 58 (1768), 68.

Ritter.
    (In 1801 he exposed muriate of silver in various parts of the spectrum and found that the action was least of all in the red, greater in the yellow, and greatest beyond the visible violet rays. Forbes, in Ency· Brit., 8 ed., 16, 594.)

Robison (John). A System of mechanical Philosophy, with notes by David Brewster. London, 1822, 4 vols., 8°. See chapter on the telescope, III, 403–522.

Rood (O. N.). Newton's Use of the Term Indigo with Reference to a Color of the Spectrum. Amer. Jour. Sci., (3) 19, 135–7; Beiblätter, 4, 460 (Abs.).

Rowland (H. A.). On recent Progress in photographing the solar Spectrum. Rept. British Assoc. (1884), 635.

Rudberg (Fr.). Dispersion de la lumière. Ann. de Chimie et de Physique, 36, 439.

—— ——. Sur la réfraction des rayons différemment colorés dans des cristaux à un ou deux axes optiques. Ann. de Chimie et de Physique, 48, 225.

Ruprecht (Rudolph). Bibliotheca chemica et pharmaceutica. Leipzig, 1858–70, 8°.

Rutherfurd (L. M.). Construction of the Spectroscope. Amer. Jour. Sci., (3) 39, (1869), 129. Note by Ditscheiner in Sitzungsber. d. Wiener Akad., 52 II, 542, 563–8.

Schwerd (F. M.). Die Beugungserscheinungen aus dem Fundamentalgesetz der Undulationstheorie analytisch entwickelt und in Bildern dargestellt. Mannheim, 1835, 8°.

Secchi (A.). Le Soleil. Exposé des principales Découvertes modernes sur la Structure de cet Astre. Paris, Gauthier-Villars, 1870. (See Nature, 13, 188.)

Seebeck (T. J.). Berlin, 1770–1831.
> Abhandlungen der Berliner Akad., 1818–19, 306; Edinburgh Jour. Sci.,
> 1 (1824), 858.

Stewart (B.). Some Points in the History of Spectrum Analysis. Nature, **21**, 35.

—— ——. Reply to Kirchhoff on the History of Spectrum Analysis. Phil. Mag., (4) **25**, 354.

Stieren (E.). Die ersten Beobachtungen über Spectralanalyse veröffentlichte Alter. Ann. Phys. u. Chem., **132**, 469.

Stokes (G. G.). Early History of Spectrum Analysis. Nature, **13**, 188–9.

—— —— ——. On the Colours of thick Plates. Cambridge Philosoph. Trans., **9** (1851), part II, 147–76.

—— —— ——. On the Composition and Resolution of Streams of polarized Light from different Sources. Cambridge Philosoph. Trans., **10** (1852), 399–416.

—— —— ——. On the Change of Refrangibility of Light. Phil. Trans. (1852), 463–562.
> (His discovery of fluorescence.)

Swan (W.). On the Prismatic Spectra of the Flames of Compounds of Carbon and Hydrogen. Edinburgh Trans., **21** (1857), 411–29; Ann. Phys. u. Chem., **100**, 306.

Tarry (H.). Report on the Researches and Experiments made by the Spectroscopic Association of Italy. (From Les Mondes of March 21, 1872.) Chem. News, **25** (1872), 179.

Thalén (Robert). Om Spektralanalys, med en Spektralkarte. Upsala Universitets Aarpkrift. Upsala, 1866, 8°.

Wollaston (Dr.), 1766–1828. A Method of examining refractive and dispersive powers by prismatic Reflection. Phil. Trans. (1802), 365–380.
> (His own account of his discovery of five fixed lines of the solar spectrum, which he said he could not explain.)

Wünsch (Christian Ernst), 1730–1810. Untersuchungen über die verschiedenen Farben des Lichtes. Leipzig, 1792, 8°, with plates.

Wurtz (A.). Histoire des Doctrines chimiques depuis Lavoisier jusqu'à nos jours. Paris, 1869, 8°.

Young (Dr. Thomas).   Elements of Natural Philosophy, Vol. 1, 786, plate 29.

> (Gives a small colored drawing of the spectrum as seen by Dr. Wollaston and himself, with the yellow line.)
>
> Life by Dr. G. Peacock, London, 1855, 8°.

Zantedeschi.   Ricerche sulla Luce, Venezia, 1846, 8°; Chap. III. (See Edinburgh Jour. Sci., n. s., 5 (1830), 76, repeating experiments of Barlocci and similar to those of Morichini.)

BOOKS.

Agnello (A.).   Eclisse totale del 22 dic. 1870.   Palermo, 1870.

Angström (A. J.).   Recherches sur le Spectre normal du Soleil.   Upsala, W. Schultz, 1868.   Avec Atlas et 6 planches.

Becquerel (Edm.).   La Lumière, ses Causes et ses Effets.   2 vols., 8°, Paris, 1867–1868, 16 fr.

Blaserna (P.).   Sulla polarizzazione della Corona solare.   Palermo, 1871, 8°.

Capron (J. R.).   Photographed Spectra.   136 photographs of spectra. London, Spon, 1877, 8°.

> (See review of, in Chem. News, 37 (1878), 118.)

Champion (P.), Pellet (H.), et Grenier.   De la Spectrométrie, Spectromètre.   Paris, 1873, 8°.

Draper (Henry).   On diffraction Spectrum Photography.   New Haven, 1873, 8°.

Grandeau (L. N.).   Instruction pratique sur l'analyse spectrale.   Paris, 1863, 8°, 3 fr.

Hirn (G. A.).   Flamme en combustion et Température du Soleil.   Paris, 1873, 8°.

Hoppe-Seyler (F.).   Handbuch der physiologisch-chemischen Analyse. 3. Auflage, Berlin, 1870, 8°.

Hough (G. W.). The total Solar Eclipse of Aug. 7, 1869. Albany, N. Y., J. Munsell, 1870, 8°.

Kirchhoff (G.). The Solar Spectrum and Spectra of the Chemical Elements. London, Macmillan, 1861–2, with plates.
(Translations of the original communications to the Academy of Sciences of Berlin.)

Lecoq de Boisbaudran (F.). Spectres Lumineux. Paris, 1874, 8°, avec atlas.

Lielegg (A.). Die Spectralanalyse. Weimar, Voigt, 1867.

Lockyer (J. N.). The Spectroscope and its Applications. London, Macmillan, 1873, 8°.

———— ——. Studies in Spectrum Analysis. London and New York, Macmillan, 1878, 8°.

Lommel (E.). The Nature of Light. New York, Appleton, 1876, 8°.

Lorscheid (J.). Die Spectralanalyse. Münster, 1870, 8°.

Mac Munn (C. A.). The Spectroscope. London, Churchill, 1880.

Proctor (R. A.). The Spectroscope. London, 1877, 8°.

Radau (R.). Le Spectre solaire. Paris, 1862, 18°.

Respighi (L.). Osservazioni spettroscopiche del Bordo e della Protuberanze Solari. Roma, 1871, 8° (with a plate).

Rood (O. N.). Modern Chromatics, with 130 illustrations. New York, Appleton, 1879.

Roscoe (H. E.). Spectrum Analysis. London, Macmillan, Fourth Edition, 1886, 8°.
(With a short bibliography of the principal works relating to the spectroscope. One of the best text-books, if not the best, on the subject.)

Ruprecht (R.). Bibliotheca chemica et pharmaceutica. Leipzig, 1858–70, 8°.

Sands (B. F.) and others. United States Naval Observatory Reports on the total Eclipse of the Sun, Aug. 7, 1869. Government Printing Office, Washington, D. C., 1869.

Schellen (H.). Die Spectralanalyse. 2 Auflage, Braunschweig, 1871, 8°.
(Translated by J. and C. Lassell, London, 1872; reviewed by Roscoe in Nature, 1, 503, and by others in Chem. News, 22, 284; 25, 80.)

Secchi (A.). Sulle ultime scoperte spettroscopiche nel Sole. Roma, Type delle Belle Arti, 1869.

————. Le Soleil. Exposé des principales Découvertes modernes sur la Structure de cet Astre. Paris, Gauthier Villars, 1870, 8°. Do. translated into German, Braunschweig, Westermann, 1872, 8°.

Simmler (R. Th.). Beiträge zur chemischen Analyse durch Spectralbeobachtungen. Chur, 1861, 8°.

Smyth (C. Piazzi). Madeira Spectroscopic. Edinburgh, W. and K. Johnston, 1881, 8°. (Spectroscopic observations made at Madeira.)

Stein (Th.). Das Licht im Dienste der wissenchaftlichen Forschung. Leipzig, 1877, 8°.

Stokes (G. G.). Mathematical and physical Papers, reprinted from the original Journals and Transactions, with additional Notes by the Author. Cambridge, University Press, 1880–1883, 2 vols., 8°.

Thalén (R.). Om Spektralanalys, exposé, med en Spektralkarte. Upsala Universitets Årsskrift, 1866, 8°.

Valentin (G.). Der Gebrauch des Spectroskops zu physiologischen und ärztlichen Zwecken. Leipzig und Heidelberg, Winter'sche Buchhandlung, 1863, 8°.

Vierordt (K.). Anwendung des Spectralapparates. Tübingen, 1871, 8°.

Vogel (H. W.). Practische Spectral-Analyse irdischer Stoffe. Nordlingen, 1877, 12°.

Watts (W. M.). Index of Spectra. London, Gillman, 1872, 8°.

Wrottesley (Lord). Applications of Spectrum Analysis. London, 1865, 8°,

Young (C. A.). The Sun. New York, 1881, 8°.

APPARATUS.

ABSORPTION SPECTROSCOPE.

Sur un nouveau spectroscope d'absorption.
>Thierry (Maurice de). Comptes Rendus, **101** (1885), 811-818; Jour. Chem. Soc., **50** (1886), 118 (Abs.).

ACTINIC BALANCE.

(See Spectro-bolometer.)

ALKALOID REACTIONS.

Alcaloïdreactionen im Spectralapparate.
>Hock (K.). Arch. f. Pharm., **19**, 858; Ber. chem. Ges., **14**, 2844 (Abs.).

ASTRONOMICAL SPECTROSCOPES.

(See Spectro-telescopes.)

AUTOMATIC SPECTROSCOPES.

A new automatic motion for the spectroscope.
>Baily (W.). Phil. Mag., (5) **4**, 100-104.

An automatic spectroscope.
>Browning (J.). Chem. News, **20** (1870), 222; **21** (1870), 201.

Automatic spectroscope.
>Proctor (R. A.). Monthly Notices Astron. Soc., **31** (1871), 47-48.

Automatic spectroscope.
>Proctor (R. A.). Monthly Notices Astron. Soc., **31** (1871), 205-208.

Automatic spectroscope for Dr. Huggins's sun observations.
>Grubb (H.). Monthly Notices Astron. Soc., **31** (1871), 86.

Automatic spectroscope.
>Reynolds (J. E.). Chem. News, **23** (1871), 118.

Universal automatic spectroscope.
>Browning (J.). Monthly Notices Astron. Soc., **32** (1872), 218.

Large automatic spectroscope.
>Browning (J.). Monthly Notices Astron. Soc., **33** (1873), 410.

Ueber Spectralapparat mit automatischer Einstellung.

Krüss (H.). Z. Instrumentenkunde, **5** (1885), 181–191, 232–244; Beiblätter, **9** (1885), 628 (Abs.).

BESSEMER–FLAME SPECTROSCOPES.

Examination of the Bessemer flame with the spectroscope.

Silliman (J. M.). Amer. Jour. Sci. (2), **50**, 297–307; Phil. Mag., **41**, 1–12; Jour. Chem. Soc. (2), **9**, 97–98 (Abs.).

Examination of the Bessemer flame with coloured glasses and with the spectroscope.

Parker (J. S.). Chem. News, **23** (1871), 25–26; Jour. Chem. Soc. (2), **9**, 98 (Abs.).

Spectroscope pour les hauts-fourneaux et pour le procédé Bessemer.

Zenger (Ch. V.). Comptes Rendus, **101** (1885), 1005; Jour. Chem. Soc., **50** (1886), 190 (Abs.).

USE OF THE BLOWPIPE.

Emploi du chalumeau à chlorhydrogène pour l'étude des spectres.

Diacon. Comptes Rendus, **56**, 653.

BOLOMETER.

(See Spectro-bolometer.)

BÖRSCH–APPARATUS.

Der Spectralapparat von Börsch zugleich Reflexions-Goniometer.

Börsch. Ann. Phys. u. Chem., **129**, 384.

COLLIMATORS.

Sur un nouveau collimateur.

Thollon (L.). Comptes Rendus, **96**, 642–643; Nature, **27**, 476 (Abs.); z. Instrumentenkunde, **3**, 180–181 (Abs.); Beiblätter, **7**, 285 (Abs.).

An easy method of adjusting the collimator of a spectroscope.

Schuster (A.). Proc. Physical Soc., **3**, 14–17; Phil. Mag., (5) **7**, 95–98; Beiblätter, 834 (Abs.).

Use of a collimating eye-piece in spectroscopy.

Liveing (G. D.) and Dewar (J.). Proc. Cambridge Phil. Soc., **4**, 836; Beiblätter, **7**, 892 (Abs.).

Construction of a compensating eye-piece.
> Proc. Royal Soc., 21, 426–442.

Zweckmässigkeit cylindrischer Linsen bei Spectralapparaten.
> Schönn (L.). Ann. Phys. u. Chem., 144, 884.

Optical densimeter for ocean water.
> Hilgard (J. E.). United States Coast Survey Rep't (1877), 108–113;
> Z. Instrumentenkunde, 1, 206–207 (Abs.); Beiblätter, 5, 658 (Abs.).

Spectroskop mit constanter Ablenkung.
> Goltzsch (H.). Carl's Repert., 18, 188–190; z. analyt. Chem., 21, 556
> (Abs.).

Ueber ein einfaches Mittel die Ablenkung oder Zerstreuung eines Licht-
strahles zu vergrössern.
> Kohlrausch (F.). Ann. Phys. u. Chem., 143, 147–149.

Die kleinste Ablenkung im Prisma.
> Lommel (E.). Ann. Phys. u. Chem., 159, 829.

Die kleinste Ablenkung im Prisma.
> Berg (F. W.). Ann. Phys. u. Chem., 158, 651.

Démonstration élémentaire des conditions du minimum de déviation d'un
rayon par le prisme.
> Hesehus (N.). Jour. soc. phys. chim. russe, 12, 226–231; Jour. de
> Phys., 10, 419–420 (Abs.); Beiblätter, 6, 227 (Abs.).

Nouvelles démonstrations des conditions du minimum de déviation d'un
rayon dans le prisme.
> Kraiewitch (K.). Jour. soc. phys. chim. russe, 16, 8–13. Notes sur
> cet article, par Wolkoff, 16, 174.

Ueber die Schwankungen in der chemischen Wirkung des Sonnenspectrums und über einen Apparat zur Messung derselben.
> Vogel (H.). Ber. chem. Ges., 7, 88–92; Jour. Chem. Soc., (2) 12, 424
> (Abs.); Amer. Jour. Sci., (3) 7, 414–415.

Das Minimum der Ablenkung eines Lichtstrahls durch ein Prisma.
> Kessler (F.). Ann. Phys. u. Chem., n. F. 15, 833–834.

DIFFRACTION SPECTROSCOPES.

(See "Gratings.")

DIRECT-VISION SPECTROSCOPES.

Nouveau spectroscope à vision directe.

> Thollon (L.). Comptes Rendus, **86**, 829-831; Beiblätter, **2**, 253-254 (Abs.).

Théorie du nouveau spectroscope à vision directe.

> Thollon (L.). Comptes Rendus, **86**, 595; Beiblätter, **2**, 253.

Nouveau prisme composé, pour spectroscope à vision directe, de très grande pouvoir dispersif.

> Thollon (L.). Comptes Rendus, **88**, 80-82; Beiblätter, **3**, 355.

Sur l'emploi de prismes à liquide dans le spectroscope à vision directe.

> Zenger (C. V.). Comptes Rendus, **92**, 1503-1504.

Le spectroscope à vision directe appliqué à l'astronomie physique.

> Zenger (C. V.). Comptes Rendus, **93**, 429-432; Beiblätter, **5**, 793 (Abs.).

Le spectroscope à vision directe, à spath calcaire.

> Zenger (C. V.). Comptes Rendus, **93**, 720-722; Beiblätter, **6**, 21 (Abs.); Z. Instrumentenkunde, **1**, 263-266.

Les observations spectroscopiques à la lumière monochromatique.

> Zenger (C. V.). Comptes Rendus, **94**, 155-156; Chem. News, **45**, 86-87 (Abs.); Jour. Chem. Soc., **42**, 677 (Abs.); Amer. Jour. Sci., (3) **23**, 322-323 (Abs.); Beiblätter, **6**, 878; Z. Instrumentenkunde, **2**, 114 (Abs.).

Spectroscope à vision directe très puissant.

> Zenger (C. V.). Comptes Rendus, **96**, 1039-1041; Nature, **27**, 596 (Abs.); Chem. News, **47**, 213 (Abs.); Beiblätter, **7**, 456-457 (Abs.); Amer. Jour. Sci., (3) **25**, 469; Z. analyt. Chem., **22**, 540-541 (Abs.).

Spectroscope à vision directe pour observation des rayons ultra-violettes.

> Zenger (C. V.). Comptes Rendus, **98**, 494.

Neues geradsichtiges Taschenspectroskop.

> Hilger (A.). Beiblätter, **1**, 124-125.

Spectroscopes à vision directe et à grande dispersion.

> Thollon (L.). Jour. de Physique, **8**, 73-77.

Note on a direct-vision spectroscope on Thollon's plan, adapted to laboratory use and capable of giving exact measurements.

> Liveing (G. D.) and Dewar (J.). Proc. Royal Soc., **28**, 482–483 ; Beiblätter, **3**, 709 (Abs.).

Ein Spectroskop à vision directe mit nur einem Prisma.

> Emsmann (H.). Ann. Phys. u. Chem., **150**, 636.

A direct-vision compound prism by Merz; with dispersion almost double that of flint glass.

> Gassiot. Proc. Royal Soc., **24**, 83.

Combinazioni spettroscopiche a visione diretta.

> Ricoó (A.). Mem. Spettr. ital., **8**, 21–84.

Ueber ein verbessertes Prisma à vision directe.

> Braun (C.). Ber. aus Ungarn, **1**, 197–200.

Note on a new form of direct-vision spectroscope.

> Liveing (G. D.) and Dewar (J.). Proc. Royal Soc., **41** (1886), 449–452.

DISPERSION APPARATUS.

Das Dispersionsparallelopiped und seine Anwendung in der Astrophysik.

> Zenger (K. W.). Sitzungsber d. Böhm. Ges. (1881), 416–429 ; Beiblätter, **6**, 286 (Abs.).

Sur un spectroskope à grande dispersion.

> Cornu (A.). Jour. de Phys., **12** (1883), 53–57 ; Amer. Jour. Sci., (3) **25**, 469.

Sur un spectroscope à grande dispersion.

> Cornu (A.). Séances de la Soc. franç. de Phys., **1882**, 165–170 ; Beiblätter, **7**, 285 (Abs.); **8**, 83 (Abs.).

Bemerkungen über die Einrichtung eines Dispersiometers.

> Mousson (A.). Ann. Phys. u. Chem., **151**, 187–145.

ECLIPSE APPARATUS.

(See "Solar and Stellar App.")

EFFICIENCY OF SPECTROSCOPES.

Efficiency of different forms of the spectroscope.

> Pickering (E. C.). Amer. Jour. Sci., **95**, 301, and (3) **22**, 397.

ELECTRIC APPÁRATUS.

Tube spectro-électrique destiné à l'observation des spectres des solutions métalliques.

> Delachanal (B.) et Mermet (A.). Comptes Rendus, **79**, 800; **81**, 726.

An arrangement of the electric arc for the study of the radiation of vapours, together with preliminary results.

> Liveing (G. D.) and Dewar (J.). Proc. Royal Soc., **34**, 119–122; Nature, **26**, 213–214 (Abs.); Beiblätter, **6**, 984–986 (Abs.); Jour. Chem. Soc., **44**, 262–268 (Abs.).

On the use of most electrodes.

> Hartley (W. N.). Chem. News, **49**, 149; Beiblätter, **8**, 581.

Apparat zur leichten Darstellung des langen electrischen Spectrums.

> Müller (J.). Ann. Phys. u. Chem., **130**, 187.

ERYTHROSCOP.

Erythroscop und Melanoskop.

> Lommel (E.). Ann. Phys. u. Chem., **143**, 483–490.

EUTHYOPTIC.

Das einfache euthyoptische Spectroskop.

> Kessler (F.). Ann. Phys. u. Chem., **151**, 507.

FINDER.

A reliable finder for a spectro-telescope.

> Winlock (Prof.). Jour. Franklin Inst., (3) **60**, 295.

FIXATOR.

Der Fixator, ein Ergänzungsapparat des Spectrometers.

> Carl's Repert., **17**, 645–651; Jour. de Phys., (2) **1**, 198–199 (Abs.).

FLAME APPARATUS.

Spectralapparat un den wärmeren oder kälteren Theile der Flammen beobachten zu können. (For Bessemer flame apparatus look above under Bessemer.)

> Salet (G.). Ber. chem. Ges., **3** (1870), 246.

FLUORESCENT EYE-PIECES.

Spectroscope à oculaire fluorescent.

    Soret (J. L.). Jour. de Phys., **3** (1874), 253.

Une spectroscope pour étudier les phénomènes de la fluorescence.

    Lamansky (S.). Jour. de Phys., **8** (1879), 411.

Some modifications of Soret's fluorescent eye-piece.

    Liveing and Dewar. Proc. Cambridge Phil. Soc., **4**, 342-343.

Spectroscope à oculaire fluorescent.

    Manet. Ann. Chim. et Phys., (5) **11**, 72.

Spectralapparat mit fluorescirendem Okular für den ultravioletten Theil des Spectrums J.

    Reye (Th.). Ann. Phys. u. Chem., **149**, 407.

Spectroscope à oculaire fluorescent.

    Soret (J. L.). Archives de Genève, (2) **49**, 338-343; Ann. Phys. u. Chem., **152**, 167-171; Jubelband, 407-411; Amer. Jour. Sci., (3) **8**, 64-65.

Spectroscope à oculaire fluorescent; seconde note.

    Soret (J. L.). Arch. de Genève, (2) **57**, 819-833; Ann. Chim. et Phys., (5) **11**, 72-86; Amer. Jour. Sci., (3) **14**, 415-416 (Abs.); Beiblätter, **1**, 190-192 (Abs.).

FULGATOR MODIFIÉ.

Nouveau tube spectro-électrique (fulgator modifié).

    Delachanal et Mermet. Comptes Rendus, **81**, 726.

GELATINE LEAVES.

Gefärbte Gelatinblättchen als Objecte für das Spectroscop.

    Lommel (E.). Ann. Phys. u. Chem., **143**, 656.

GRATINGS.

Preliminary notice of the results accomplished in the manufacture and theory of gratings for optical purposes.

    Rowland (H. A.). Johns Hopkins Univ. Circular (1882), 248-249; Phil. Mag., (5) **13**, 469-474; Nature, **26**, 211-213; Amer. Jour. Sci., (3) **24**, 63 (Abs.); Observatory (1882), 224-228; Z. Instrumentenkunde, **2**, 304 (Abs.).

On concave gratings for optical purposes.

> Rowland (H. A.). Amer. Jour. Sci., (3) **26**, 87–98; Phil. Mag., (5)
> **16**, 197–210; Beiblätter, **7**, 862–868 (Abs.); Z. Instrumentenkunde,
> **4**, 185–186 (Abs.); Jour. de Phys., (2) **3**, 184 (Abs.).

Curved diffraction gratings.

> Glazebrook (R. T.). Proc. Physical Soc., **5**, 243–258; Phil. Mag., (5)
> **15**, 414–428; Amer. Jour. Sci., (3) **26**, 67 (Abs.); Beiblätter, **8**, 84
> (Abs.); Jour. de Phys., (2) **3**, 152–154 (Abs.).
>
> Remarks on the above by Rowland (H. A.). Amer. Jour. Sci., (3)
> **26**, 214; Phil. Mag., (15) **16**, 210; Beiblätter, **8**, 84 (Abs.); Jour.
> de Phys., (2) **3**, 184–185 (Abs.).

Concave gratings for giving a diffraction spectrum.

> Rowland (H. A.). Nature, **27**, 95.

The spectra formed by curved diffraction gratings.

> Baily (W.). Proc. Physical Soc., **5**, 181–185; Phil. Mag., (5) **15**, 183–
> 187; Beiblätter, **7**, 465–566 (Abs.); Jour. de Phys., (2) **3**, 152–154;
> Chem. News, **47** (1883), 54.

Notes on diffraction gratings.

> · Blake (J. M.). Amer. Jour. Sci., (3) **8**, 33–39.

Optische Experimentaluntersuchungen über Beugungsgitter.

> Quincke (G.). Ann. Phys. u. Chem., **146**, 1–65.

Note on the use of a diffraction grating as a substitute for the train of
prisms in a solar spectroscope.

> Young (C. A.). Amer. Jour. Sci., (3) **5**, 472–478; Phil. Mag., (4)
> **46**, 87–88; Ann. Phys. u. Chem., **152**, 868 (Abs.).

Preliminary note on the reproduction of diffraction gratings by means of
photography.

> Strutt (J. W.). Proc. Royal Soc., **20**, 414–417; Phil. Mag., (4) **44**,
> 392–394; Amer. Jour. Sci., (3) **5**, 216 (Abs.); Ann. Phys. u. Chem.,
> **152**, 175–176 (Abs.).

On the manufacture and theory of diffraction gratings.

> Rayleigh (Lord). Phil. Mag., (4) **47**, 81–93, 193–205.

On copying diffraction gratings.

> Rayleigh (Lord). Phil. Mag., (5) **11**, 196–205.

On the determination of the coefficient of expansion of a diffraction grating by means of the spectrum.

> Medenhall (T. C.). Amer. Jour. Sci. (3) **21**, 230–232.

Use of the reflecting grating in eclipse photography.

Lockyer (J. N.). Proc. Royal Soc., 27, 107–108.

Sur les réseaux métalliques de M. Rowland.

Mascart. Soc. franç. de Phys. (1882), 232–238; Jour. de Phys., (2) 2, 5–11; Beiblätter, 7, 466–468 (Abs.).

Sur la théorie des réseaux courbes.

Sokoloff (A.). Jour. soc. phys. chim. russe, 15, 293–305.

On a theorem relating to curved diffraction gratings.

Baily (W.). Phil. Mag., (5) 22 (1886), 47–49.

HAND–SPECTROSCOPE.

Handspectroskop.

Simmler. Jour. prackt. chem., 90, 299; Ann. Phys. u. Chem., 120, 623.

HELPS.

Ein neuer Hülfsapparat zur Spectralanalyse.

Schultz (H.). Pfluger's Arch. f. Physiol., 28, 197–199; Ber. chem. Ges., 15, 2754 b (Abs.); Beiblätter, 6, 674 (Abs.).

Ueber einige physikalische Versuche und Hülfseinrichtungen.

Z. Instrumentenkunde, 3, 888–892; Beiblätter, 8, 220 (Abs.).

INDEX.

Selbstleuchtender Index im Spectroskop.

Sundell (A. F.). Astronom. Nachr., 102, 90; Beiblätter, 6, 876–877 (Abs.); Z. Instrumenten., 2, 422 (Abs.).

INTERFERENCE APPARATUS.

Sur les phénomènes d'interférence produits par les réseaux parallèles, interférence-spectromètre.

Crova (A.). Comptes Rendus, 72, 855–858, 74, 932–936; Ann. Chim. et Phys., (5) 1, 407–432.

Sur l'application du spectroscope à l'observation des phénomènes d'interférence.

Mascart. Jour. de Phys., 1 (1872), 177.

KOLORIMETER.

Dr. von Konkoly's Spectralapparat in Verbindung mit einem Kolorimeter.

> Gothard (E. von). Centralzeitung für Optik und Mechanik, **4**, 241–248.

LAMPS.

Ueber Lampen für monochromatisches Licht.

> Laspeyres (H.). Z. Instrumenten., **2**, 96–99; Beiblätter, **6**, 480.

Un illuminateur spectral.

> Le Roux (F. P.). Comptes Rendus, **76**, 960, 998–1000; Chem. News, **27** (1873), 238.

Illumination des corps opaques.

> Lallemand (A.). Comptes Rendus, **69**, 192; **78**, 1272.

Spectralilluminator.

> Jahresber. d. Chem. (1873), 147.

Illumination of spectroscope micrometers.

> Konkoly (N. von). Monthly Notices Astronom. Soc., **44**, 250.

End-on in place of transverse illumination in private spectroscopy.

> Smyth (Piazzi). Chem. News, **39** (1879), 145, 166, 188; Nature, **19**, 400 (Abs.).

Des minima produits, dans une spectre calorifique, par l'appareil réfringent et la lampe qui servent à la formation de ce spectre.

> Aymonnet et Maquenne. Comptes Rendus, **87**, 494.

Spectre calorifique du Soleil et de la lampe à platine incandescent Bourbouze.

> Mouton. Comptes Rendus, **89**, 295.

On an improvement of the Bunsen burner for spectrum analysis.

> Kingdon (F.). Chem. News, **30**, 259.

Sur l'emploi de la lumière Drummond.

> Debray (H.). Ann. Chim. et Phys., (3) **65**, 331.

Note on the Littrow form of spectroscope.

> Brackett (C. F.). Amer. Jour. Sci., (3) **24**, 60–61; Beiblätter, **6**, 875–876 (Abs.).

The monochromatic lamp.

> Brewster (Sir D.). Trans. Edinburgh Royal Soc., 1822.

Ueber das Spectrum der Sell'schen Schwefelkohlenstofflampe.

Vogel (H. W.). Ber. chem. Ges., **8**, 96–98.

Relation between radiant energy and radiation in the spectrum of incandescence lamps.

Abney (W. de W.) and Festing (R.). Proc. Royal Soc., **37** (1884), 157–173.

Ein einfacher Brenner für monochromatisches Licht.

Noack. Z. zur Förderung des physischen Unterrichts, **2**, 67–69; Beiblätter, **9** (1885), 739 (Abs.).

Natriumlampe für Polarizationsapparate.

Landolt (H.). Z. Instrumentenkunde, **4** (1884), 390; Beiblätter, **8**, 339 (Abs.).

FOR MAGNETIC SPECTRA.

Fixing and exhibiting magnetic spectra.

Mayer (A. M.). Jour. Franklin Inst., **91**, 355.

MEASURING APPARATUS.

Eine vergleichbare Spectralscale.

Weinhold (A.). Ann. Phys. u. Chem., **138**, 417, 434; Jahresber. d. Chemie (1869), 175.

Glass reading-scale for direct-vision spectroscopes.

Proctor (H. R.). Chem. News, **27** (1878), 149; Nature, **6**, 473.

Measurement of faint spectra.

Proctor (H. R.). Nature, **6**, 534.

Spectroscopic scale.

Capron's Photographed Spectra. London, 1877, p. 17.

Measuring scales for pocket spectroscopes.

Herschel (A. S.). Nature, **18**, 300–301; Beiblätter, **2**, 560–561 (Abs.).

New form of measuring apparatus for a laboratory spectroscope.

Reynolds (J. E.). Scientific Proc. Dublin Soc., new ser., **1**, 5–9; Phil. Mag., (5) **5**, 106–110; Chem. News, **37** (1878), 115–116.

Messung des Brechungsexponenten während des Unterrichtes.

Kurz (A.). Carl's Repert., **18**, 190–192.

Mesure des indices de réfraction des liquides à l'aide des lentilles formées des mêmes.

Piltchikoff. Jour. soc. phys. chim. russe, **13**, 390–410; Beiblätter, **7**, 189–190 (Abs.); Jour. de Phys. (2) **1**, 578–579 (Abs.).

Eine Interferenz-Scala für das Spectroskop.

     Müller (J.). Dingler's Jour., **199**, 138–145.

Combination der Interferenz-Scala mit der photographischen Spectral-Scala.

     Müller (J.). Dingler's Jour., **199**, 268–271.

FOR METALLIC SPECTRA.

Apparat zur Objectivdarstellung der Metallspectren.

     Edelmann (Th.). Ann. Phys. u. Chem., **149**, 119–122; Chem. Central-blatt (1872), 691; Jour. Chem. Soc., (2) **11**, 461 (Abs.).

METEOROLOGICAL.

A meteorological spectroscope.

     Donelly (Col. J. F.). Nature, **26**, 501; Beiblätter, **7**, 25 (Abs.); Jour. de Phys., (2) **3**, 44, (Abs.).

     (See Rain-Band Spectroscope, below.)

SPECTRO–MICROMETERS.

Illumination of spectroscope micrometers.

     Konkoly (N. von). Monthly Notices Astronom. Soc., **44**, 250.

A convenient eye-piece micrometer for the spectroscope.

     Rood (O. N.). Amer. Jour. Sci., (3) **6**, 44–45; Phil. Mag., (4) **46**, 176.

Direct-vision micrometer for pocket spectroscopes.

     Proctor (H. R.). Chem. News, **27** (1878), 150.

A new form of micrometer for use in spectroscopic analysis.

     Watts (W. M.). Proc. Physical Soc., **1**, 160–164; Phil. Mag., (4) **50**, 81–85; Ann. Phys. u. Chem., **156**, 813–818; Chem. News, **32** (1875), 14.

MICRO–SPECTROSCOPES. (SPECTRUM–MICROSCOPES.)

Some technical applications of the spectrum-microscope.

     Sorby (H. C.). Quar. Jour. Microscop. Sci., **9** (1869), 358–383; Dingler's Jour., **198**, 243–254, 334–348.

A new and improved microscope spectrum apparatus.

     Sorby (H. C.). Monthly Microscop. Jour., **13**, 198–208.

A new micro-spectroscope, and on a new method of printing a description of the spectra seen with the spectrum microscope.

     Sorby (H. C.). Chem. News, **15**, 220.

Use of the micro-spectroscope in the discovery of blood-stains.
> Herepath (W. Bird). Chem. News, 17, 113, 128.

Spectrum analysis as applied to microscopic observation.
> Suffolk (W. T.). Chem. News, 29 (1874), 195.

Binoculares Spectrum-Mikroscop.
> Jahresber. d. Chemie, (1869), 175.

New arrangement of a binocular spectrum-microscope.
> Crookes (W.). Proc. Royal Soc., 17, 448.

Ueber ein Polari-Spectrum-Mikroscrop, mit Bemerkungen über das Spectrumocular.
> Rollett (A.). Z. Instrumentenkunde, 1, 366–372; Beiblätter, 6, 229–230 (Abs.); Z. analyt. Chemie, 21, 554–555 (Abs.).

Mikrochemische Reactionsmethoden im Dienste der technischen Microscopic.
> Tschirch (A.). Generalversammlung d. deutsch. Apotheker Ver. 1888; Archiv f. Pharm., (3) 20, 801–812; Jour. Chem. Soc., 44, 376–378 (Abs.).

MINERALOGICAL SPECTROSCOPE.

The spectroscope applied to mint-assaying.
> Outerbridge (A. E.). Jour. Franklin Inst., 98, 276; Jahresber. d. Chemie, (1868), 180.

MIRRORS.

Sur la transparence actinique de quelques milieux et en particulier sur la transparence actinique des miroirs de Foucault et leur application en photographie.
> Chardonnet (de). Jour. de Phys., (2) 1, 305–312; Comptes Rendus, 94, 1171.

Miroir tremblant pour la recomposition des couleurs du spectre.
> Luvini (J.). Les Mondes, 43, 427–429; Beiblätter, 1, 556 (Abs.).

Miroir tournant pour la recomposition de la lumière spectrale.
> Lestrade (Lavaut de). Les Mondes, 44, 416–417.

Neues Spiegelprisma mit konstanten Ablenkungswinkeln. Absteck ganzer und halber rechter Winkel mit den Wollaston'schen Spiegelprisma
> Bauernfeind (O. M.). Ann. Phys. u. Chem., 134, 169–172.

Un nouveau spectroscope.

> Govi (S. G.). Chem. News, **52** (1885), 201 (Abs.); Comptes Rendus, **101** (1885).

Ueber ein neues Spectroskop.

> Gothard (E. von). Ber. aus. Ungarn, **2** (1884), 263–265; Beiblätter, **11** (1887), 87 (Abs.).

OPTOMETER.

Sur un optomètre spectroscopique.

> Zenger (C. V.). Comptes Rendus, **101** (1885), 1008; Amer. Jour. Sci., (3) **31**, 60.

OVERLAPPING SPECTROSCOPE.

An overlapping spectroscope.

> Love (J.). British Assoc. Rept. (1881), 564; Beiblätter, **8**.

OXYHYDROGEN APPARATUS.

Production of spectra by the oxyhydrogen flame.

> Marvin (T. H.). Phil. Mag., (5) **1**, 67–68; Jour. Chem. Soc., **2** (1876), 156 (Abs.).

PHOSPHORESCENT EYE–PIECE.

Spectroscop mit phosphorescirendem Ocular.

> Lommel (E.). Ann. Phys. u. Chem., n. F. **20**, 847.

PHOSPHOROGRAPHIES.

Sur les phosphorographies du spectre solaire.

> Becquerel (E.). Jour. de Phys., **11** (1882), 189.

Phosphorographies du spectre solaire infra-rouge.

> Becquerel (H.). Comptes Rendus, **96** (1883); Amer. Jour. Sci., (3) **25**, 280.

Phosphorograph of the spectrum.

> Draper. Amer. Jour. Sci., (3) **21**, 171.

Phosphorographie, angewandt auf die Photographie des Unsichtbaren.

> Zenger (K. V.). Comtes Rendus, **103** (1886), 454–456; Beiblätter, **11** (1887), 94 (Abs.).

PHOTOGRAPHIC SPECTROSCOPY.

Notice imprimée sur les effects chimiques des radiations et sur l'emploi qu'en a fait M. Daguerre pour fixer les images de la chambre noire.

Biot. Comptes Rendus, **9**, 200.

Application aux opérations photographiques des propriétés reconnus par M. Ed. Becquerel dans ce qu'il nomme les rayons continuateurs.

Gaudin. Comptes Rendus, **12**, 862.

Action des rayons rouges sur les placques daguerriennes.

Foucault et Fizeau. Comptes Rendus, **23**, 679.

Observations sur les expériences de M. M. Foucault et Fizeau.

Becquerel (Ed.). Comptes Rendus, **23**, 800.

Remarques. Foucault (L.). Do., 856.

Des actions que les diverses radiations solaires exercent sur les couches d'iodure, de chlorure ou de bromure d'argent.

Claudet. Comptes Rendus, **25**, 554.

Note sur ce Mémoire. Becquerel (Ed.). Do., 594.

Note sur les transformations successives de l'image photographique par la prolongation de l'action lumineuse.

Janssen (J.). Comptes Rendus, **91**, 199.

Beschreibung eines höchst einfachen Apparatus um das Spectrum zu photographiren.

Vogel (H. W.). Ann. Phys. u. Chem., **154**, 806.

Ueber die Hülfsmittel, photographische Schichten für grüne, gelbe und rothe Strahlen empfindlich zu machen.

Vogel (H. W.). Ber. chem. Ges., **17**, 1196–1208; Jour. Chem. Soc., **46**, 1081 (Abs.); Beiblätter, **8**, 583–585 (Abs.).

Early contributions to spectrum-photography and photo-chemistry.

Draper (J. W.). Nature, **10**, 243–244.

Spectrum photography.

Lockyer (J. N.). Nature, **10**, 109, 254.

Photographie du spectre chimique.

Prazmowski. Comptes Rendus, **79**, 108.

Theory of absorption-bands in the spectrum, and its bearing in photography.
>   Amory (Dr. Rob't).   Proc. Amer. Acad., **13**, 216.

Dunkle Linien in dem photographirten Spectrum weit über dem sichtbaren Theil hinaus.
>   Müller (J.).   Ann. Phys. u. Chem., **97**, 185.

Physics in photography.
>   Abney (W. de W.).   Nature, **18**, 489–491, 528–531, 543–546.

Method of fixing, photographing, and exhibiting the magnetic spectra.
>   Mayer (A. M.).   Chem. News, **23** (1871), 266.

Reversal of the metallic lines as seen in over-exposed photographs of spectra.
>   Hartley (W. N.).   Proc. Royal Soc., **34**, 84.

Reversal of the developed photographic image.
>   Abney (W. de W.).   Phil. Mag., (5) **10**, 200–208.

Photographische Spectral-Beobachtungen im rothen und indischen Meere.
>   Vogel (H. W.).   Ann. Phys. u. Chem., **156**, 819–825.

Delicacy of spectrum photography.
>   Hartley (W. N.).   Proc. Royal Soc., **36** (1885), 421–422; Jour. Chem. Soc., **48** (1885), 466 (Abs.).

Ueber neue Fortschritte in dem farbenempfindlichen photographischen Verfahren.
>   Vogel (H. W.).   Sitzungsber. preuss. Akad., **51** (1886), 1205–1208; Photogr. Mitt., **22**, 295; Beiblätter, **11** (1887), 255.

Ueber einige geeignete praktische Methoden zur Photographie des Spectrums in seinen verschiedenen Bezirken mit sensibilisirten Bromsilberplatten.
>   Eder (J. M.).   Monatschr. f. Chemie, **7** (1886), 429–454; Beiblätter, **11** (1887), 39 (Abs.); Jour. Chem. Soc., **52** (1887), 93 (Abs.).

PHOTOMETERS.

Ein neues Photometer.
>   Glan (P.).   Ann. Phys. u. Chem., n. F. **1**, 351.

Photometrische Untersuchungen.
>   Ketteler (E.) und Pulfrich (C.).   Ann. Phys. u. Chem., n. F. **15**, 337–378; Amer. Jour. Sci., (3) **23**, 486–487 (Abs.).

Études photométriques.

Cornu (A.). Jour. de Phys., **10**, 189–198; Beiblätter, **6**, 229 (Abs.).

Ein Photometer zu schulhygienischen Zwecken.

Petruschewski (Th.). Jour. soc. phys. chim. russe, **16**, (2) 295–308, 1884; Beiblätter, **9** (1885), 248 (Abs.).

POLARIZATION SPECTROSCOPES.

A rotary polarization spectroscope of great dispersion.

Tait (P. G.). Nature, **22**, 360–361; Beiblätter, **4**, 725 (Abs.).

Ein Polarizationsapparat aus Magnesiumplatincyanur.

Lommel (E.). Ann. Phys. u. Chem., n. F. **13**, 347.

PRISMS.

Absorption of light by prisms.

Robinson (T. R.). Observatory (1882), 58–54; Beiblätter, **6**, 589 (Abs.).

Projection du foyer du prisme.

Crova (A.). Jour. de Phys., (2) **1**, 84–86.

Étude des aberrations des prismes et de leur influence sur les observations spectroscopiques.

Crova (A.). Ann. Chim. et Phys., (5) **22**, 518–543.

Bemerkungen über Prismen.

Radau (R.). Ann. Phys. u. Chem., **118**, 452.

Déplacement des raies du spectre sous l'action de la température du prisme.

Blaserna (P.). Arch. de Genève, (2) **41**, 429–430; Ann. Phys. u. Chem., **143**, 655–656; Jour. Chem. Soc., (2) **10**, 118 (Abs.); Phil. Mag., (4) **43**, 239–240.

A direct-vision compound prism by Merz, with dispersion almost double that of ordinary flint glass.

Mr. Gassiot. Proc. Royal Soc., **24**, 33.

Note on the use of compound prisms.

Browning (J.). Monthly Notices Astronom. Soc., **31**, 203–205.

Auflösung scheinbar einfacher Linien durch Vermehrung der Prismen.

Merz (Sigismund). Ann. Phys. u. Chem., **117**, 655.

The best form of compound prism for the spectrum microscope.

  Sorby (H. C.). Nature, **4**, 511–512.

Ueber ein verbessertes Prisma à vision directe.

  Braun (C.). Ber. aus Ungarn, **1**, 197–200.

Ein Spectroscop à vision directe mit nur einem Prisma.

  Emsmann (H.). Ann. Phys. u. Chem., **150**, 686.

Geradsichtiges Prisma.

  Fuchs (F.). Z. Instrumentenkunde, **1**, 849–853 ; Z. analyt. Chemie., **21**, 555.

Nouveau modèle de prisme pour spectroscope à vision directe.

  Hofmann (J. G.). Comptes Rendus, **79**, 581.

Geradsichtige Prismen.

  Riccó (A.). Z. Instrumentenkunde, **2**, 105; Z. analyt. Chem., **21**, 555

  (Abs.) ; Beiblätter, **6**, 794 (Abs.).

Minimum du pouvoir de resolution d'un prisme.

  Thollon (L.). Comptes Rendus, **92**, 128–130.

The magnifying power of the half-prism as a means of obtaining great dispersion, and on the general theory of the half-prism spectroscope.

  Christie (W. H. M.). Proc. Royal Soc., **26**, 8–40; Beiblätter, **1**, 556–561 (Abs.).

New form of spectroscope with half-prisms.

  Chem. News, **35** (1875), 161.

Use of prisms of flint glass.

  Rood (O. N.). Amer. Jour. Sci., **85**, 856.

Ueber die anomale Dispersion spitzer Prismen.

  Lang (V. von). Ann. Phys. u. Chem., **143**, 269.

Nicht alle Quarzprismen verlängern das Spectrum am ultra-violetten Ende.

  Salm-Horst (Der Fürst). Ann. Phys. u. Chem., **109**, 158.

Use of carbon bisulphide in prisms.

  Draper (H.). Amer. Jour. Sci., (3) **29**, 269–277, 1885; Jour. Chem. Soc., **48**, 853 (Abs.), 1885; Jour. de Phys., (2) **5**, 182 (Abs.), 1886.

Ueber die Anwendung von Schwefelkohlenstoffprismen zu spectroscopischen Beobachtungen von hoher Präcision.

Hasselberg (B.). Ann. Phys. u. Chem., (2) **27** (1886), 415–436.

Neues Flüssigkeitsprisma für Spectralapparate.

Wernicke (W.). Z. Instrumentenkunde, **1**, 853–857; Beiblätter, **6**, 94–95 (Abs.); Z. analyt. Chemie, **21**, 555.

PROJECTION OF THE SPECTRUM.

Projection du foyer du prisme.

Crova (A.). Jour. de Phys., **11** (1882), 84.

Projection of the Fraunhofer lines of diffraction and prismatic spectra on a screen.

Draper (J. C.). Amer. Jour. Sci., (8) **9**, 22–24; Phil. Mag., (4) **49**, 142–4.

Nouvelle méthode pour projecter les spectres.

Moigno. Les Mondes, **43**, 554–5; Beiblätter, **1**, 555.

PROTUBERANCE SPECTROSCOPE.

Protuberanz Spectroscop mit excentrischer bogenförmiger Spaltvorrichtung.

Brunn (J.). Z. Instrumentenkunde, **1**, 281–282; Beiblätter, **6**, 280 (Abs.).

QUANTITATIVE APPARATUS.

Quantitative Analyse durch Spectralbeobachtung, Apparat.

Hennig (R.). Ann. Phys. u. Chem., **149**, 350.

Zur quantitativen Spectralanalyse.

Krüss (H.). Carl's Repert., **2**, 17–22.

RAIN–BAND SPECTROSCOPE.

Rain-band Spectroscope.

Bell (L.). Amer. Jour. Sci., (8) **30**, 347.

REFLECTOR.

Anwendung eines Reflectors bei Spectraluntersuchungen.

Fleck. Jour. prackt. Chemie, n. F. **3** (1870), 352; Jour. Chem. Soc., (2) **9**, 857 (Abs.).

REFRACTOMETERS.

Sur un réfractomètre destiné à la mesure des indices et de la dispersion des corps solides.

> Soret (C.). Comptes Rendus, **95**, 517–520; Beiblätter, **6**, 870–872 (Abs.); Z. Instrumenten., **2**, 414–415 (Abs.).

Sur l'emploi d'un verre biréfringent dans certaines observations d'analyse spectrale.

> Cruls. Comptes Rendus, **96**, 1298–1294; Nature, **28**, 48 (Abs.); Beiblätter, **7**, 529 (Abs.).

Interference phenomena in a new form of refractometer.

> Michelson (A. A.). Amer. Jour. Sci., (3) **23**, 395–400; Phil. Mag., (5) **13**, 286–242; Beiblätter, **7**, 584–535 (Abs.).

Appareils refringents en sel gemme.

> Dessains (P.). Comptes Rendus, **97**, 689, 782; Beiblätter, **7**, 858 (Abs.).

A new refractometer for measuring the mean refractive index of plates of glass and lenses by the employment of Newton's rings.

> Royston-Pigott (G. W.). Proc. Royal Soc., **24**, 893–899.

REGISTERING SPECTROSCOPE.

A registering spectroscope.

> Huggins (W.). Proc. Royal Soc., **19**, 817–818; Phil. Mag., (4) **41**, 544–546; Ann. Chim. et Phys., (4) **26**, 275–276; Chem. News, **23** (1871), 98.

REVERSION SPECTROSCOPES.

Ein neues Reversionsspectroscop.

> Zöllner (F.). Ber. d. Sächs. Ges. d. Wiss., **23**, 800–806; Ann. Phys. u. Chem., **144**, 449–456; Phil. Mag., (4) **43**, 47–52; Jahresber. d. Chemie (1869), 175.

Ein neuer Reversionsspectralapparat.

> Konkoly (N. von). Centralzeitung f. Optik u. Mechanik, **4**, 122–124; Beiblätter, **7**, 595; Ber. aus Ungarn, **1**, 128–183.

Reversion spectroscope.

> Langley (S. P.). Comptes Rendus (1884), 1145–1147.

On a method of estimating the thickness of Young's Reversing Layer.

> Pulsifer (W. H.). Amer. Jour. Sci., (3) **17**, 303.

A new form of reversible spectroscope.

> Stevens (W. L.). Amer. Jour. Sci., (3) **23**, 226–229.

RIGID SPECTROSCOPES.

Description of a rigid spectroscope; constructed to ascertain whether the position of the known and well-defined lines of a spectrum is constant while the coefficient of terrestrial gravity under which the observations are taken is made to vary.

Gassiot (J. P.). Proc. Royal Soc., 14, 820.

On the observations made with a rigid spectroscope by Captain Mayne and Mr. Connor.

Gassiot (J. P.). Proc. Royal Soc., 16, 6.

ROTARY SPECTROSCOPE.

Ueber einen rotirenden Spectralapparat.

Lohse (O.). Z. Instrumentenkunde, 1, 22-25; Beiblätter, 5, 278.

SCALES.

(See "Measuring Apparatus.")

SCREENS.

Die Beugungserscheinungen geradlinig begrenzter Schirme.

Lommel (E.). Abhandl. d. bayr. Akad., (2) 15, 529-664, 1886; Beiblätter, 11 (1887), 42-46 (Abs.).

APPARATUS FOR SECONDARY SPECTRA.

On a secondary spectrum of very large size, with a construction for secondary spectra.

Rood (O. N.). Amer. Jour. Sci., (8) 6, 172-180.

Du spectre secondaire et de son influence sur la vision dans les instruments d'optique.

Foucault (Léon). Ann. Chim. et Phys., (5) 15, 288.

SELENACTINOMETER.

Un Selénactinomètre.

Morize (H.). Comptes Rendus, 100, 271-272; Beiblätter, 9, 256.

SLITS FOR SPECTROSCOPES.

Sur un spectroscope à fente inclinée.

Garbe (G.). Comptes Rendus, 96, 836; Jour. de Phys., 12 (1883), 818.

Die Anwendung des Vierordt'schen Doppelspaltes in der Spectralanalyse.

Dietrich (W.).   Beiblätter, **5**, 488–441.

Protuberanzspectroscop mit excentrischer, bogenförmiger Spaltvorrichtung.

Brunn (J.).   Z. Instrumenten., **1**, 281; Beiblätter, **6**, 230.

Spectralspalt mit symmetrischer Bewegung der Schneiden.

Krüss (H.).   Carl's Repert., **18**, 217–228; Z. analyt. Chemie, **21**, 182–191; Beiblätter, **6**, 286 (Abs.); Jour. Chem. Soc., **42**, 1229 (Abs.); Z. Instrumenten., **3**, 62–68.

Spectroscope with slide, approved by Tyndall and others.

Hofmann.   Chem. News, **26** (1872), 180.

Slit for the spectroscope.

Tucker (Alex. E.).   Chem. News, **41** (1880), 79.

SPECTRO–BOLOMETER.

Use of the spectro-bolometer.

Langley (S. P.).   Amer. Jour. Sci., (3) **21**, 187; **24**, 895; **25**, 170; **27**, 169; **30**, 477.

SPECTROGRAPH.

Beschreibung eines Spectrographen mit Flüssigkeitsprisma.

Lohse (O.).   Z. Instrumenten., **5** (1884), 11–13; Beiblätter, **9** (1885), 167 (Abs.).

SPECTROMETERS.

Description d'un spectromètre.

Zantedeschi.   Comptes Rendus, **54**, 208.

Description d'un nouveau spectromètre à vision directe rendu plus simple et moins dispendieux.

Valz.   Comptes Rendus, **57**, 69, 141, 298.

On a spectrometer and universal goniometer, adapted to the ordinary wants of a laboratory.

Liveing (G. D.).   Proc. Cambridge Phil. Soc., **4**, 848.

On a new form of spectrometer.

Draper (J. W.).   Amer. Jour. Sci., (3) **18**, 30–34; Phil. Mag., (5) **7**, 818–316; Beiblätter, **3**, 621.

Interferenzspectrometer.

Fuchs (F.).   Z. Instrumenten., **1**, 326–329; Beiblätter, **6**, 228.

Das Lang'sche Spectrometer.
 Miller (F.). Carl's Repert., **16**, 250–251.

Der Fixator, ein Ergänzungsapparat des Spectrometers.
 Ketteler (E.). Carl's Repert., **17**, 645–651.

A Spectrometer.
 Browning (J.). Monthly Notices Astronom. Soc., **33**, 411.

De la spectrométrie, spectromètre.
 Champion (P.), Pellet (H.), et Grenier (M.). Comptes Rendus, **76**, 707–711; Jour. Chem. Soc., (2) **11**, 984 (Abs.).

SPECTROPHOTOMETERS.

Ueber ein Spectrophotometer.
 Zahn (von). Ber. d. naturforsch. Ges. in Leipzig, **5**, 1–4.

Ein Spectrophotometer.
 Fuchs (F.). Z. Instrumenten., **1**, 349–353; Beiblätter, **6**, 228.

Ein neues Spectrophotometer.
 Hüfner (G.). J. prackt. Chemie, n. F. **16** (1877), 290; Chem. News, **37** (1878), 81; Carl's Repert., **15**, 116–118.

On a spectrophotometer.
 Glazebrook (R. T.). Proc. Cambridge Phil. Soc., **4**, 304–308; Beiblätter, **8**, 211–212 (Abs.).

Étude sur les spectrophotomètres.
 Crova (A.). Comptes Rendus, **92**, 86–87; Phil. Mag., (5) **11**, 155–156.

Description d'un spectrophotomètre.
 Crova (A.). Ann. Chim. et Phys., (5) **29**, 556–573.

Das neue Spectrophotometer von Crova, verglichen mit dem von Glan, nebst einem Vorschlag zur weiteren Verbesserung beider Apparate.
 Zenker (W.). Z. Instrumenten., **4**, 83–87; Beiblätter, **8**, 499.

Ueber die Umwandlung meines Photometers in ein Spectrophotometer.
 Wild (H.). Ann. Phys. u. Chem., n. F. **20**, 452–468; Nature, **29**, 258 (Abs.); Jour. de Phys., (2) **3**, 142–143 (Abs.).

Ein Spectrophotometer.
 Wild (H.). Dingler's Jour., **252**, 462–465.

SPECTROPOLARISCOPE.

A spectropolariscope for sugar analysis.
 Levison (W. G.). Amer. Jour. Sci., **124**, 469.

3 T

SPECTROSCOPES (MISCELLANEOUS).

Construction of the spectroscope.

> Rutherfurd (L. M.).   Amer. Jour. Sci., (3) **39** (1869), 129.
> Note by Ditscheiner in Sitzungsber. Wiener Akad., **52** II, 542, 563–568.

Construction of the spectroscope.

> Cooke (J. P., Jr.).   Amer. Jour. Sci., **90**, 305.

Description of a large spectroscope.

> Gibbs (Wolcott).   Amer. Jour. Sci., (2) **25**, 110.

Spectral-Apparat.

> Kirchhoff (G.) und Bunsen (R.).   Ann. Phys. u. Chem., **110**, 162;
> Jour. prakt. Chem., **85**, 65, 74.

Spectral-Apparat.

> Mousson (A.).   Ann. Phys. u. Chem., **112**, 428.

Ursache der mangelnden Proportionalität in den Abständen bestimmter Streifen bei verschiedenen Apparaten.

> Gottschalk (F.).   Ann. Phys. u. Chem., **121**, 64–96.

Notiz zur Theorie der Spectralapparate.

> Ditscheiner (L.).   Ann. Phys. u. Chem., **129**, 336.

Convenient form of spectroscope for use in a laboratory.

> Browning (J.).   Chem. News, **22** (1870), 229.

Improvement of the spectroscope.

> Grubb (T.).   Chem. News, **29** (1874), 222.

On a quartz and Iceland spar spectroscope corrected for chromatic aberration.

> Stone (W. H.).   Chem. News, **41**, 91.

Note accompagnant le présentation de trois nouveaux spectroscopes.

> Janssen (J.).   Comptes Rendus, **55**, 576.

Un appareil destiné à réproduire les expériences d'optique, relatives à la réfraction, à la réflexion de la lumière polarisée, à la mesure des indices et à la spectroscopie.

> Lutz.   Comptes Rendus, **84**, 201.

Eine Verbesserung an Spectralapparaten.

> Miller (F.).   Z. Instrumenten., **2**, 29–30; Beiblätter, **6**, 231.

Ein sehr einfacher und wirksamer Spectralapparat.

> Konkoly (N. von). Centralzeitung f. Optik u. Mechanik, **4**, 76–77; Beiblätter, **7**, 456 (Abs.); Z. Instrumenten., **3**, 324 (Abs.); Ber. aus Ungarn, **1**, 184.

Vorschlag zur Construction eines neuen Spectralapparates.

> Lippich (F.). Z. Instrumenten., **4**, 1–8; Beiblätter, **8**, 300–302 (Abs.).

Neuere Apparate für die Wollaston'sche Methode zur Bestimmung von Lichtbrechungsverhältnissen.

> Liebich (T.). Z. Instrumentenkunde, **4**, 185–189.

Nouveau spectroscope.

> Thollon (L.). Jour. de Phys., **7**, 141–148.

Spectroscop-Apparate.

> Jahresber. d. Chemie, (1861) 41, (1862) 27, (1863) 114, (1864) 115, (1865) 94, (1866) 78, (1867) 105, (1868) 180, 132, (1869) 175, (1870) 1062, (1872) 948, (1873) 146, 147, (1874) 152, (1876) 142.

Spectralapparat.

> Mitscherlich. Jour. prakt. Chem., **86**, 13.

Arcobaleno in mare e modificazione allo spettroscopio descritto nel Vol. V.

> Riccò (A.). Mem. spettr. ital., **8**, 87.

Nouveau spectroscope.

> Stoney. Moniteur scientifique (3) **6**, 657.

Apparate zur Untersuchung der Farbenempfindungen.

> Glan (P.). Archiv. f. Physiol., **24**, 307–308; Beiblätter, **5**, 445 (Abs.).

A new spectroscope.

> Zenger (C. V.). Phil. Mag., (4) **46**, 439–445.

An improvement in the construction of the spectroscope.

> Madan (H. G.). Phil. Mag., (4) **48**, 118.

A home-made spectroscope.

> Furniss (J. J.). Pop. Sci. Monthly, **15**, 808.

Description of a large spectroscope.

> Gassiot (J. P.). Proc. Royal Soc., **12** (1863), 536.

The improvement of the spectroscope.

> Grubb (T.). Proc. Royal Soc., **22**, 308–309; Phil. Mag., (4) **48**, 532–534; Chem. News, **29**, 222–223; note by G. G. Stokes, Proc. Royal Soc., **22**, 309–310, and Phil. Mag., (4) **48**, 534.

Neue Einrichtung des Spectroscops.

> Littrow (Otto von). Sitzungsber. Wiener Akad., **46** II, 521; **48** II, 26–82; note by Prof. C. F. Brackett in Amer. Jour. Sci., **124**, 60.

SPECTRO—TELESCOPES.

Ein Spectrotelescop.

> Glan (P.). Ann. Phys. u. Chem., n. F. **9**, 492.

Description of a hand spectrum-telescope.

> Huggings (W.). Proc. Royal Soc., **16**, 241; Ann. Phys. u. Chem., **136**, 167.

Spectrum-telescop.

> Jahresber. d. Chemie (1868), 188.

A reliable finder for a spectro-telescope.

> Winlock (J.). Jour. Franklin Inst., (8) **60**, 295.

Ueber das spectroscopische Reversionsfernrohr.

> Zöllner (F.). Ber. Sächs. Acad. Wiss., **24**, 129–184; Phil. Mag., (4) **43**, 47; **44**, 417–421; Ann. Phys. u. Chem., **147**, 617–628; Comptes Rendus, **69**, 421.

A tele-spectroscope for solar observations.

> Browning (J.). Monthly Notices Astronom. Soc., **32**, 214–215.

Appareil destiné à observer les raies noires du spectre solaire.

> Dujardin (F.). Comptes Rendus, **8**, 253.

Improvements in a solar spectroscope made by Mr. Grubb for Prof. Young.

> Erck (W.). Monthly Notices Astronom. Soc., **38**, 331–332.

Spectroscopes furnished by the Royal Society to Mr. Hennessey for observing the solar eclipse of 1868 at Mussoorie, in India.

> Proc. Royal Soc., **16**, 169.

An eclipse spectroscope.

> Lockyer (J. N.). Nature, **18**, 224.

Neue Methode die Sonne spectroscopisch zu beobachten.

> Secchi (A.). Ann. Phys u. Chem., **143**, 154; Amer. Jour. Sci., (3) **1**, 463–464.

Sur un nouveau moyen d'observer les éclipses et les passages de Vénus.

> Secchi (A.). Comptes Rendus, **73**, 984–985; Monthly Notices Astronom. Soc., **31**, 202.

Sur l'emploi de la lunette horizontale pour les observations de la spectroscopie solaire.

> Thollon (L.). Comptes Rendus, **96**, 1200–1202; Nature, **28**, 24; Beiblätter, **7**, 456 (Abs.).

Apparatus for recording the position of lines in the spectrum, especially adapted to solar eclipses.

> Winlock (J.). Proc. Amer. Acad., **8**, 299.

Ein Spectroscop für Cometen-und Fixstern-Beobachtungen.

> Gothardt (E. von). Centralzeitung für Optik u. Mechanik, **4**, 121; Beiblätter, **7**, 595 (Abs.).

A star spectroscope.

> Gould (B. A.). Proc. Amer. Acad., **8**, 499.

A small universal stellar spectroscope.

> Merz (S.). Phil. Mag., (4) **41**, 129–132.

The spectroscope and the transit of Venus.

> Nature, **11**, 171.

Spectroscopie stellaire.

> Secchi (A.). Comptes Rendus, **65**, 889.

Secchi met sous les yeux de l'Académie l'appareil dont il s'est servi pour ses recherches.

> Comptes Rendus, **64**, 738.

Un nouveau spectroscope stellaire.

> Thollon (L.). Comptes Rendus, **89**, 749–752; Beiblätter, **4**, 860–861 (Abs.).

Ueber ein neués Spectroscop, nebst Beiträgen zur Spectralanalyse der Gestirne.

> Zöllner (F.). Ann. Phys. u. Chem., **138**, 82, 85; Phil. Mag., (4) **38**, 860; Amer. Jour. Sci., **99**, 58.

Nouveau spectroscope et recherches spectroscopiques de M. Zöllner; rapport verbal sur ces publications.

> Faye. Comptes Rendus, **69**, 689.

Ein einfaches Ocularspectroscop für Sterne.

> Zöllner (F.). Ann. Phys. u. Chem., **152**, 503; Phil. Mag., (4) **48**, 156–157.

Nouveau spectroscope stellaire.

> Zenger (Ch. V.). Comptes Rendus, **101** (1885), 616.

TUBES.

Sur les tubes lumineux à électrodes extérieures.

> Alvergniat. Comptes Rendus, **73**, 561; Jour. Chem. Soc., (2) **9**, 1141 (Abs.).

Tube spectro-électrique destiné à l'observation des spectres de solutions métalliques.

> Delachanal (B.) et Mermet (A.). Comptes Rendus, **79**, 800; Ann. Chim. et Phys., (5) **3**, 485.

Nouveau tube spectro-électrique (fulgator modifié).

> Delachanal et Mermet. Comptes Rendus, **81**, 726; Bull. Soc. chim., (2) **25**, 194-197; Jour. Chem. Soc., **2** (1876), 85 (Abs.).

Ein einfaches Stativ für Geissler'sche Spectralröhren.

> Gothardt (E. von). Z. Instrumenten., **3**, 320-321; Centralzeitung f. Optik u. Mechanik, **4**, 146-147; Beiblätter, **8**, 216.

End-on gas vacuum-tubes in spectroscopy.

> Smyth (C. Piazzi). Nature, **19**, 458; Beiblätter, **3**, 604 (Abs.).

End-on tubes brought to bear upon the carbon and carbo-hydrogen question.

> Smyth (C. Piazzi). Nature, **20**, 75-76.

Tube for observing the spectra of solutions.

> Nature, **13**, 75.

Spectralröhren mit longitudinaler Durchsicht.

> Zahn (W. von). Ann. Phys. u. Chem., n. F. **8**, 675.

ULTRA-VIOLET APPARATUS.

Spectroscope pour la partie ultra-violette du spectre.

> Cornu (A.). Les Mondes, **49**, 16-17; Beiblätter, **3**, 501.

Spectroscope destiné à l'observation des radiations ultra-violettes.

> Cornu (A.). Jour. de Phys., **8**, 185-193; Beiblätter, **4**, 84 (Abs.).

UNIVERSAL-SPECTROSCOPES.

Ein neues Universalstativ für die Benützung des Taschenspectroskopes.

> Lepel (F. von). Ber. chem. Ges., **12**, 263-266.

Ein Universalstativ für die Benützung des Taschenspectroskopes.

> Vogel (H. W.). Ber. chem. Ges., **10**, 1428-1432; Jour. Chem. Soc., **2** (1877), 915 (Abs.).

Neues Universalspectroskop für quantitative und qualitative chemische Analyse.

Krüss (G.). Ber. chem. Ges., **19** (1885), 2789–2745; Jour. Chem. Soc., **52**, 179 (Abs.), 1887; Amer. Jour. Sci., (8) **33** (1887).

WIDTH IN APPARATUS.

Bei der kleinsten Breite des Spectrums haben die Linien die geringste Krummung in dem Spectralapparat.

Ditscheiner (L.). Ann. Phys. u. Chem., **129**, 887.

ADDENDA.

On iquids of high dispersive powers for prisms.

Gibbs (Wolcott). Amer. Jour. Sci., vol. 4, 1870.

Appareil destiné à l'étude des intensités lumineuses et chromatiques des couleurs spectrales et de leurs mélanges.

Parinaud et Duboscq. Jour. de Phys., (2) **4** (1885), 271–8.

Sur in nouvel appareil dit " hema-spectroscope."

Thierry (M. de). Comptes Rendus, **100** (1885), 1244.

Sur in nouveau spectroscope d'absorption.

Thierry (M. de). Comptes Rendus, **101**, (1885), 811.

Veraischte Mittheilungen, betreffend Spectralapparate.

Vogel (H. C.). Z. Instrumentenkunde, **1**, 19–22; Beiblätter, **5**, 279 (Abs.).

Sur in nouveau spectroscope stellaire.

Zenger (Ch. V.). Comptes Rendus, **101** (1885), 616.

Sur in optomètre spectroscopique.

Zenger (Ch. V.). Comptes Rendus, **101** (1885), 1008.

Spetroscope pour les hautes fourneaux et le procédé Bessemer.

Zenger (Ch. V.). Comptes Rendus, **101** (1885), 1005.

SPECTRUM ANALYSIS.

a, GENERAL.

On the production of coloured spectra by light.

Abney (W. de W.).　Proc. Royal Soc., **29** (1879), 190; Chem. News, **39** (1879), 282.

The production of monochromatic light, or a mixture of colours on a screen.

Abney (W. de W.).　Phil. Mag., (5) **20** (1885), 172–174.

Mathematische Theorie der Spectralerscheinungen.

Akin (C. H.).　Sitzungsber. Wiener Akad., **53** I, 892; **53** II, 574.

Welchen Stoffen die Fraunhofer'schen Linien angehören.

Angström (A. J.).　Ann. Phys. u. Chem., **117**, 296–302; Proc. Roal Soc., **19**, 120.

Spectra of non-metallic bodies.

Angström and Thalèn.　Chem. News, **36** (1877), 111.

Spectres de quelques corps composés dans les mélanges gazeux en équilibre.

Berthelot et Richard.　Ann. Chim. et Phys., (4) **18**, 191; Bull. loc. chim. Paris, **13**, 109.

Nouvelles remarques sur la nature des éléments chimiques.

Berthelot.　Comptes Rendus, **77**, 1847–52, 1857, 1899–1403.

Certain spectral images produced by a rotating vacuum-tube.

Bidwell (Shelford).　Nature, **32** (1885), 80.

Photochemical researches.

Bunsen (R.) and Roscoe (H. E.).　Rept. British Assoc. (1856), I, 6‹

Spectralanalytische Untersuchungen.

Bunsen (R.).　Ann. Phys. u. Chem., **155**, 230–252, 366–384; P.1. Mag., (4) **50**, 417–480, 527–539.

Spectrum Analysis.

Carpenter (J.).　Once a Week, **8**, 708.

Untersuchungen über die optischen Eigenschaften von fein vertheiltn Körpern.

Christiansen (C.).　Ann. Phys. u. Chem., (2) **24** (1885), 439–446.

Spectren der chemischen Elemente und ihrer Verbindungen.

> Ciamician (G. L.). Sitzungsber. Wiener Akad., **76** II, 499; Ber. chem. Ges., **14**, 1101a.

Spectroskopische Untersuchungen.

> Ciamician (G. L.). Sitzungsber. Wiener Akad., **79** II, 8; Amer. Jour. Sci., **1**, 301; Chem. News, **40**, 285; **43**, 211, 270.

The spectroscope and evolution.

> Clarke (F. W.). Pop. Sci. Monthly, **2**, 320.

Lecture experiments in chemical analysis.

> Clemenshaw (E.). Nature, **31** (1885), 329; Phil. Mag., (5) **19** (1885), 365-368; Jour. Chem. Soc., **48**, 1085 (Abs.); note on the above, Chem. News, **51**, 57, 189.

Sur les raies spectrales spontanément renversables et l'analogie de leurs lois de répartition et d'intensité avec celles des raies de l'hydrogène.

> Cornu (A.). Jour. de Phys., (2) **5** (1886), 93-100.

Distinction between spectral lines of solar and terrestrial origin.

> Cornu (A.). Phil. Mag., (5) **22** (1887), 458-463; Jour. Chem. Soc., **52**, 818 (Abs.).

Radiant matter spectroscopy and residual glow.

> Crookes (W.). Chem. News, **53** (1885), 75, 133; **54** (1886), 28, 40, 54, 63, 75; **55** (1887), 107, 119, 131; Ber. chem. Ges., **16**, R. 1689a; note par Damien (B. C.), Jour. de Phys., (2) **4** (1885), 333.

Genesis of the elements.

> Crookes (W.). Chem. News, **55** (1887), 83, 99.

Production normale des trois systèmes de franges des rayons rectilignes.

> Croullebois. Comptes Rendus, **92**, 1009.

Notice sur la constitution de l'univers. Première Partie, Analyse spectrale.

> Delaunay. Ann. des Longitudes, 1869.

Sur quelques procédés de spectroscopie pratique.

> Demarçay (Eug.). Comptes Rendus, **99** (1885), 1022, 1069-71.

Loi de répartition des raies et des bandes; analogie avec la loi de succession de sons d'un corps solide.

> Deslandres. Comptes Rendus, **103** (1887), 972-976; Chem. News, **55** (1887), 204 (Abs.).

De spectral analyse. Academisch Proefschrift.

> Dibbits (H. C.), Rotterdam, 1863, with plates.

Over spectroscopische vergelikingen, betrekking hebbende tot de samen-
stelling van verschillende lichtbronnen en hoofdzalijk tot den licht
en kleurenzin.

> Donders.   Proc. Verb. Akad. Wetensch., Amsterdam, 1882–3, No. 10,
> 4–6.

The spectroscope and its revelations.

> Draper (H.).   Galaxy, **1**, 313.

Essai d'analyse spectrale.

> Dubrunfaut.   Bull. Soc. chim. Paris, n. s. **13**, 412; Comptes Rendus,
> **70**, 448.

Chemical Changes produced by Sunlight.

> Duclaux (E.).   Comptes Rendus, **103** (1887), 881–2.

Comparative Actions of Heat and Solar Radiation.

> Duclaux (E.).   Comptes Rendus, **104** (1887), 294–7.

Recherches spectrographiques de la scource normale de lumière et de son
emploi à la mesure photochimique de la sensibilité lumineuse.

> Eder (J. M.).   Wiener, Anzeigen (1885), 93; note par Gripon (E.),
> Jour. de Phys., (2) **5** (1886), 241, and note by Abney (W. de W.),
> Chem. News, **49**, 57.   [Chiefly interesting to photographers.]

Position du foyer des rayons de lumière monochromatique qui, issus d'un
même point, ont traversé un prisme à vision directe.

> Exner (K.).   Wiener Anzeigen (1885); Jour. de Phys., (2) **5** (1886),
> 287.

Les vibrations de la matière et les ondes de l'éther dans les combinaisons
photochimiques.

> Favé.   Comptes Rendus, **86**, 560–565.

Influence du magnétisme sur les caractères des lignes spectrales.

> Fievez (Ch.).   Mém. Acad. Bruxelles, **9** (1885), No. 3; Chem. News,
> **52** (1885), 802.

Bestimmung des Brechungs-und Farbenzerstreuungs-Vermögens verschiedener Glasarten.

> Fraunhofer (Jos.).   Denkschr. d. k. Akad. d. Wiss., München, **V**
> (1814–15), 193–226, mit drey Kupfertafeln, München, 1817, 4°.

Mischung von Spectralfarben.

> Frey (M. von) und Kries (J. von).   Archiv f. Physiol. (1881), 336–353;
> Jour. de Phys., (2) **1**, 513–514 (Abs.).

Spectrum analysis.

> Gassiot (J. P.).   Proc. Royal Soc., **12**, 536.

Spectre rotatoire.
> Govi (G.).   Comptes Rendus, **91**, 517.

Note on the theoretical explanation of Fraunhofer's lines.
> Hartshorne (H.).   Jour. Franklin Inst., **75**, 38–43; **105**, 38; Les
> Mondes, **45**, 517–522; Beiblätter, **2**, 561.

On the methods and recent progress of spectrum analysis.
> Herschel (A. S.).   Chem. News, **19**, 157.

Die Fraunhofer'schen Linien auf grossen Höhen dieselben wie in der
Ebne.
> Heusser (J. C.).   Ann. Phys. u. Chem., **91**, 319.

Der Gang der Lichtstrahlen durch ein Spectroskop.
> Hoorweg (J. L.).   Ann. Phys. u. Chem., **154**, 423.

On the spectra of some of the chemical elements, with maps.
> Huggins (W.).   Phil. Trans. (1884), 139; Proc. Royal Soc., **13**, 43.

Le prix Lalande decerné à M. Huggins.
> Comptes Rendus, **75**, 1305.

On some recent spectroscopic researches.
> Huggins (W.).   Quar. Jour. Sci., April, 1869.

Chemische Wirkung der verschiedenen Theile des Spectrums.
> Jahresber. d. Chemie. **1**, 197, 221; **2**, 156; **3**, 154; **4**, 152, 201; **4**, 152,
> 201; **5**, 124, 125, 126, 131, 211; **6**, 167; **7**, 137; **8**, 123; **12**, 643;
> **13**, 598; **14**, 27; (1870), 930; (1872), 146; (1873), 152; (1874), 152,
> 958.

Leçons sur l'analyse spectrale.
> Jamin.   Jour. de Pharm., (3) **42**, 9.

Chemische Analyse durch Spectralbeobachtungen.
> Kirchhoff (G.) und Bunsen (R.).   Ann. Phys. u. Chem., **110**, 161–187;
> **113**, 337–379; Phil. Mag., (4) **20**, 89.

Spectroscopic method for determining chemical action in solutions containing two or more colored salts.
> Krüss (G.).   Nature, **26**, 568.

Analyse spectrale simplifiée.
> Laborde (l'abbé).   Comptes Rendus, **60**, 53.

On certain remarkable groups in the lower spectrum.
> Langley (S. P.).   Proc. Amer. Acad., **14**, 92.

Nouvelle méthode spectroscopique.

> Langley (S. P.). Comptes Rendus, **84**, 1145–47; Beiblätter, **1**, 471–2.

Recomposition de la lumière spectrale.

> Lavaut de Lastrade. Les Mondes, **43**, 828–830.

Spectroscopic Notes.

> Leach (J. H.). Nature, **6**, 125; J. Franklin Inst., **93**, 418.

Remarques sur quelques particularités observées dans des recherches d'analyse spectrale.

> Lecoq de Boisbaudran (F.). Comptes Rendus, **69**, 1189; **76**, 1263–1265; Jour. Chem. Soc., (2) **11**, 1257–1258 (Abs.).

Théorie des spectres.

> Lecoq de Boisbaudran (F.). Comptes Rendus, **82**, 1264–1266; Jour. Chem. Soc., **2** (1876), 470 (Abs.).

Note on "Spectroscopic Papers."

> Liveing (G. D.) and Dewar (J.). Proc. Royal Soc., **29**, 166–168; Beiblätter, **4**, 88 (Abs.).

On the identity of the spectral lines of different elements.

> Liveing (G. D.) and Dewar (J.). Proc. Royal Soc., **32**, 225; Beiblätter, **5**, 741.

Studies in Spectrum Analysis.

> Liveing (G. D.) and Dewar (J.). Proc. Cambridge Phil. Soc., **3**, 208–209; Nature, **19**, 163–164.

Preliminary note on the compound nature of the line spectra of elementary bodies.

> Lockyer (J. N.). Proc. Royal Soc., **24**, 352–354; Phil. Mag., (5) **2**, 229–281; Ann. Chim. et Phys., (5) **25**, 190; Jahresber. d. Chemie, **14**, 45.

The spectroscope and its applications.

> Lockyer (J. N.). Nature, **7**, 125–466; **8**, 10, 89, 104.

Some recent methods in spectroscopy.

> Lockyer (J. N.). Chem. News, **33**, 29.

On a new method of spectrum observation.

> Lockyer (J. N.). Proc. Royal Soc., **30**, 22–31; Chem. News, **41**, 84–87; Amer. Jour. Sci., (3) **19**, 303–311; Beiblätter, **4**, 361 (Abs.); Ber. chem. Ges., **13**, 988–9 (Abs.).

On the necessity for a new departure in spectrum analysis.

> Lockyer (J. N.). Nature, **21**, 5–8; Beiblätter, **4**, 363 (Abs.).

Recomposition of the component colours of white light.
> Loudon (J.). Phil. Mag., (5) **1**, 170–171.

Das Stokes'sche Gesetz.
> Lubarsch (O.). Ann. Phys. u. Chem., n. F. **9**, 665.

Recomposition de la lumière spectrale.
> Luvini (J.). Les Mondes, **44**, 97–99.

Recherches sur la comparaison photométrique des scources diversement colorées, et en particulier sur la comparaison des divers parties d'une même spectre.
> Macé de Lépinay (J.) et Nicati (W.). Bull. soc. franç. de Phys. (1883), 11–23; Jour. de Phys., (2) **2**, 64–76; Ann. Phys. u. Chem., n. F. **22** (1884), 567.

Applications des spectres cannelées de Fizeau et Foucault.
> Macé de Lépinay (J.). Jour. de Phys., (2) **4** (1885), 261–271.

The logical spectrum.
> Macfarlane (A.). Phil. Mag. (5) **19**, 286.

Spectre chimique rendu visible avec ses raies cannelées.
> Matthiesen. Comptes Rendus, **16**, 1281.

Lectures on spectrum analysis, 1862.
> Miller (W. A.). Pharmaceutical Jour., (2) **3**, 899; Chem. News, **5**, 201.

Recent spectrum discoveries, 1863.
> Miller (W. A.). Jour. Franklin Inst., **76**, 29.

Exeter Lecture, 1869.
> Miller (W. A.). Popular Sci. Rev., Oct., 1869.

Beitrag zur Spectralanalyse.
> Mitscherlich (Alex.). Ann. Phys. u. Chem., **116**, 499–504; Ann. Chim. et Phys., (8) **69**, 169; Phil. Mag., (4) **28**, 169.

Sur l'analyse spectrale.
> Moigno (Fr.). Cosmos, **22**, 23, 52, 75.

Spectrum Analysis.
> Morton (H.). Jour. Franklin Inst., (3) **58**, 56, 186.

Die Spectren der chemischen Verbindungen.
> Moser (J.). Ann. Phys. u. Chem., **160**, 177–190; Phil. Mag., (5) **4**, 444–449 (Abs.); Nature, **16**, 193–194 (Abs.).

Résumé de nos connaissances actuelles sur le spectre.
> Mousson (A.). Archives de Genève (1861).

Sur le mélange des couleurs.
> Moutier (J.). Bull. Soc. Philom., (7) **7**, 19–21; Carl's Repert., **19**, 672–674.

On certain spectral images produced by a rotating vacuum-tube.
> Muirhead (Dr. Henry). Nature, **32** (1885), 55.

Present state of spectrum analysis.
> Nature, **22**, 523.

Upon an optical method for the measurement of high temperatures.
> · Nichols (E. L.). Amer. Jour. Sci., (3) **19**, 42–49.

Mutual attraction of spectral lines.
> Peirce (C. S.). Nature, **21**, 108; Beiblätter, **4**, 278 (Abs.)

Die Spectren der chemischen Verbindungen.
> Plücker. Ann. Phys. u. Chem., **105**, 78.

Spectrum Analysis.
> Pritchard (C.). Contemporary Review, **11**, 481

Lettre relative à l'analyse spectrale.
> Regimbeau. Comptes Rendus, **54**, 921.

Die Méthode des Spectrophors.
> Reinke (J.). Ann. Phys. u. Chem., (2) **27** (1886), 444–448.

Preliminary Report of the Committee appointed to construct and print Catalogues of Spectral Rays arranged upon a Scale of Wave-numbers.
> Rept. British Assoc., 1872; later Reports of same Committee, Repts. British Assoc., 1873 and 1874.

Report of the Committee consisting of Professor Dewar, Dr. Williamson, Dr. Marshall Watts, Captain Abney, Mr. Stoney, Prof. W. N. Hartley, Prof. McLeod, Prof. Carey Foster, Prof. A. K. Huntington, Prof. Emerson Reynolds, Prof. Reinold, Prof. Liveing, Lord Rayleigh, Dr. Arthur Schuster, and Mr. W. Chandler Roberts (Secretary), appointed for the purpose of reporting upon the Present State of our Knowledge of Spectrum Analysis.
> Reports of the British Association (1881), 317–422; (1884), 295–350.

Report of the Committee consisting of Professor Sir H. E. Roscoe, Mr. J. N. Lockyer, Professors Dewar, Wolcott Gibbs, Liveing, Schuster, and W. N. Hartley, Captain Abney, and Dr. Marshall Watts (Secretary), appointed for the purpose of preparing a new series of Wave-length Tables of the Spectra of the Elements. (Gives the wave-lengths of the elements and of certain compounds, " so far as they are known to the committee or have proved accessible.")

> Report of the British Association, (1884) 351–446, (1885) 288–322, (1886) 167–204.

Sur quelques phénomènes spectroscopiques singuliers.

> Riccò (A.). Comptes Rendus, **102** (1886), 851–853.

Secondary Spectra.

> Rood (O. N.). Amer. Jour. Sci., **106**, 172.

Spectrum Analysis.

> Roscoe (H. E.). Cornhill Mag., **6**, 109.

Lectures on Spectrum Analysis, delivered at the Royal Institution of Great Britain, 1861, 1862.

> Roscoe (H. E.). Chem. News, **4**, 118; **5**, 218, 261, 287.

Six Lectures on Spectrum Analysis, delivered in 1868, before the Society of Apothecaries of London.

> Roscoe (H. E.). London, 1869 (published in book form by Macmillan).

Address to the Chemical Section of the British Association; Remarks on the Spectroscope and Spectrum Analysis.

> Roscoe (Prof. Sir H. E.). Rept. British Assoc. (1884), 664.

Principles of spectrum analysis.

> Rowney (T.). Jour. Franklin Inst., **75**, 81.

Recherches spectroscopiques.

> Salet (G.). Bull. Soc. chim. Paris, n. s. **16**, 195.

Teachings of modern spectroscopy.

> Schuster (A.). Popular Science Monthly, **19**, 468.

Résumé des résultats de l'analyse spectrale.

> Secchi (A.). N. Arch. Phil. Nat., **23**, 145.

Beitrag zur chemischen Analyse durch Spectralbeobachtungen.

> Simmler (R. Th.). Ann. Phys. u. Chem., **115**, 242, 425.

Madeira spectroscopic.

> Smyth (C. Piazzi), Edinburgh, 1881–1882 (book).

Vorschläge zur Herstellung übereinstimmender Angaben.
>Steinheil. Ann. Phys. u. Chem., **122**, 167.

The Janssen-Lockyer Method of Spectrum Analysis.
>Stewart (B.). Nature, **7**, 301–302, 381–382.

Spectrum Analysis.
>Stewart (B.). Nature, **21**, 35.

On a simple mode of eliminating errors of adjustment in delicate observations of compared spectra.
>Stokes (G. G.). Proc. Royal Soc., **31**, 470–473; Beiblätter, **5**, 360–361 (Abs.).

On a remarkable phenomenon of crystalline reflection.
>Stokes (G. G.). Nature, **31** (1885); 565–568.

On a method of destroying the effects of slight errors of adjustment in experiments of change of refrangibility due to relative motions in the line of sight.
>Stone (E. J.). Proc. Royal Soc., **31**, 881.

Sur la récomposition de la lumière blanche avec l'aide des couleurs du spectre.
>Stroumbo. Comptes Rendus, **103** (1886), 737–8.

Prismatic Spectra.
>Talbot (H. Fox). Phil. Mag., **9** (1836), 8.

Notices spectroscopiques.
>Thenard (P.). Comptes Rendus, **91**, 887; Beiblätter, **5**, 44 (Abs.).

Eine neue Methode für spectralanalytische Untersuchungen.
>Timiriasef. Soc. phys. chim. russe, Mar. 27, 1872; Ber. chem. Ges., **5**, 828–829 (Abs.); Jour. Chem. Soc., (2) **10**, 1118 (Abs.).

Eine Lichteinheit.
>Trowbridge (J.). Proc. Amer. Acad. (1885), 494–499; Beiblätter, **9** (1885), 789 (Abs.).

Effect of resistance in modifying spectra.
>Tyndall (J.). Nature, **7**, 884.

Ueber die Beziehungen zwischen Lichtabsorption und Chemismus.
>Vogel (H. V.). Monatsber. Berliner Akad. (1875), 80–88; Pharmaceutical Jour. Trans., (3) **6**, 464–465; Scientific American, 1876.

Ueber einige Farbenwahrnehmungen und über Photographie in natürlichen Färben.

> Vogel (H. W.). Ann. Phys. u. Chem., (2) **28** (1886), 130–135; Jour.
> Chem. Soc., **50** (1886), 749 (Abs.).

General methods of observing and mapping spectra.

> Watts (W. Marshall). Rept. British Ass. (1881), 817.

On a means to determine the pressure at the surface of the Sun and stars,
and some spectroscopic remarks.

> Wiedemann (E.). Phil. Mag., (5) **10**, 123–125; Proc. Phys. Soc., **4**,
> 81–84.

Darstellung eines Spectrums mit einer Fraunhofer'schen Linie.

> Wüllner (A.). Ann. Phys. u. Chem., **135**, 174.

Spectroscopic Notes.

> Young (C. A.). Nature, **2**, 338; **3**, 110; **5**, 85–88; Phil. Mag., (5)
> **16**, 460–463; Beiblätter, **8**, 221 (Abs.); Amer. Jour. Sci., (8) **26**,
> 333–336; Jour. Franklin Inst., **60**, 331–340; **88**, 416; **90**, 64, 331;
> **92**, 348; **94**, 349; Chem. News, **22**, 218.

Ueber eine neue spectrometrische Methode.

> Zenger (K. W.). Sitzungsber. Prager Ges. (1877), 20–40; Beiblätter,
> **3**, 187–188 (Abs.).

b, QUALITATIVE ANALYSIS.

On the use of the prism in qualitative analysis.

> Gladstone (J. H.). Jour. Chem. Soc., **10** (1858), 79.

On a definite method of qualitative analysis of animal and vegetable col-
ouring-matters by means of the spectrum microscope.

> Sorby (H. C.). Proc. Royal Soc., **15**, 433.

c, QUANTITATIVE ANALYSIS.

Ueber quantitative Bestimmung des Lithiums mit dem Spectral-Apparat.

> Ballmann (H.). Z. analyt. Chem., **14**, 297–301; Jour. Chem. Soc., **2**
> (1876), 550 (Abs.).

De la spectrométrie.

> Champion (P.), Pellet (H.), et Grenier (M.). Comptes Rendus, **76**,
> 707–711; Jour. Chem. Soc., (2) **11**, 934 (Abs.).
> Note par M. J. Janssen. Comptes Rendus, **76**, 711–718; Jour. Chem.
> Soc., (2) **11**, 1258 (Abs.).

Use of the spectroscope in quantitative analysis.

> Gibbs (Wolcott). Proc. Amer. Acad., **10**, 401, 417.

4 T

De la loi d'absorption des radiations de toute espèce à travers les corps, et de son emploi dans l'analyse spectrale quantitative.

> Govi (G.). Comptes Rendus, **85**, 1046–1049, 1100–1103; Phil. Mag., (5) **5**, 78–80; Jour. Chem. Soc., **34**, 190–191 (Abs.); Beiblätter, **2**, 842–343 (Abs.).

Researches on spectrum photography in relation to new methods of quantitative chemical analysis.

> Hartley (W. N.). Proc. Royal Soc., **34**, 81–84; Ber. chem. Ges., **15**, 2924–5 (Abs.); Jour. Chem. Soc., **44**, 263–4 (Abs.); Beiblätter, **7**, 109–110 (Abs.); Z. analyt. Chem., **22**, 539–540 (Abs.); Phil. Trans., **175** (1884), 49–62.

> The same, continued. Proc. Royal Soc., **36**, 421–2; Chem. News, **49**, 128 (Abs.); Beiblätter, **8**, 705 (Abs.).

Ueber quantitative Analyse durch Spectralbeobachtung.

> Hennig (R.). Ann. Phys. u. Chem., **149**, 349–353; Jour. Chem. Soc., (2) **12**, 495 (Abs.).

Ueber quantitative Spectralbeobachtung.

> Hufner (G.). Jour. prakt. Chem., (2) **16**, 290.

Quantitative Spectralanalyse.

> Jahresber. d. Chemie, (1872) 873, (1873) 147, 173, (1875) 901.

Analyse spectrale quantitative.

> Janssen (J.). Comptes Rendus, **71**, 626.

Zur quantitativen Spectralanalyse.

> Krüss (H.). Carl's Repert. analyt. Chem., **2**, 17–22.

Quantitative Spectralanalyse.

> Krüss (H.). Ber. chem. Ges., **18**, 983–6; Jour. Chem. Soc., **48** (1885), 835 (Abs.).

Quantitative spectroscopic experiments.

> Liveing (G. D.) and Dewar (J.). Proc. Royal Soc., **29**, 482–489; Beiblätter, **4**, 367 (Abs.).

Quantitative analysis of certain alloys by means of the spectroscope.

> Lockyer (J. N.). Proc. Royal Soc., **21**, 507–8; Phil. Trans., **164** (1874), 495–499; Phil. Mag., (4) **47**, 811–812 (Abs.); Ber. chem Ges., **6**, 1426 (Abs.); Jour. Chem. Soc., (2) **12**, 495 (Abs.).

Quantitative Spectralanalyse, insbesondere zu denjenigen des Blutes.

> Noorden (C. v.). Ber. chem. Ges., **13** (1880), 439; Z. physiolog. Chem., **4**, 9–35.

Quantitative Bestimmung von Farbstoffen durch den Spectralapparat.

> Preyer (W.). Ber. chem. Ges., **4**, 404.

Analyse quantitative de la lumière blanche.
> Rood (O. N.). Les Mondes, **48**, 610–611.

Emploi du spectroscope pour la détermination quantitative des matières colorantes.
> Schiff (H.). Bull. Soc. chim. Paris, n. s. **16**, 97.

Beiträge zur quantitativen Spectralanalyse.
> Settegast (H.). Ann. Phys. u. Chem., n. F. **7**, 242–271; Jour. Chem. Soc., **36**, 828–9 (Abs.).

Quantitative Bestimmung von Farbstoffen durch den Spectralapparat.
> Vierordt (K.). Ber. chem. Ges., **4**, 327, 457, 519.

Zur quantitativen Spectralanalyse.
> Vierordt (K.). Ber. chem. Ges., **5**, 34–38; Ann. Phys. u. Chem., n. F. **3**, 357.

Die Anwendung des Spectralapparates zur Photometrie der Absorptionsspectren und zur quantitativen chemischen Analyse.
> Vierordt (Dr. Karl). Tübingen, 1873, 8°.

Die Anwendung der quantitativen Spectralanalyse bei den Titrirmethoden.
> Vierordt (K.). Ann. Phys. u. Chem., **177**, 31–45; Amer. Jour. Sci., (3) **10**, 216–7 (Abs.).

Beschreibung einiger quantitativen Spectralanalyse.
> Wolff (C. H.). Ber. chem. Ges., **12**, 128; Z. analyt. Chem., **18**, 38–49.

Anwendung eines Spectrophotometers zur quantitativen Spectralanalyse.
> (Von Lahn). Ber. d. naturforsch. Ges. in Leipzig, **5**, 1–4.

ABSORPTION SPECTRA.

On the photographic method of registering absorption spectra, and its application to solar physics.

Abney (W. de W.). Proc. Phys. Soc., **3**, 43–46; Phil. Mag., (5) **7**, 313–316; Beiblätter, **3**, 621.

Photographic records of absorption spectra.

Abney (W. de W.). Chem. News, **39** (1879), 132.

Absorption spectra of organic bodies.

Abney (Capt.) and Festing (Col.). Chem. News, **43** (1881), 126.

Absorption-spectra thermograms.

Abney (W. de W.) and Festing (R.). Proc. Royal Soc., **38**, 77–83; Jour. Chem. Soc., **48** (1885), 1175 (Abs.).

Transverse absorption of light.

Ackroyd (W.). Chem. News, **36**, 159–161.

Selective absorption of light.

Ackroyd (W.). Proc. Physical Soc., **2**, 110–118; Phil. Mag., (5) **2**, 423–430; Beiblätter, **1**, 350–2 (Abs.).

Note on the absorption of sea-water.

Aitken (J.). Proc. Royal Soc. Edinburgh, **11**, 637; Beiblätter, **7**, 372 (Abs.).

Theory of absorption bands in the spectrum, and its bearing in photography and chemistry.

Amory (Dr. Robert). Proc. Amer. Acad., **13**, 216.

Pouvoirs absorbants des corps pour la chaleur ; analyse spectroscopique.

Aymonnet. Comptes Rendus, **83**, 971.

Sur les variations des spectres d'absorption, et des spectres d'émission par phosphorescence d'un même corps.

Becquerel (H.). Comptes Rendus, **102** (1886), 106–110.

Sur les lois de l'absorption de la lumière dans les cristaux et sur une méthode nouvelle permettant de distinguer dans un cristal certaines bands d'absorption appartenant à des corps différents.

Becquerel (H.). Comptes Rendus, **103** (1887), 165–169.

Absorption spectrum of nitrogen peroxide.

> Bell (L.). Amer. Chem. Jour., **7**, 82–84; Jour. Chem. Soc., **48** (1885), 949 (Abs.).

A new form of absorption cell.

> Bostwick. Amer. Jour. Sci., (3) **30**, 452.

Ueber das Absorptionsspectrum des übermangansauren Kalis und seine Benützung bei chemisch-analytischen Arbeiten.

> Brücke (E.). Chemisches Centralblatt, (3) **8** (1877), 189–148; Jour. Chem. Soc., **34**, 242–248 (Abs.).

Das Absorptionsspectrum des Didyms.

> Bührig (H.). Jour. prakt. Chem., (2) **12**, 209–215; Amer. Jour. Sci., (3) **11**, 142 (Abs.).

Sur les spectres d'absorption de l'ozone et de l'acide pernitrique.

> Chappuis (J.). Comptes Rendus, **94**, 946–948; Jour. Chem. Soc., **42**, 1017 (Abs.); Beiblätter, **6**, 488 (Abs.); Amer. Jour. Sci., (3) **24**, 58–59 (Abs.).

Ueber die Veränderlichkeit der Lage der Absorptionsstreifen.

> Claes (F.). Ann. Phys. u. Chem., n. F. **3**, 889–414.

Sur la loi de répartition suivant l'altitude de la substance absorbant dans l'atmosphère; les radiations solaires ultra-violettes.

> Cornu (A.). Comptes Rendus, **90**, 940–946; Beiblätter, **4**, 727.

Sur l'observation comparative des raies telluriques et métalliques comme moyen d'évaleur les pouvoirs absorbants de l'atmosphère.

> Cornu (A.). Soc. franç. de Phys. (1882), 241–247; Jour. de Phys., (2) **2**, 58–63; Z. Instrumenten., **3**, 290 (Abs.).

Sur l'intensité calorifique de la radiation solaire et son absorption par l'atmosphère terrestre.

> Crova (A.). Comptes Rendus, **81**, 1205–1207.

Effect of various dyes on the behavior of silver bromide towards the solar spectrum; connection between absorption and photographic sensitiveness.

> Eder (J. M.). Monatsschr. f. Chemie, **6**, 927–953; Jour. Chem. Soc., **50**, 405 (Abs.).

Connection between absorption and photographic sensitiveness.

> Eder (J. M.). Monatsschr. f. Chemie, **7**, 331–350; Jour. Chem. Soc., **50** (1886), 958 (Abs.).

Salpetersaure Nickellösung als Absorptionspäparat.

> Emsmann (H.). Ann. Phys. u. Chem., Ergänzungsband 6 (1874), 334–5; Phil. Mag., (4) 46, 829–330; Jour. Chem. Soc., (2) 12, 118.

Sur les raies d'absorption produites dans le spectre par les solutions des acides hypoazotiques, hypochloriques et chloreux.

> Gernez (D.). Comptes Rendus, 74, 465–468; Jour. Chem. Soc., (2) 10, 280 (Abs.); Ber. chem. Ges., 5, 218 (Abs.).

Note sur le prétendu spectre d'absorption special de l'acide azoteux.

> Gernez (D.). Bull. Soc. Philom., (7) 5, 42.

Sur les spectres d'absorption des vapeurs de sélénium, de protochlorure et de bromure de sélénium, de tellure, de protochlorure et de bromure de tellure, protobromure d'iode et d'alizarine.

> Gernez (D.). Comptes Rendus, 74, 1190–1192; Jour. Chem. Soc., (2) 10, 665 (Abs.); Phil. Mag., (4) 43, 473–475; Amer. Jour. Sci., (3) 4, 59–60.

Sur les spectres d'absorption de quelques matières colorantes.

> Girard (Ch.) et Pabst. Comptes Rendus, 101 (1885), 157–160; Jour. Chem. Soc., 48, 1098 (Abs.).

Ueber den Einfluss der Dichtigkeit eines Körpers auf die Menge des von ihm absorbirten Lichtes.

> Glan (P.). Ann. Phys. u. Chem., n. F. 3, 54–82.

Sur la mesure de l'intensité des raies d'absorption et des raies obscures du spectre solaire.

> Gouy. Comptes Rendus, 89, 1033–4; Beiblätter, 4, 869–870 (Abs.).

On the action of heat on the absorption spectra and chemical constitution of saline solutions.

> Hartley (W. N.). Proc. Royal Soc., 23, 872–873 (Abs.); Ber. chem. Ges., 8, 765 (Abs.); Phil. Mag., (5) 1, 244–245.

On the absorption spectrum of ozone.

> Hartley (W. N.). Jour. Chem. Soc., 39, 57 60; Ber. chem. Ges., 14, 672 (Abs.); Beiblätter, 5, 505–506 (Abs.).

On the absorption of solar rays by atmospheric ozone.   Part I.

> Hartley (W. H.). Jour. Chem. Soc., 39, 111–128; Ber. chem. Ges., 14, 1890 (Abs.).

Researches on the relation between the molecular structure of carbon compounds and their absorption spectra.

> Hartley (W. N.). Jour. chem. Soc., 39, 153–168; 41, 45–49; 47, 685–757; 51, 152–202; Beiblätter, 6, 375–6 (Abs.); Nature, 32 (1885), 93–4.

Die Oxydationsproducte der Gallenfarbstoffe und ihre Absorptionsstreifen.

Heynsius (A.) und Campbell (G. F.). Archiv. f. Physiol., **4**, 497–547; Jour. Chem. Soc., (2) **10**, 807–808 (Abs.).

Absorptionsspectra.

Jahresber. d. Chemie (1875), 124.

Photometrie des Absorptionsspectrums der Blutkörperchen.

Jessen (E.). Zeitschr. f. Biologie, **17**, 251–272; Ber. chem. Ges., **15**, 952 (Abs.).

On the absorption of radiant heat by carbon dioxide.

Keeler (J. E.). Amer. Jour. Sci., (3) **28**, 190–198; Nature, **31**, 46.

Zusammenhang zwischen Absorption und Dispersion.

Ketteler (E.). Ann. Phys. u. Chem., **160**, 478.

Notiz, betreffend die Dispersionscurve der Mittel mit mehr als einem Absorptionsstreifen.

Ketteler (E.). Ann. Phys. u. Chem., n. F. **1**, 840–851.

Experimentaluntersuchung über den Zusammenhang zwischen Refraction und Absorption des Lichtes.

Ketteler (E.). Ann. Phys. u. Chem., n. F. **12**, 481–519.

Ueber den Zusammenhang zwischen Emission und Absorption von Licht und Wärme.

Kirchhoff (G.). Monatsber. d. Berliner Akad., 27 Oct., 1859; Phil. Mag., (4) **19**, 168.

(This contains the statement of the Law of Exchanges, and the first announcement of the discovery of the cause of Fraunhofer's lines.— *Roscoe.*)

Ueber das Verhältniss zwischen dem Emissionsvermögen und dem Absorptionsvermögen der Körper für Wärme und Licht.

Kirchhoff (G.). Ann. Phys. u. Chem., **109**, 275, 299; Phil. Mag., (4) **20**, 1.

(This paper contains a discussion of the Mathematical Theory of the Law of Exchanges, and is followed by a postscript on the history of the subject.—*Roscoe.*)

Beziehungen zwischen der Zusammensetzung und den Absorptionsspectren organischer Verbindungen.

Krüss (J.) und Oecomenides (S.). Ber. chem. Ges., **16**, 2051–56; **18**, 1426–33; Jour. Chem. Soc., **44**, 1041–2 (Abs.); **48**, 949; Beiblätter, **7**, 807–9 (Abs.).

Ueber das Absorptionsspectrum der flüssigen Untersalpetersäure.

Kundt (A.). Ann. Phys. u. Chem., **141**, 157–159; Jour. Chem. Soc., (2) **9**, 185 (Abs.); Z. analyt. Chem., (2) **7**, 64 (Abs.).

Ueber einige Bezeihungen zwischen der Dispersion und Absorption des Lichtes.

Kundt (A.). Ann. Phys. u. Chem., Jubelband, 615–624.

Ueber den Einfluss des Lösungsmittels auf die Absorptionsspectra gelöster absorbirenden Medien.

Kundt (A.). Sitzungsber. d. Münchener Akad. 1877, 234–262; Ann. Phys. u. Chem., n. F. 4, 34–54.

Die Absorptionsstreifen in Prismen von Schwefelkohlenstoff, Flintglass und Steinsalz entsprechend.

Lamansky (S.). Ann. Phys. u. Chem., 146, 213–215.

Zur Kenntniss der Absorptionsspectra.

Landauer (J.). Ber. chem. Ges., 11, 1772–1775; 14, 391–394; Jour. Chem. Soc., 36, 101 (Abs.); 40, 591 (Abs.); Beiblätter, 3, 195–6 (Abs.); 5, 441 (Abs.).

The selective absorption of solar energy.

Langley (S. P.). Amer. Jour. Sci., (3) 25, 169–196; Ann. Phys u. Chem., n. F. 19, 226–244, 384–400; Phil. Mag., (5) 15, 153–183; Ann. Chim. et Phys., (5) 29, 497–542; Z. Instrumentenkunde, 4, 27–32 (Abs.); Jour. de Phys., (2) 2, 371–374 (Abs.); Jour. Franklin Inst., 88, 157–8 (Abs.).
Note on the above by Koyl (C. H.). Johns Hopkins Univ. Cir., 2, 145–6; Phil. Mag., (5) 16, 317–318; Beiblätter, 7, 899.

On the amount of atmospheric absorption.

Langley (S. P.). Amer. Jour. Sci., (3) 28 (1885), 163, 242; Phil. Mag., (5) 18, 289–307; Jour. Chem. Soc., 28 (1885), 819 (Abs.).

Absorption dunkler Wärmestrahlen durch Gasen und Dämpfen.

Lecher und Pernter. Sitzungsber. d. Wiener Akad., 82 II, 265; Phil. Mag., Jan., 1881; Amer. Jour. Sci., (3) 21, 236.

Ueber die Absorption der Sonnenstrahlung durch die Kohlensäure unserer Atmosphäre.

Lecher (E.). Sitzungber. d. Wiener Akad., 82 II, 851–863.

Ueber Ausstrahlung und Absorption.

Lecher (E.). Sitzungsber. d. Wiener Akad., 85 II, 441–490; Ann. Phys. u. Chem., n. F. 17, 477–518 (Abs.).

Ueber die Aenderung der Absorptionsspectra einiger Farbstoffe in verschiedenen Lösungsmitteln.

Lepel (F. von). Ber. chem. Ges., 11, 1146–1151; Jour. Chem. Soc., 34 925 (Abs.); Beiblätter, 3, 860.

On the absorption of great thicknesses of metallic and metalloidal vapours.
Note 1, of Spectroscopic Notes.

> Lockyer (J. N.). Proc. Royal Soc., **22**, 371.

On a new class of absorption phenomena.

> Lockyer (J. N.). Proc. Royal Soc., **22**, 878.

On the absorption spectra of metals volatilized by the oxyhydrogen flame.

> Lockyer (J. N.) and Roberts (W. C.). Proc. Royal Soc., **23**, 344–349;
> Phil. Mag., (5) **1**, 284–289; Jour. Chem. Soc., **2** (1876), 156 (Abs.).

Emploi de la gélatine pour montrer l'absorption dans le spectre.

> Lommel (E.). Ann. Chim. et Phys., (4) **26**, 279.

Theorie der Absorption und Fluorescenz.

> Lommel (E.). Ann. Phys. u. Chem., n. F. **3**, 251–283.

Sur la théorie de l'absorption atmosphérique de la radiation solaire.

> Maurer (J.). Archives de Genève, (3) **9**, 874–891.

Absorption des Lichtes durch gefärbten Flüssigkeiten.

> Melde (F.). Ann. Phys. u. Chem., **124**, 91; **126**, 264.

Absorption spectra of brucine, morphine, strychnine, veratrine and santonine in concentrated acids.

> Meyer (A.). Archives Pharmaceutical Soc., (3) **13**, 413–416; Jour.
> Chem. Soc., **36**, 269.

Absorption spectra of anthrapurpurin.

> Perkin (W. H.). Jour. Chem. Soc., (2) **11**, 433.

New way of observing absorption spectra.

> Phipson (T. L.). Chem. News, **31** (1875), 255.

M. Chautard's classification of the absorption band of chlorophyll.

> Pocklington (H.). Pharmaceutical Trans., (3) **4**, 61–68.

Ueber die Absorptionsspectra der Chlorophyllfarbstoffe.

> Pringsheim. Monatsber. d. Berliner Akad. (1874), 628–659.

Photometrische Untersuchungen über die Absorption des Lichtes in isotropen und anisotropen Medien.

> Pulfrich (O.). Ann. Phys. u. Chem., n. F. **14**, 177–218; Amer. Jour.
> Sci., (3) **23**, 50 (Abs.); Jour. de Phys., (2) **1**, 285–286.

On the absorption bands in the visible spectrum produced by certain colourless liquids.

> Russell (W. J.) and Lapraik (W.). Jour. Chem. Soc., **39** (1881), 168–
> 173; Nature, **22**, 368–70; Beiblätter, **5**, 44–45; Amer. Jour. Sci., (3)
> **21**, 500–501 (Abs.).

Sur le spectre d'absorption de la vapeur du soufre.

    Salet (G.). Comptes Rendus, **74**, 865–866 ; Jour. Chem. Soc., (2) **10**, 882 (Abs.); Ber. chem. Ges., **5**, 323 (Abs.).

Ueber die Absorptionsstreifen des Blattgrüns.

    Schönn (L.). Ann. Phys. u. Chem., **145**, 166–167; Arch. de Genève, (2) **43**, 282–283.

Ueber die Absorption des Lichtes durch Flüssigkeiten.

    Schönn (J. L.). Ann. Phys. u. Chem., n. F. **6**, 267–270.

Ueber die Absorption des Lichtes durch Wasser, Steinöl, Ammoniak, Alcohol und Glycerin.

    Schönn (J. L.). Ann. Phys. u. Chem., Ergänzungsband **8** (1878), 670–5; Jour. Chem. Soc., **34**, 693.

Ueber die Lichtempfindlichkeit der Silberhaloidsalze und den Zusammenhang von optischer und chemischer Lichtabsorption.

    Schulz–Sellack (C.). Ann. Phys. u. Chem., **143**, 161–171; Ber. chem. Ges., **4**, 210–211 (Abs.); Jour. Chem. Soc., (2) **9**, 802–808 (Abs.); Phil. Mag., (4) **41**, 549–550 (Abs.).

Sur les spectres d'absorption ultra-violets des différents liquides.

    Soret (J. L.). Arch. de Genève, (2) **60**, 298–300; Beiblätter, **2**, 30–31 (Abs.), 410–411 (Abs.).

Recherches sur l'absorption des rayons ultra-violets par diverses substances ; spectres d'absorption des terres de la gadolinite et du didyme.

    Soret (J. L.). Arch. de Genève, (2) **63**, 89–112; Comptes Rendus, **86**, 1062–1064; Beiblätter, **3**, 196–197 (Abs.).

Sur les spectres d'absorption du didyme et de quelques autres substances extraits de la samarskite.

    Soret (J. L.). Comptes Rendus, **88**, 422–424.

Recherches sur l'absorption des rayons ultra-violets par diverses substances ; nouvelle étude des spectres d'absorption des métaux terreaux.

    Soret (J. L.). Arch. de Genève, (3) **4**, 261–292; Beiblätter, **5**, 124–125 (Abs.).

Absorption des rayons ultra-violets.

    Soret (J. L.). Arch. de Genève, (3) **4**, 877–880 ; remarques par M. A. Rilliet, do., 880–1.

Recherches sur l'absorption des rayons ultra-violets par diverses substances.

    Soret (J. L.). Arch. de Genève, (3) **10**, 429–494.

Spectre d'absorption du sang dans la partie violette et ultra-violette.

> Soret (J. L.). Comptes Rendus, **97**, 1269–70; Jour. Chem. Soc., **46**, 881.

Absorption der unsichtbaren Strahlen durch Alkalien, Glukoside, u. s. w.

> Stokes (G. G.). Ann. Phys. u. Chem., **123**, 48.

Ueber eine Methode zur Untersuchung der Absorption des Lichtes durch gefärbte Lösungen.

> Tumlirz (O.). Wiener Anzeigen (1882), 165–6; Beiblätter, **7**, 895–6; Chem. News, **49**, 201.

Observations of absorbing vapours upon the Sun.

> Trouvelot (E. L.). Monthly Notices Astronom. Soc., **39**, 874.

Die graphische Darstellung der Absorptionsspectren.

> Vierordt (K.). Ann. Phys. u. Chem., **151**, 119–124.

Ueber die Absorption der chemisch wirksamen Strahlen in der Atmosphäre der Sonne.

> Vogel (H. C.). Ber. d. Sächs. Ges. d. Wiss., **24**, 135–141; Ann. Phys. u. Chem., **148**, 161–168; Phil. Mag., (4) **45**, 345–350; Jour. Chem. Soc., (2) **11**, 712 (Abs.).
> Note on this by A. Schuster in Phil. Mag., (4) **45**, 350.

Ueber die Beziehung zwischen chemischer Wirkung des Sonnenspektrums, der Absorption und anomalen Dispersion.

> Vogel (H.). Ber. chem. Ges., **7**, 976–979; Jour. Chem. Soc., (2) **12**, 1121–1122.

Ueber die Beziehungen zwischen Lichtabsorption und Chemismus.

> Vogel (H.). Monatsber. d. Berliner Akad. (1875), 82–83.

Spectral-photometrische Untersuchungen insbesondere zur Bestimmung der Absorption der die Sonne umgebenden Gashülle.

> Vogel (H. C.). Monatsber. d. Berliner Akad. (1877), 104–142.

Absorptionsspectrum des Granats und Rubins.

> Vogel (H. W.). Ber. chem. Ges., **10** (1877), 373.

Untersuchungen über Absorptionsspectra.

> Vogel (H. W.). Monatsber. d. Berliner Akad. (1878), 409–431.

Ueber Verschiedenheit der Absorptionsspectra eines und desselben Stoffs.

> Vogel (H. W.). Ber. chem. Ges., **11**, 913–920, 1363–71; Jour. Chem. Soc., **36**, 189 (Abs.); Beiblätter, **2**, 699–702 (Abs.); note on the above by J. Moser. Ber. chem. Ges., **11**, 1416 and 1562; Bull. Soc. chim. Paris, n. ser., **32** (1879), 52.

Ueber den Zusammenhang zwischen dem Absorptionsspectrum und der sensibilisirenden Wirkung von Farbstoffen.

> Vogel (H. W.).   Ann. Phys. u. Chem., (2) 26, 527–30.

Ueber die Absorption und Brechung des Lichtes in metallisch undurchsichtigen Körpern.

> Wernicke (W.).   Monatsber. d. Berliner Akad. (1874), 728–787; Ann. Phys. u. Chem., 155, 87–95.

Untersuchungen über die bei der Beugung des Lichtes auftretenden Absorptionserscheinungen.

> Wien (Willy).   Ann. Phys. u. Chem., (2) 28 (1886), 117–130.

Einige neuen Absorptionsspectren.

> Wolff (O. H.).   Carl's Repert., 2, 55–56; Z. analyt. Chem., 22, 96–7; Chem. News, 47, 178 (Abs.).

Ueber die Absorptionsspectren verschiedener Ultramarinsorten.

> Wunder (J.).   Ber. chem. Ges., 9, 295–299; Jour. Chem. Soc., 1 (1876), 864–5.
> Bemerkungen, von R. Hoffmann.   Ber. chem. Ges., 9, 494–5.

(For the absorption spectra of particular substances look under those substances.)

ALCALIES AND ALCALOIDS.

Nachweis der Spectralanalyse der Alcalien.
>Belohoubek. Jour. prackt. Chem., **99**, 285.

Absorption spectra of the alcaloids.
>Hartley (W. N.). Chem. News, **51** (1885), 135; Phil. Trans. (1885), Part II, 9; Proc. Royal Soc., **38**, 1-4 and 191-198; Jour. Chem. Soc., **48** (1885), 1174 (Abs.).

Spectralreactionen der Alcaloïde.
>Hock (C.). Ber. chem. Ges., **14** (1881), 2844*b* (Abs.); Arch. f. Pharm., **19**, 858-9; Comptes Rendus, **93**, 849-51; Jour. Chem. Soc., **42**, 849 (Abs.); Beiblätter, **6**, 282 (Abs.).

Spectra der Alkalien.
>Kirchhoff und Bunsen. Jour. prakt. Chem., **80**, 449.

Zur Lehre von den Fäulnissalkaloïden.
>Poehl (A.). Ber. chem. Ges., **16**, 1975-1988.

Absorptionsspectra der Alkalichromate und der Chromsäure.
>Sabatier (P.). Beiblätter, **11** (1887), 223.

Absorption der unsichtbaren Strahlen durch Alkaloïde, Glukoside, u. s. w.
>Stokes (G. G.). Ann. Phys. u. Chem., **123**, 48.

Ueber die Lichtempfindlichkeit der Silberhaloïdsalze unter alkalischer Entwickelung.
>Vogel (H.). Ber. chem. Ges., **6**, 88-92.

Spectra der Alkalien.
>Wolf und Diacon. Jour. prakt. Chem., **88**, 67.

ALUMINIUM.

Phosphorescence de l'alumine.

> Becquerel (E.). Comptes Rendus, **103** (1886), 1224; **104** (1887), 334–5; Amer. Jour. Sci., (3) **33**, 808 (Abs.); Jour. Chem. Soc., **52**, 409 (Abs.); Chem. News, **55** (1887), 99.

Aluminium spark spectrum, photographed.

> Capron (J. R.). Photographed Spectra, London, 1877, p. 19, 40, 47.

Renversement des raies spectrales de l'aluminium.

> Cornu (A.). Comptes Rendus, **73**, 332.

Détermination des longueurs d'onde des radiations très-réfrangibles de l'aluminium, etc.

> Cornu (A.). Jour. de Phys., **10**, 425–431; Arch. de Genève, (3) **2**, 119–126; Beiblätter, **4**, 34–35 (Abs.).

Crimson line of phosphorescent alumina.

> Crookes (W.). Proc. Royal Soc., **42** (1887), 25–30; Nature, **35** (1887), 310; Amer. Jour. Sci., (3) **33**, 304 (Abs.); Chem. News, **55** (1887), 25.

Action des fluorures sur l'alumine.

> Frémy et Verneuil. Comptes Rendus, **103** (1887), 738–40.

Specific refraction and dispersion of the alums.

> Gladstone (J. H.). Phil. Mag., (5) **20**, 162–168; Jour. Chem. Soc., **50** (1886), 293 (Abs.).

Spectre continu de l'alumine.

> Gouy. Comptes Rendus, **86**, 878.

Distribution of heat in the spectra of various scources of radiation; white oxide of aluminium, etc.

> Jacques (W. W.). Proc. Amer. Acad., **14**, 142.

Spectrum von Aluminium.

> Jahresber. d. Chemie (1872), 145.

Aluminium métallique, étincelle.

> Lecoq de Boisbaudran (F.). Spectres Lumineux, Paris, 1874, p. 102, planche XV.

Sur la fluorescence rouge de l'alumine.

Lecoq de Boisbaudran (F.). Comptes Rendus, **103**, 478–482, 554–556, 1107; **104**, 830–834; Jour. Chem. Soc., **52** (1887), 191, 409 (Abs.).

Remarques par M. Edm. Becquerel. Comptes Rendus, **104**, 834–86 et 824–26.

Phosphorescence de l'alumine.

Lecoq de Boisbaudran (F.). Comptes Rendus, **103** (1887), 1224–1227; Jour. Chem. Soc., **52** (1887), 191 (Abs.).

Indice du quartz pour les raies de l'alumine.

Sarasin (Ed.). Comptes Rendus, **85**, 1230.

Spectre de l'aluminium dans l'arc voltaïque.

Secchi (A.). Comptes Rendus, **77**, 178.

Indices de réfraction des aluns.

Soret (C.). Comptes Rendus, **101**, 156–157; Jour. Chem. Soc., **48** (1885), 1097 (Abs.).

Réaction très-sensible de l'alumine.

Vogel (H. W.). Bull. Soc. chim. Paris, n. sér. **28**, 475–8.

ANTIMONY.

Antimony Spark Spectrum.
> Capron's Photographed Spectra, London, 1877, p. 19, 34.

L'antimoine n'a donné aucune apparence de renversement.
> Cornu (A.). Comptes Rendus, **73**, 832.

Protochlorure d'antimoine, en solution, étincelle.
> Lecoq de Boisbaudran. Spectres Lumineux, Paris, 1874, p. 150, planche 23.

Spectrum of antimony at elevated temperatures.
> Lockyer (J. N.). Chemical News, **30**, 98.

ARSENIC.

Arsenic spark spectrum, photographed.

Capron's Photographed Spectra, London, 1877, p. 18.

Spectrum of arsenic.

Huntington (O. W.). Proc. Amer. Akad., (2) **9**, 85–88; Amer. Jour. Sci., (8) **22**, 214–217; Beiblätter, **5**, 868 (Abs.).

The spectrum of arsenic at elevated temperatures.

Lockyer (J. N.). Chem. News, **30**, 98.

Sur l'origine de l'arsénic et de la lithine dans les eaux sulfatées calciques

Schlagdenhauffen. Jour. de Pharm., (5) **6**, 457–468; Jour. Chem. Soc., **44**, 302 (Abs.).

ASTRONOMICAL.

a, GENERAL.

Spectroscopic Researches.
> D'Arrest. Nature, **17**, 311.

Notes on some recent astronomical experiments at high elevations on the Andes.
> Copeland (R.). Nature, **28**, 606; Beiblätter, **8**, 220–221 (Abs.).

Spectroscopic observations made at the Earl of Crawford's observatory, Dun Echt.
> Copeland (R.). Monthly Notices Astronom. Soc., **45**, 90.

Recherches spectroscopiques sur quelques étoiles non encore étudieés.
> Cruls (L.). Comptes Rendus, **91**, 486–7; Beiblätter, **5**, 180–1.

Intorno alle strie degli stellari.
> Donati. Il nuovo Cimento, **15**, 292.

Rapport sur un mémoire et plusieurs notes de M. Janssen concernant l'analyse prismatique de la lumière solaire et de celle de quelques étoiles.
> Fizeau. Comptes Rendus, **58**, 795.

Recherches sur les spectres des gaz dans leur rapports avec la constitution du Soleil, des étoiles et des nébuleuses.
> Franckland et Lockyer. Comptes Rendus, **68**, 1519.

Astrophysical observations made during the year 1882 at the Herény Observatory, Hungary.
> Gothard (E. von). Monthly Notices Astronomical Soc., **43**, 420–424; Math.-naturwiss. Ber. aus Ungarn, **1**, 207–9.

Spectroscopic observations at the Royal Observatory, Greenwich.
> Christie (W. H. M.). Nature, **28**, 136–9; **30**, 147–8.

Ditto.
> Airy (G. B.). Monthly Notices Astronom. Soc., **36**, 27–37; **37**, 22–36; Beiblätter, **11**, 95 (Abs.).

Beiträge zur Untersuchung der Sternbewegungen und der Lichtbewegung durch Spectral-Messungen.
> Homann (Hans). Inaugural.-Diss., Berlin, 1885; Beiblätter, **11** (1887), 146.

Spectrum analysis applied to the heavenly bodies.

> Huggins (W.). Rept. British Assoc., 1866; do., 1868; Chem. News, 19, 187.

Spectra of some of the fixed stars. [The first complete and accurate investigation of the stellar spectra.—*Roscoe.*]

> Huggins (W.) and Miller (W. A.). Phil. Trans. (1864), 413; Phil. Mag., June, 1866; Proc. Royal Soc., 12, 444; 13, 242.

Lecture on the physical and chemical constitution of the fixed stars and nebulæ.

> Huggins (W.). Chem. News, 11, 270.

Further observations of the Sun and of some of the stars and nebulæ; with an attempt to discover therefrom whether these bodies are moving towards or from the earth.

> Huggins (W.). Proc. Royal Soc., 16, 382.

Note on the heat of the stars.

> Huggins (W.). Proc. Royal Soc., 17, 309.

Spectren von Gestirne.

> Jahresber. d. Chemie, (1856) 140, (1862) 26 u. 27, (1863) 107, 108 u. 110, (1864) 115, (1865) 92, (1866) 78, (1867) 107, (1870) 176.

Remarques sur la note du père Secchi relative aux spectres prismatiques des corps célestes.

> Janssen. Comptes Rendus, 57, 215.

Nouvelle lettre annonçante la présence de la vapeur d'eau dans les planètes et les étoiles.

> Janssen. Comptes Rendus, 68, 876.

Sur quelques spectres stellaires remarquables par les caractères optiques de la vapeur d'eau.

> Janssen. Comptes Rendus, 68, 1545.

Les méthodes en astronomie physique.

> Janssen. Ann. du Bureau des Longitudes (1883), 779–812; Beiblätter, 7, 823–4 (Abs.).

Note sur divers points de physique céleste.

> Janssen. Comptes Rendus, 96, 527–529; Nature, 475 (Abs.).

Testimony of the spectroscope to the nebular hypothesis.

> Kirkwood (D.). Amer. Jour. Sci., (3) 2, 155; Phil. Mag., (4) 42, 399.

Astrophysiche Beobachtungen.

> Konkoly (N. von). Math.-naturwiss. Ber. aus Ungarn, 1, 126–127.

Untersuchungen über das Spectrum der Fixsterne.
>   Lamont. Jahrb. d. Sternwarte bei München (1868), 90.

The Mt. Whitney Expedition.
>   Langley (S. P.). Nature, 26, 814–817.

Note on the bright lines in the spectra of stars.
>   Lockyer (J. N.). Proc. Royal Soc., 27, 50.

Spectrum der Fixsterne.
>   Merz (S.). Ann. Phys. u. Chem., 117, 654.

A course of four lectures on spectrum analysis, with its applications to astronomy; delivered at the Royal Institution of Great Britain in May and June, 1867.
>   Miller (W. A.). Chem. News, 15, 259, 276; 16, 8, 20, 47, 71.

Spectrum analysis of the Sun and other heavenly bodies.
>   Miller (W. A.). Pop. Sci. Monthly, 8, 385.

Stars with peculiar spectra, discovered at the astronomical observatory of Harvard College.
>   Pickering (E. C.). Astronom. Nachr., 101, 73–74; Beiblätter, 6, 106 (Abs.).

The spectroscope in astronomical observation.
>   Proctor (R. A.). Pop. Sci. Rev., 8, 141.

The measurement of stellar spectra.
>   Rutherfurd (L. M.). Amer. Jour. Sci., (3) 35, 71.

Sur les spectres prismatiques des corps célestes.
>   Secchi (A.). Comptes Rendus, 57, 71.
>   Remarques par M. Janssen, do., 215.

Analyse spectrale de la lumière de quelques étoiles.
>   Secchi (A.). Comptes Rendus, 63, 824, 864.

Nouvelles recherches sur l'analyse de la lumière spectrale des étoiles.
>   Secchi (A.). Comptes Rendus, 63, 621.

Sur les spectres de quelques étoiles.
>   Secchi (A.). Comptes Rendus, 64, 345.

Nouvelle note sur les spectres stellaires.
>   Secchi (A.). Comptes Rendus, 64, 774.

Note accompagnant la présentation d'un exemplaire de son mémoire "Sur les Spectres stellaires" imprimé dans les publications de la Societé des Quarante de Modène.

> Secchi (A.). Comptes Rendus, **65**, 562.

Note sur les spectres stellaires.

> Secchi (A.). Comptes Rendus, **67**, 878.

Étude spectrale des divers rayons du Soleil et rapprochements entre les spectres obtenus et ceux de certaines étoiles.

> Secchi (A.). Comptes Rendus, **68**, 959.

Note sur l'intervention probable des gaz composés dans les caractères spectroscopiques de la lumière de certaines étoiles ou de diverses régions du Soleil.

> Secchi (A.). Comptes Rendus, **68**, 1086.

Nouvelles remarques sur les spectres fournis par divers types d'étoiles.

> Secchi (A.). Comptes Rendus, **71**, 252; Ann. Phys. u. Chem., **131**, 156.

Les spectres stellaires.

> Secchi (A.). Comptes Rendus, **75**, 655.

Spettri prismatici delle Stelle fisse.

> Secchi (A.). Atti della Soc. Ital., Roma, 1868.

Stellar Spectrometry.

> Secchi (A.). Chemical News, **18**, 168.

Bright lines in stellar spectra.

> Sherman. Amer. Jour. Sci., (8) **30**, 378, 475; note by Maunder (E. W.), Monthly Notices, **46** (1885), 282-4; reply to note, do., **47** (1886), 14.

Colour in practical astronomy, spectroscopically examined.

> Smyth (Piazzi). Trans. Royal Soc. Edinburgh, **28**, 779-843; Beiblätter, **4**, 548.

Physical constitution of the Sun and stars.

> Stoney (G. J.). Proc. Royal Soc., **16**, 25; **17**, 1.

Spectroscopic observations with the great Melbourne telescope.

> Sueur (A. Le). Proc. Royal Soc., **18**, 242.

Spectroscopic observations of various stars.

> Sueur (A. Le. Proc. Royal Soc., **19**, 18.

Ueber die Spectra der weissen Fixsterne.

    Vogel (H. V.). Monatsber. Berliner Akad. (1880), 192–198; Beiblätter,
    4, 786 (Abs.); Photographic News, Feb. 20, 1880; Nature, 21, 410.

Einige spectralanalytische Untersuchungen an Sternen, ausgeführt mit
dem grossen Refractor der Wiener Sternwarte.

    Vogel (H. W.). Sitzungsber. d. Wiener Akad., 88 II, 791–815; Bei-
    blätter, 8, 508–511 (Abs.).

Spectroscopie stellaire.

    Wolf et Rayet. Comptes Rendus, 65, 292.

Analyse spectrale de la lumière de quelques étoiles.

    Wolf. Comptes Rendus, 68, 1470.

Ursache der ungleichen Intensität der dunklen Linien im Spectrum der
Sonne und der Fixsterne.

    Zöllner (F.). Ann. Phys. u. Chem., 141, 873.

b, COMETS.

1, *Spectra of Comets in general.*

La matière radiante et les comètes.

    Begouen. Revue scientifique, 30, 297.

Remarques sur la lumière propre des comètes.

    Berthelot. Ann. Chim. et Phys., (5) 27, 282–3; Jour. Chem. Soc.,
    44, 261 (Abs.).

Comets; their composition, purpose and effect upon the earth.

    Boss (L.). Observatory (1882), 215–221.

Sur l'analyse spectrale appliquée aux comètes.

    Faye. Comptes Rendus, 93, 861.

Sur les queues des comètes.

    Flammarion. Comptes Rendus, 93, 186.

On Comets.

    Huggins (W.). Proc. Royal Institution, 10, 1–11; Ann. Chim. et
    Phys., (5) 27, 408–425.

Ueber die chemische Constitution der Cometen, verglichen mit der der
Meteore.

    Konkoly (N. von). Math.-naturwiss. Ber. aus Ungarn, 1, 185–189.

Observations sur la réfraction cométaire.

    Meyer (W.). Arch. de Genève, (3) 8, 526–535; Beiblätter, 7, 141–142
    (Abs.); Jour. de Phys., (2) 2, 387–8.

Sur la polarization de la lumière des comètes.
> Prazmowski. Comptes Rendus, **93**, 262.

Sur la lumière des comètes.
> Respighi. Comptes Rendus, **93**, 439–440 ; Phil. Mag., (5) **12**, 800–807 ;
> Beiblätter, **5**, 745 (Abs.).

Observations sur le spectre des comètes.
> Secchi (A.). Comptes Rendus, **78**, 1467.

Cometary Theory.
> Tyndall (J.). Phil. Mag., (4) **37**, 241.

Ueber die Spectra der Cometen.
> Vogel (H.). Astronom. Nachr., **80**, 183–188 ; Ann. Phys. u. Chem.,
> **149**, 400–408 ; Nature, **9**, 198.

2, *Particular Comets.*

(In the order of their last known dates.)

Comet c, 1859 (*Donati's*).

c, 1859, Donati's Comet. Comparaison du spectre produit par la lumière
de la comète de Donati et par celle d'Arcturus.
> Porro. Comptes Rendus, **47**, 873.

Comet a, 1866.

Spectrum of Comet *a*, 1866.
> Huggins (W.). Proc. Royal Soc., **15**, 5.

Comet b, 1867.

Spectrum of Comet *b*, 1867.
> Huggins (W.). Monthly Notices Astronom. Soc., **17**, 288.

Comet b, 1868.

Spectrum of Comet *b*, 1868.
> Huggins (W.). Proc. Royal Soc., **16**, 481.

Comet a, 1871.

Spectrum of Comet *a*, 1871.
> Huggins (W.). Chem. News, **23**, 265.

Comet c, 1873.

Spectre de la comète c, 1873.
> Wolf (C.) et Rayet (G.). Comptes Rendus, **77**, 529.

Comet d, 1873.

Spectre de la comète d, 1873.
> Rayet (G.) et André. Comptes Rendus, **77**, 564.

Comet c, 1874 (Coggia's).

Observations spectroscopiques de la queue de la comète de Coggia.
> Barthélemy (A.). Comptes Rendus, **79**, 313, 578.

Spectrum of Coggia's Comet.
> Huggins (W.). Proc. Royal Soc., **23**, 154–159.

Coggia's Comet, its physical condition and structure. Physical theory
of comets.
> Norton (W. A.). Amer. Jour. Sci., (3) **15**, 161–77.

Note sur le spectre de la comète de Coggia (c, 1874).
> Rayet (G.). Comptes Rendus, **78**, 1650–2; Amer. Jour. Sci., (3) **8**,
> 156 (Abs.).

Spectre de la comète de Coggia.
> Secchi (A.). Comptes Rendus, **79**, 20, 284.

Observations spectroscopiques sur la comète de Coggia.
> Wolf et Rayet. Comptes Rendus, **79**, 870–1.

Comet b, 1877 (Winnecke's).

On the spectrum of Comet b, 1877 (Winnecke's).
> Airy (G. B.). Monthly Notices Astronom. Soc., **37**, 469, 470.

The spectra of comets b and c, 1877.
> Lindsay (Lord). Monthly Notices Astronom. Soc., **37**, 480.

Spectre de la comète de Winnecke.
> Secchi (A.). Comptes Rendus, **66**, 1299, 1386.

Lumière de la comète de Winnecke.
> Wolf et Rayet. Comptes Rendus, **71**, 49.

Comet c, 1877 (Swift-Borelly).

On the spectra of comets b and c, 1877.
> Lindsay (Lord). Monthly Notices Astronom. Soc., **37**, 480.

Observations du spectre de la comète Borelly.
> Secchi (A.). Comptes Rendus, **84**, 427, 1289.

Ueber das Spectrum des von Borelly am 20, August entdeckten Cometen, sowie über das des hellen von Henry am 23 August aufgefundenen Cometen.
> Vogel (H.). Astronom. Nachr. **82**, 217-20; Amer. Jour. Sci., (8) **6**, 398 (Abs.).

Observations des comètes b (Winnecke) et c (Swift-Borelly), 1877.
> Wolf. Comptes Rendus, **84**, 929-81, 1289-92.

Comet a, 1878 (Brorsen's).

Spectrum of Brorsen's Comet, observed at Greenwich.
> Airy (G. B.). Monthly Notices Astronomical Soc., **39**, 428-80.

Spectrum of Brorsen's Comet.
> Backhouse (T. W.). Nature, **20**, 28.

Spectrum des Brorsen'schen Cometen.
> Brédischin (T.). Astronom. Nachr., **95**, 15-16.

Spectrum of Brorsen's Comet.
> Christie (W. H. M.). Nature, **20**, 5, 75; Amer. Jour. Sci., (8) **17** 496-7.

Spectrum of Brorsen's Comet.
> Huggins (W.). Proc. Royal Soc., **16**, 386; Nature, **19**, 579.

Vorläufige Anzeige über das Spectrum des Brorsen'schen Cometen.
> Konkoly (N. von). Astronom. Nachr., **94**, 385-6; **95**, 193-6.

Observations of Brorsen's Comet.
> Lindsay (Lord). Monthly Notices Astronom. Soc., **39**, 430.

Spectre de la comète de Brorsen.
> Secchi (A.). Comptes Rendus, **66**, 881.

Spectrum of Brorsen's Comet.
> Watts (W. M.). Nature, **20**, 27-8, 94.

Spectrum of Brorsen's Comet.
> Young (C. A.). Amer. Jour. Sci., (8) **17**, 373-5; Nature, **19**, 559; Phil. Mag., (5) **8**, 178-9.

Comet d, 1879 (Palisa's).

Spectroscopische Beobachtung des Cometen Palisa.
> Konkoly (N. von). Astronom. Nachr., **96**, 39-42.

Observations of the spectrum of comet *d*, 1879.

 Lindsay (Lord). Monthly Notices Astronom. Soc., **40**, 22–5.

Comet *d*, 1880 (Hartwig's). Spectrum of.

 Christie (W. H. M.). Monthly Notices Astronom. Soc., **41**, 52–3;
 Nature, **22**, 557; Beiblätter, **5**, 129.

Comet *b*, 1881.

Observations of comet *b*, 1881.

 Backhouse (T. W.). Monthly Notices Astronom. Soc., **42**, 413–21.

Spectra of comets *b* and *c*, 1881.

 Capron (J. R.). Nature, **24**, 430–1.

Spectra of comets *b* and *c*, 1881.

 Greenwich Observatory Reports, Monthly Notices Astronom. Soc., **42**,
 14–19.

Note on the observations of comet *b*, 1881, made at the United States
Naval Observatory.

 Harkness (W.). Amer. Jour. Sci., (3) **22**, 137–9.

Spectroscopische Beobachtungen der Cometen *b* und *c*, 1881.

 Hasselberg (B.). Bull. Acad. St. Petersburg, **27**, 417–25.

Preliminary notes on the photographic spectrum of comet *b*, 1881.

 Huggins (W.). Proc. Royal Soc., **33**, 1; Chem. News, **44**, 183; Rept.
 British Assoc. (1881), 320; Comptes Rendus, **92**, 1488; **93**, 26.

Note sur la photographie de la comète *b*, 1881, obtenu à l'observatoire de
Meudon.

 Janssen (J.). Jour. de Phys., (2) **1**, 441–9.

Spectroscopische Beobachtungen der Cometen *b* und *c*, 1881, angestellt
in O'Gyalla, Ungarn.

 Konkoly (N. von). Naturforscher, **14**, 321, 328, 331.

Physical observations of comet *b*, 1881, made at Forrest Lodge, Mares-
field.

 Noble (W.). Monthly Notices Astronom. Soc., **42**, 47–49.

Spectrum of comet *b*, 1881.

 Seabroke (G. M.). Nature, **24**, 201, 431.

Observations spectroscopiques sur la comète *b*, 1881.

 Thollon (L.). Comptes Rendus, **93**, 37, 259, 388; Nature, **24**, 224.

Ueber die Spectra der Cometen *b* und *c*, 1881.

Vogel (H. C.). Astronom. Nach., **100**, 801-4; Beiblätter, **5**, 867 (Abs.).

Observations de la comète *b*, 1881.

Wolf (C.). Comptes Rendus, **93**, 86.

Spectroscopic observations upon the comet *b*, 1881.

Young (C. A.). Amer. J. Sci., (3) **22**, 185-7; Beiblätter, **5**, 663-4 (Abs.).

Comet c, 1881.

Note on the spectrum of comet *c*, 1881, as seen with a Browning's miniature spectroscope on the 4½ telescope.

Backhouse (T. W.). Monthly Notices Astronom. Soc., **42**, 43.

Note on photographs of the spectrum of the comet of June, 1881.

Draper (H.). Amer. Jour. Sci., (3) **22**, 184-5; Chem. News, **44**, 75-6; Mem. Spettr. ital., **10**, 150-1; Jour. de Phys., (2) **1**, 153 (Abs.).

Spectra of comets *b* and *c*, 1881.

Greenwich Observatory, Monthly Notices Astronom. Soc., **42**, 14-19.

Spectroscopische Beobachtungen der Cometen *b* und *c*, 1881.

Hasselberg (B.). Bull. Acad. St. Petersburg, **27**, 417-25.

Spectroscopische Beobachtungen der Cometen *b* und *c*, 1881, angestellt am astrophysikalischen Observatorium in O'Gyalla (Ungarn).

Konkoly (N. von). Naturforscher, **14**, 321, 328, 331.

Études spectroscopiques sur les comètes *b* et *c*, 1881.

Thollon (L.). Comptes Rendus, **93**, 883.

Ueber die Spectra der Cometen *b* und *c*, 1881.

Vogel (H. C.). Astronomische Nachr., **100**, 801-4; Beiblätter, **5**, 867.

Spectrum of Schaeberle's Comet.

Capron (J. R.). Nature, **24**, 430-1. (See also Tacchini, in Comptes Rendus, **93**, 261.)

Telbutt's Comet, origination of its proper light.

Smyth (C. Piazzi). Nature, **24**, 430.

Comet a, 1882 (Wells's).

Spectrum of comet *a*, 1882 (Wells's).

Backhouse (T. W.). Nature, **26**, 56; Beiblätter, **6**, 678.

Les vapeurs du sodium dans la comète de Wells.

> Bredichin (T.). Astronom. Nachr., **102**, 207; Beiblätter, **6**, 678 (Abs.).

Ueber das Spectrum des Cometen Wells.

> Dunér (N. C.). Astronom. Nachr., **102**, 159, 169; Monthly Notices Astronom. Soc., **42**, 412–13; Beiblätter, **6**, 678 (Abs.).

Spectroscopic observations of comet a, 1882 (Wells).

> Greenwich Observatory Rept., Monthly Notices Astronom. Soc., **42**, 251, 410–12.

Ueber das Spectrum des Cometen a, 1882 (Wells).

> Hasselberg (B.). Astronom. Nachr., **102**, 259–64; Beiblätter, **6**, 744 (Abs.); Nature, **26**, 344 (Abs.).

On the photographic spectrum of comet a, 1882 (Wells).

> Huggins (W.). Proc. Royal Soc., **34**, 148–150; Nature, **26**, 179 (Abs.); Beiblätter, **6**, 679 (Abs.); Amer. Jour. Sci., (3) **24**, 402–3; Comptes Rendus, **94**, 1689–91.

Spectroscopische Beobachtungen des Cometen Wells, angestellt am astrophysikalischen Observatorium in O'Gyalla (Ungarn).

> Konkoly (N. von). Naturforscher, **15**, 245; Beiblätter, **6**, 678 (Abs.).

On the spectrum of comet a, 1882 (Wells), observed at the Royal Observatory of Greenwich.

> Maunder. Monthly Notices Astronom. Soc., **42**, 251, 410–12; Mem. Spettr. ital., **11**, 79.

Spettro della Cometa Wells osservato à Palermo.

> Riccò (A.). Mem. Spettr. ital., **11**, 76.

Cometa Wells, Spettro osservato all'Equatore Merz del R. Osservatorio del Collegio romano.

> Tacchini (R.). Mem. Spettr. ital., **11**, 77–8; Comptes Rendus, **94**, 1081–3.

Ueber das Spectrum des Cometen Wells.

> Vogel (H. C.). Astronom. Nachr., **102**, 159, 199–202; Beiblätter, **6**, 678 (Abs.).

Su di una particolaritá luminosa rimarcata a Palermo nella coda della cometa (Wells).

> Zona (T.). Mem. Spettr. ital., **11**, 76–7; Beiblätter, **6**, 679 (Abs.).

Comet b, 1882 (Cruls).

Analyse spectrale de la grande comète australe.

> Cruls. Comptes Rendus, **95**, 825.

Beobachtungen des grossen September Cometen, 1882, am astrophysical-ischen Observatorium zu Herény, Ungarn.

> Gothard (E. von). Astronom. Nachr., **103**, 877–80; Beiblätter, **7**, 116 (Abs.).

Spectroscopische Beobachtungen des grossen September Cometen, 1882 II.

> Gothard (E. von). Astronom. Nachr., **105**, 811–14.

Sur le déplacement des raies du sodium observé dans le spectre de la grande comète de 1882.

> Gouy et Thollon. Comptes Rendus, **96**, 371–2; Nature, **27**, 880 (Abs.); Amer. Jour. Sci., (8) **25**, 309; Beiblätter, **7**, 298 (Abs.).

Zur Spectroscopie des grossen September Cometen, 1882.

> Hasselberg (B.). Astronom. Nachr., **104**, 13–16; Beiblätter, **7**, 298 (Abs.).

Beobachtung des grossen September Cometen auf der Sternwarte in O'Gyalla (Ungarn).

> Konkoly (N. von). Astronom. Nachr., **104**, 45–8; Monthly Notices Astronom. Soc., **43**, 56–7; Beiblätter, **7**, 298.

Osservazioni astrofisiche della grande cometa di settembre, 1882.

> Riccò (A.). Astronom. Nachr., **103**, 281–4; Beiblätter, **7**, 28 (Abs.).

Osservazioni spettroscopiche della cometa Cruls fatte collo spettroscopio di Clesh applicato al refrattore di Om. 25 nell'Osservatorio di Palermo.

> Riccò (A.). Mem. Spettr. ital., **11**, Sept. 15–17.

Observations of the great comet b, 1882, made at Sydney Observatory.

> Russell (H. C.). Monthly Notices Astronom. Soc., **43**, 81.

Sur une comète observée à Nice.

> Thollon et Gouy. Comptes Rendus, **95**, 555–7; Beiblätter, **7**, 116 (Abs.).

Observations spectroscopiques sur la grande comète (Cruls).

> Thollon et Gouy. Comptes Rendus, **95**, 712–14; Nature, **27**, 24 (Abs.); Beiblätter, **7**, 28–9 (Abs.).

Sur le déplacement des raies du sodium observé dans le spectre de la grande comète de 1882.

> Thollon et Gouy. Comptes Rendus, **96**, 371.

Beobachtungen des grossen September Cometen, 1882.

> Vogel (H. C.). Astronom. Nachr., **103**, 279–282; Beiblätter, **7**, 28 (Abs.).
> (See also Tacchini, in Comptes Rendus, **93**, 261.)

Comet *a*, 1883 (Brooks-Swift). Beobachtung des Cometen *a*, 1883 (Brooks-Swift).

> Gothard (E. von). Astronom. Nachr., **105**, 185–6.

Spectroscopic Observations of Comet *a*, 1883 (Brooks-Swift).

> Konkoly (N. von). Monthly Notices Astronom. Soc., **43**, 328–9.

Finlay's Comet. Sulla spettro della cometa Finlay, Settembre, 1883.

> Hasselberg (B.). Mem. Spettr. ital., **11**, no. 11, 1–3; Beiblätter, **7**, 298 (Abs.).

Comet *a*, 1884 (*Pons-Brooks*).

Aspect de la comète Pons-Brooks, le 13 Janvier, 1884.

> Cruls (L.). Comptes Rendus, **98**, 898.

Spectroscopische Beobachtungen des Cometen *a*, 1884 (Pons-Brooks).

> Gothard (E. von). Astronom. Nachr., **109**, 99–106.

Spectrum of Comet *b*, 1883 (Pons-Brooks).

> Greenwich Observatory Rept., Monthly Notices Astronom. Soc., **44**, 62–3.

Spectroscopische Beobachtungen des Cometen Pons-Brooks.

> Hasselberg (B.). Astronom. Nachr., **108**, 55–56.

Vorläufige spectroscopische Beobachtung des Cometen Pons-Brooks.

> Konkoly (N. von). Astronom. Nachr., **107**, 41–2; Observatory, **6**, 333–4; Amer. Jour. Sci., (3) **27**, 76–7: Beiblätter, **8**, 33 (Abs.); Monthly Notices Astronom. Soc., **44**, 251–3.

Spectroscopische Beobachtungen des Cometen Pons-Brooks.

> Kövesligethy (R. v.). Astronom. Nachr., **108**, 169–174.

Observations spectroscopiques sur la comète Pons-Brooks.

> Perrotin. Comptes Rendus, **98**, 344.

Spectre de la comète Pons-Brooks, à l'observatoire de Bordeaux.

> Rayet (G.). Comptes Rendus, **97**, 1352; **98**, 348.

Sullo spettro della cometa Pons-Brooks.

> Riccò (A.). Mem. Spettr. ital., **13**, 39–40.

Observations spectroscopiques faites à Nice sur la comète Pons-Brooks.

> Thollon (L.). Comptes Rendus, **98**, 33; Beiblätter, **8**, 221.

Étude spectroscopique de la comète Pons-Brooks, faite au réflecteur de Om. 50 de l'Observatoire d'Alger.

> Trépied (C.). Comptes Rendus, **97**, 1540–1; Nature, **19**, 255 (Abs.).

Sur le spectre de la comète Pons-Brooks.
> Trépied (C.). Comptes Rendus, **98**, 82–3.

Variation singulière de la comète Pons-Brooks.
> Trépied (C). Comptes Rendus, **98**, 614.

Einige Beobachtungen über den Cometen Pons-Brooks, insbesondere über das Spectrum desselben.
> Vogel (H. C.). Astronom. Nachr., **108**, 21–6.

Observations of Comet Pons-Brooks.
> Young (C. A.). Astronom. Nachr., **108**, 305–8.

Encke's Comet.

Note on the spectrum of Encke's Comet.
> Huggins (W.). Proc. Royal Soc., **20**, 45; Comptes Rendus, **73**, 1297–1301.

Sur le spectre de la comète Encke.
> Tacchini (P.). Comptes Rendus, **93**, 949; Beiblätter, **6**, 106.

Spectre de la comète de Tempel.
> Secchi (A.). Comptes Rendus, **62**, 210.

Spectrum of comet *c*, 1886.
> Sherman. Amer. Jour. Sci., (3) **32**, 1

C, DISPLACEMENT OF STELLAR SPECTRA.

Effect of a star's rotation on its spectrum.
> Abney (W. de W.). Monthly Notices Astronom. Soc., **37**, 278.

Spectroscopic results for the motions of stars in the line of sight, obtained at the Royal Observatory, Greenwich.
> Airy (G. B.). Monthly Notices Astronom. Soc., **36**, 218; **38**, 493; **41**, 109; **42**, 230; **43**, 80; **44**, 89; **45**, 330; **46**, 126; **47**, 101.

Note on the displacement of lines in the spectra of stars.
> Christie (W. H. M.). Monthly Notices Astronom. Soc., **36**, 813–817.

Remarques sur le déplacement des raies du spectre par le mouvement du corps lumineux ou de l'observateur.
> Fizeau. Comptes Rendus, **69**, 743; **70**, 1062.

Sur un travail de M. l'abbé Spée concernant le déplacement des raies des spectres d'étoiles.
> Houzeau et Montigny. Bull. de l'Acad. de Belgique, **47**, 818–324.

Sur le déplacement des raies dans les spectres des étoiles produits par leur mouvement dans l'épace.

> Huggins (W.).　Comptes Rendus, **82**, 1291–1298 ; Phil. Mag., (5) **2**, 72–74.

On a method of finding the parallax of double stars, and on the displacement of the lines of the spectrum of a planet.

> Niven (C.).　Monthly Notices Astronom. Soc., **34**, 339–847.

Spectroscopic observations of the motions of stars in the line of sight, made at the Temple Observatory, Rugby.

> Seabroke (G. M.).　Monthly Notices Astronom. Soc., **39**, 450–453 ; **47** (1887), 93.

Sur le déplacement des raies dans les spectres des étoiles produit par leurs mouvements dans l'épace.

> Secchi (A.).　Comptes Rendus, **82**, 761, 812.

Nouvelles remarques sur question du déplacement des raies spectrales, dû au mouvement propre des astres.

> Secchi (A.).　Comptes Rendus, **83**, 117.

d, FIXED STARS.

1, In general.

Lecture on the physical and chemical constitution of the fixed stars and nebulæ.

> Huggins (W.).　Chem. News, **11**, 270.

Spectra of some of the fixed stars.

> Huggins (W.) and Miller (W. A.).　Phil. Trans. (1864), 418 ; Phil. Mag., June, 1866 ; Proc. Royal Soc., **12**, 444 ; **13**, 242.

Untersuchungen über das Spectrum der Fixsterne.

> Lamont.　Jahrbuch d. Sternwarte bei München (1868), 90.

Spectrum der Fixsterne.

> Merz (S.).　Ann. Phys. u. Chem., **117**, 654.

Spettri prismatici delle stelle fisse.

> Secchi (A.).　Atti della Soc. Ital., Roma, 1868.

2, Particular fixed stars.

Spectrum of Novæ Andromedæ.

> Sherman.　Amer. Jour. Sci., (8) **30**, 878.

Observations of the spectrum of a new star in Andromeda at Greenwich·
> Maunder (E. W.). Monthly Notices Astronom. Soc., **46** (1885), 19–21.

Outburst in Andromeda.
> Perry (S. J.). Monthly Notices Astronom. Soc., **46** (1885–0), 22.

Note sur le spectre d'Antarès.
> Secchi (A.). Comptes Rendus, **69**, 163.

Spectrum of η Argo with bright lines.
> Sueur (A. Le). Nature, **1**, 517.

Spectroscopische Beobachtung von γ Cassiopeiæ.
> Konkoly (N. von). Astronom. Nachr., **107**, 61–2; Beiblätter, **8**, 221.

Beobachtungen der hellen Linien in dem Spectrum von γ Cassiopeiæ.
> Gothard (E. von). Astronom. Nachr., **106**, 298; **108**, 233; Beiblätter,
> **7**, 862 (Abs.).

Spectrum of a new star in Corona Borealis.
> Huggins (W.) and Miller (W. A.). Proc. Royal Soc., **15**, 146.

On the spectrum of the new star in Cygnus.
> Backhouse (J. W.). Monthly Notices Astronom. Soc., **39**, 34–87;
> Nature, **15**, 295–6.

The new star in Cygnus.
> Becquerel (E.). Monthly Notices Astronom. Soc., **37**, 200–202; Amer.
> Jour. Sci., (8) **13**, 895–97.

The new star in Cygnus.
> Copeland (R.). Astronom. Nachr., **89**, 37–40, 68; **90**, 351–2; Nature,
> **15**, 315–16; Amer. Jour. Sci., (8) **15**, 76–77.

Sur le spectre de l'étoile nouvelle de la constellation du Cygne.
> Cornu (A.). Comptes Rendus, **83**, 1172–1174; Nature, **15**, 158.

Spectrum of Nova Cygni.
> Nature, **16**, 400–403.

Étude spectroscopique de la nouvelle étoile signalée par M. Schmidt.
> Secchi (A.). Comptes Rendus, **84**, 107, 290.

Der neue Stern in Cygnus.
> Vogel (H.). Astronom. Nachr., **89**, 37–40, 68; **90**, 851; Nature, **15**,
> 315; Amer. Jour. Sci., (8) **15**, 76.

Spectrum of the star Ll 13412.
> Pickering (E. C.). Nature, **23**, 604; Beiblätter, **5**, 511 (Abs.).

6 T

Photographs of the spectra of α Lyra and of Venus.

>   Draper (H.). Amer. Jour. Sci., (3) **13**, 95; Nature, **15**, 218; Phil.
>   Mag., (5) **3**, 288.

Beobachtungen der hallen Linien in dem Spectrum von β Lyræ.

>   Gothard (E. von). Astronom. Nachr., **108**, 233.

Lettre accompagnant l'envoi d'une figure du spectre d'α d'Orion.

>   Secchi (A.). Comptes Rendus, **62**, 591; Monthly Notices Astronom.
>   Soc., **26**, 214.

Spectrum of the variable star α Orionis.

>   Huggins (W.) and Miller (W. A.). Monthly Notices Astronom. Soc.,
>   **26**, 215.

Sur le spectre de l'étoile α d'Orion.

>   Janssen (J.). Comptes Rendus, **57**, 1008.

Spectrum of a new star in Orion.

>   Copeland (R.). Monthly Notices, **46**, 109–114.
>   Note by Maunder, do., 284–6.

Observations on the spectrum of Nova Orionis at Greenwich.

>   Maunder (E. W.). Monthly Notices Astronom. Soc., **46** (1885–6), 114–
>   115.

Disappearance of ε Piscium at its occultation of Jan. 4, 1865, with con-
clusions as to the non-existence of a lunar atmosphere.

>   Huggins (W.). Monthly Notices, **25**, 60; Chem. News, **11**, 175.

Sur le spectre de Sirius.

>   Janssen (J.). Comptes Rendus, **57**, 1008.

Note sur les spectres des trois étoiles de Wolf.

>   Secchi (A.). Comptes Rendus, **69**, 39, 163, 1053.

Sur trois petites étoiles.

>   Wolf et Rayet. Comptes Rendus, August, 1867.

e, MEASUREMENTS OF STELLAR SPECTRA.

Measurements of stellar lines.

>   Airy (G. B.). Monthly Notices Astronom. Soc., **23**, 190.

Stellar spectrometry.

>   Report of the British Assoc., 1868.

Measurement of stellar spectra.

>   Rutherfurd (L. M.). Amer. Jour. Sci., **35**, 71.

Measurement of a few stellar lines.

>   Secchi (A.). Astronom. Nachr., 8. März, 1868.

f. SPECTRA OF METEORS.

Spectra of the meteors of November 13–14, 1866.
> Browning (J.). Phil. Mag., (4) **33**, 234.

Presence of lithium in meteorites.
> Bunsen. Phil. Mag., (4) **23**, 474.

Meteoric Arc Spectrum.
> Capron (J. R.). Photographed Spectra, London, 1877, p. **82**, 88.

Spectra of shooting stars.
> Herschel (A. S.). Nature, **9**, 142–3.

Progress of meteor spectroscopy.
> Herschel (A. S.). Nature, **24**, 507–8; Beiblätter, **5**, 871.

Spectroscopische Beobachtungen der Meteorite.
> Konkoly (N. von). Astronom. Nachr., **95**, 283–6; Monthly Notices Astronom. Soc., **33**, 575–6; Nature, **20**, 521–2 (Abs.).

Ueber die chemische Constitution der Planeten verglichen mit der der Meteore.
> Konkoly (N. von). Math.-naturwiss. Ber. aus Ungarn, **1**, 135–9.

A catalogue of observations of luminous meteors,
> by Baden Powell from 1848 till 1859, by Glaisher till 1867, and by others till 1882; all in the Reports of the British Assoc. for those years.

Note sur les spectres stellaires, et sur les étoiles filantes.
> Secchi (A.). Comptes Rendus, **65**, 979; **75**, 606–618.

Sur les diverses circonstances de l'apparition d'un bolide aux environs de Rome et sur les spectres stellaires.
> Secchi (A.). Comptes Rendus, **75**, 655–9.

L'existence d'essaines d'étoiles filantes à proximité du globe terrestre.
> Silbermann (J.). Comptes Rendus, **74**, 553–7, 638–642.

Spectroscopic examination of gases from meteoric iron.
> Wright (A. W.). Amer. Jour. Sci., (3) **9**, 294–302; Jour. Chem. Soc. (1876), **1**, 27–8 (Abs.).

Preliminary note on an examination of gases of the meteorite of Feb. 12, 1875.
> Wright (A. W.). Amer. Jour. Sci., (3) **9**, 459–60; Jour. Chem. Soc. (1876), **1**, 352 (Abs.).

g, NEBULÆ.

1, *In general.*

Recherches sur l'intensité relative des raies spectrales des nébuleuses.

Fiévez (O.). Bull. de l'Acad. de Belgique, (2) **49**, 107–113; Phil. Mag., (5) **9**, 309–312; Beiblätter, **4**, 461–2.

Recherches sur les spectres des gaz dans leurs rapports avec la constitution du Soleil, des étoiles et des nébuleuses.

Franckland et Lockyer. Comptes Rendus, **68**, 1519.

Spectra of the nebulæ.

Huggins (W.). Phil. Trans. (1864), 437.

Further observations on the spectra of some of the nebulæ.

Huggins (W.). Phil. Trans. (1866), 381–387; Proc. Royal Soc., **15**, 17.

On the motions of some of the nebulæ towards or from the Earth.

Huggins (W.). Proc. Royal Soc., **22**, 251–4; Amer. Jour. Sci., (3) **8**, 75–77; Phil. Mag., (4) **48**, 471–4.

Note on the bright lines in the spectra of stars and nebulæ.

Lockyer (J. N.). Proc. Royal Soc., **27**, 50.

New planetary nebulæ.

Pickering (E. O.). Amer. Jour. Sci., (3) **20**, 303–305; Beiblätter, **5**, 180 (Abs.).

Spettro di alcune nebulose.

Secchi (A.). Naturforscher (Berliner), **1**, 279; **2**, 279, 356; Mem. Spettr. ital.; **1**, 38.

2, *Spectra of particular nebulæ.*

Nebula of Argo.

Le Sueur. Proc. Royal Soc., **18**, 245.

The nebula in Cygnus.

Winnecke. Monthly Notices Astronom. Soc., **40**, 92.

On the inferences to be drawn from the appearance of bright lines in the spectra of irresolvable nebulæ.

Huggins (W.). Proc. Royal Soc., **26**, 179–181.

On a cause for the appearance of bright lines in the spectra of irresolvable star-clusters.

Stone (E. J.). Proc. Royal Soc., **26**, 156–7, 517–19; Monthly Notices Astronom. Soc., **38**, 106–8.

On photographs of the nebula in Orion and of its spectrum.

> Draper (H.). Amer. Jour. Sci., (3) **23**, 889; Monthly Notices Astronom. Soc., **42**, 867–8; Nature, **26**, .88; Comptes Rendus, **94**, 1243.

Spectrum of the Great Nebula in the Sword-Handle of Orion.

> Huggins (W.). Proc. Royal Soc., **14**, 39.

On the spectrum of the Great Nebula in Orion, and on the motions of some stars towards or from the earth.

> Huggins (W.). Proc. Royal Soc., **20**, 879–894; Phil. Mag., (4) **45**, 133–147; Nature, **6**, 231–235; Amer. Jour. Sci., (3) **5**, 75–78; Monthly Notices Astronom. Soc., **32**, 859–862; Comptes Rendus, **94**, 685.

Photographic spectrum of the Great Nebula in Orion.

> Huggins (W.). Nature, **25**, 489; Ann. Chim. et Phys., (5) **28**, 282; Proc. Royal Soc., **33**, 425; Amer. Jour. Sci., (3) **23**, 355–6.

Lumière spectrale de la nébuleuse d'Orion.

> Secchi (A.). Comptes Rendus, **60**, 543.

Observations of the Nebula of Orion, made with the great Melbourne Telescope.

> Sueur (A. Le). Proc. Royal Soc., **18**, 242.

New planetary nebulæ.

> Pickering (E. C.). Amer. Jour. Sci., (3) **20**, 803–5; Beiblätter, **5**, 130 (Abs.).

Neue Linien im Spectrum planetischer Nebel.

> Zöllner (F.). Ann. Phys. u. Chem., **144**, 451.

Spectra of southern nebulæ.

> Herschel (Lieut. John). Proc. Royal Soc., **16**, 416, 417, 451; **17**, 58 61, 303.

Note on the Rev. T. W. Webb's new nebula.

> Lindsay (Lord). Monthly Notices Astronom. Soc., **40**, 91; Beiblätter, **4**, 614 (Abs.).

Ueber das Spectrum des von Webb entdeckten Nebels im Schwan.

> Vogel (H. C.). Astronom. Nachr., **96**, 287; Beiblätter, **4**, 468 (Abs.); Monthly Notices Astronom. Soc., **40**, 294.

h, PHOTOGRAPHY OF STELLAR SPECTRA.

Researches upon the photography of stellar and planetary spectra.

> Draper (H.). Proc. Amer. Acad., n. s. **11**, 231–261; Amer. Jour. Sci., (3) **18**, 419–425; Nature, **21**, 83–85; Beiblätter, **4**, 874.

Note on the photographic spectra of stars.

> Huggins (W.). Proc. Royal Soc., **25**, 445; **30**, 20; Nature, **21**, 269–270; Phil. Trans., **171**, 669–690; Beiblätter, 467–468 (Abs.).

Note préliminaire sur les photographies des spectres stellaires.

> Huggins (W.). Comptes Rendus, **83**, 1229.

Sur les spectres photographiques des étoiles.

> Huggins (W.). Comptes Rendus, **90**, 70–73; Amer. Jour. Sci., (3) **19**, 817.

Investigations in stellar photography.

> Pickering (E. C.). Memoirs Amer. Acad., **11** (1886), 179–226; Beiblätter, **11** (1887), 115 (Abs.).

Report on the present state of celestial photography in England.

> Rue (Warren de la). Rep'ts British Assoc. for 1859 and 1861.

Études astrophotographiques.

> Zenger (C. V.). Comptes Rendus, **97**, 552–555; Beiblätter, **7**, 860–862 (Abs.).

i, SPECTRA OF PLANETS.

1, *In general.*

On some points connected with the chemical constituents of the solar system.

> Gladstone (J. H.). Phil. Mag., (5) **4**, 879–385; Jour. Chem. Soc., **34**, 189 (Abs.).

Ueber die chemische Constitution der Planeten verglichen mit der der Meteore.

> Konkoly (N. von). Math.-naturwiss. Ber. aus Ungarn, **1**, 185–189.

On the displacement of the lines of the spectrum of a planet.

> Niven (C.). Monthly Notices Astronom. Soc., **34**, 839–847.

Sur les raies atmosphériques des planètes.

> Secchi (A.). Comptes Rendus, **59**, 182.

Untersuchungen über die Spectra der Planeten.

> Vogel (H. C.). Ann. Phys. u. Chem., **158**, 461–472.

2, *Spectra of particular planets.*

On a photograph of Jupiter's spectrum showing evidence of intrinsic light from that planet.

> Draper (H.). Monthly Notices Astronom. Soc., **40**, 433–435; Amer. Jour. Sci., (3) **20**, 118–120.

Note on the spectrum of the red spot on Jupiter.

> Lindsay (Lord). Monthly Notices Astronom. Soc., **40**, 87–88; Bei-blätter, **4**, 614 (Abs.).

Observation du spectre de Jupiter.

> Secchi (A.). Comptes Rendus, **59**, 809.

Spectroscopic observations of Jupiter, made with the great Melbourne telescope.

> Sueur (A. Le). Proc. Royal Soc., **18**, 242.

Physical observations of Mars.

> Airy (G. B.). Monthly Notices Astronom. Soc., **38**, 34–38.

Spectrum of Mars.

> Huggins (W.). Monthly Notices Astronom. Soc., **27**, 178; Jour. Franklin Inst., **84**, 261.

Note on the spectrum of the eclipsed Moon.

> Noble (W.). Monthly Notices Astronom. Soc., **38**, 34.

Sur l'application de l'analyse spectrale à la question de l'atmosphère lunaire.

> Janssen (J.). Comptes Rendus, **56**, 962.

Lettre sur le spectre de la planète Neptune et sur quelques faits d'analyse spectrale.

> Secchi (A.). Comptes Rendus, **69**, 1050.

Raies du spectre du planète Saturne.

> Secchi (A.). Comptes Rendus, **60**, 1167; Phil. Mag., (4) **30**, 78.

Spectrum of Uranus.

> Huggins (W.). Chem. News, **23**, 265; Proc. Royal Soc., **19**, 488–491; Phil. Mag., (4) **42**, 223–226; Nature, **4**, 88; Amer. Jour. Sci., (3) **2**, 188.

Résultats fournis par l'analyse spectrale de la lumière d'Uranus.

> Secchi (A.). Comptes Rendus, **68**, 761.

The Transit of Venus.

> Cacciatore. Nature, **27**, 180.

Osservazioni del passagio di Venere sul disco solare fatte in Italia nel 6 Dicembre 1882.

> Crova (A.). Mem. Spettr. ital., **11**, Dic. 1–23; Beiblätter, **7**, 375 (Abs.).

Photographs of the spectrum of Venus, Dec., 1876.

> Draper (H.).   Nature, **15**, 218; Amer. Jour. Sci., (3) **13**, 95; Phil.
> Mag., (5) **3**, 238.

Observations of the transit of Venus, Dec. 6, 1882, made at Mells, ten
miles south of Bath.

> Horner (Maurer):   Mon. Not. Astronom. Soc., **43**, 276.

Note sur l'observation du passage de la planète Vénus sur le Soleil.

> Janssen (J.).   Comptes Rendus, **96**, 288–92; Beiblätter, **7**, 875.

Observation of the transit of Venus, Dec. 6, 1882, made at the Allegheny
Observatory.

> Langley (S. P.).   Mon. Not. Astronom. Soc., **41**, 71.

The spectroscope and the transit of Venus.

> Nature, **11**, 171; **27**, 156–157.

Nouveau moyen d'observer les éclipses et les passages de Vénus.

> Secchi (A.).   Comptes Rendus, **73**, 984.

Essai pendant une éclipse solaire, de la nouvelle méthode spectroscopique
proposée pour le prochain passage de Vénus.

> Secchi (A.).   Comptes Rendus, **76**, 1327.

Observations du passage de Vénus à l'Observatoire royal du Collège
romain.

> Tacchini (P.).   Comptes Rendus, **95**, 1209–1211.

Observation du passage de Vénus, à Avila, Espagne.

> Thollon (L.).   Comptes Rendus, **95**, 1340–42.

Observations of the transit of Venus, Dec. 6, 1882, made at Princeton,
N. J., and South Hadley, Mass.

> Young (C. A.).   Amer. Jour. Sci., (3) **25**, 321–29.

j, SOLAR SPECTRUM.

1, *Solar spectrum in general.*

Influence of water in the atmosphere on the solar spectrum.

> Abney and Festing.   Proc. Royal Soc., **35**, 328–341; Beiblätter, **8**, 507
> (Abs.).

Lecture on solar physics.

> Abney (W. de W.).   Nature, **25**, 162–166, 187–191, 252–257.

Sunlight and skylight at high altitudes.
> Abney (W. de W.). Nature, **26**, 586; Beiblätter, **7**, 28 (Abs.); Jour.
> de Phys., (2) **3**, 47–48 (Abs.).

The solar spectrum, from λ 7150 to λ 10000.
> Abney (W. de W.). Phil. Trans. (1886), Part II, XIII.

Remarques sur quelques raies du spectre solaire.
> Angström (A. J.) Comptes Rendus, **63**, 647; Phil. Mag., (4) **23**,
> 76; **24**, 1.
> Remarques de M. Janssen. Comptes Rendus, **63**, 728.

Ueber die Fraunhofer'schen Linien im Sonnenspectrum.
> Angström (A. J.). Ann. Phys. u. Chem., **117**, 290.

Mémoire sur la constitution du spectre solaire.
> Becquerel (E.). Comptes Rendus, **14**, 901–3.

Des effets produits sur les corps par les rayons solaires.
> Becquerel (E.). Comptes Rendus, **17**, 882.

Constitution physique du Soleil.
> Boillot (A.). Comptes Rendus, **72**, 728.

Mémoire sur le spectre solaire.
> Brenta. Comptes Rendus, **11**, 766.

On the lines of the solar spectrum, and on those produced by the Earth's
atmosphere, and by the action of nitrous acid gas.
> Brewster (Sir D.). Phil. Mag., (3) **8**, 384; Proc. Royal Soc., **10**, 339
> (Abs.); Comptes Rendus, **30**, 578.

On the lines of the solar spectrum, with a map of the solar spectrum, giving the absorption lines of the Earth's atmosphere.
> Brewster and Gladstone. Phil. Trans. (1860), 149.

Catalogue of the oscillation-frequencies of solar rays.
> British Association Rep't for 1878.

Ueber die Fraunhofer'schen Linien im Sonnenspectrum, wie sie sich dem
unbewaffneten Auge zeigen.
> Broch (O. J.). Ann. Phys. u. Chem., Ergänzungsband, **3**, 311.

Constitution physique du Soleil.
> Chacornac. Comptes Rendus, **60**, 170.

Sur la distribution de l'intensité lumineuse et de l'intensité visuelle dans
le spectre solaire.
> Charpentier (Aug.). Comptes Rendus, **101** (1885), 182–188.

Spectral estimates of the Sun's distance.

> Chase (P. E.). Proc. Amer. Philosoph. Soc., 18, 227.

Sur le spectre normal du Soleil.

> Cornu (A.). Ann. de l'Ecole normale, (2) 3, 421–484; Arch. de Genève, (2) 52, 62–3 (Abs.).

Constitution du Soleil ; reponse à M. Janssen.

> Cornu (A.). Comptes Rendus, 73, 545.

Sur quelques conséquences de la constitution du spectre solaire.

> Cornu (A.). Comptes Rendus, 86, 530.

Considération sur les couleurs du spectre solaire.

> Dalet. Comptes Rendus, 28, 278.

Action du spectre solaire sur les sels haloïdes d'argent, accroissement de leur sensibilité dans certaines parties du spectre par l'adjonction de matières colorantes et autres.

> Eder (J. M.). Jour. de Phys., (2) 4 (1885), 185.

Constitution physique du Soleil.

> Faye. Comptes Rendus, 60, 89, 138, 168.

Résultats concernant la constitution physique du Soleil, obtenus soit par l'analyse spectrale, soit par l'étude mécanique de la rotation.

> Faye. Comptes Rendus. 68, 1189.

Analyse spectrale du Soleil.

> Faye. Comptes Rendus, 74, 921.

Sur la théorie physique du Soleil proposée par M. Vicaire.

> Faye. Comptes Rendus, 77, 298–301.

Sur la constitution physique et mécanique du Soleil.

> Faye. Comptes Rendus, 96, 855–861.

Sur une objection de M. Tacchini relative à la théorie du Soleil dans les "Memorie dei Spettroscopisti italiana."

> Faye. Comptes Rendus, 96, 811–816.

Réponse à une note do M. Thollon sur l'interprétation d'une phénomène de spectroscopie solaire.

> Faye. Comptes Rendus, 97, 779–782.

Studien über den Ursprung der Fraunhofer'schen Linien in ihrer Beziehung zur Constitution der Sonne.

> Fievez (Ch.). Bull. de l'Acad. de Belgique, (3) 12 (1886), 25–32; Beiblätter, 11 (1887), 94 (Abs.).

Rapport sur un Mémoire et plusieurs Notes de M. Janssen concernant l'analyse prismatique de la lumière solaire.
> Fizeau. Comptes Rendus, **58**, 795.

Spectroscopische Beobachtungen der Sonne.
> Franckland u. Lockyer. Ber. chem. Ges., **2**, 742.

On some points connected with the chemical constituents of the solar system.
> Gladstone (J. H.). Phil. Mag., (5) **4**, 879–885; Jour. Chem. Soc., **34**, 189 (Abs.).

Solar Chemistry.
> H. (G.). Nature, **24**, 581-2.

Spectrum of the Sun; spectra of the limb and centre of the Sun.
> Hastings (C. S.). Amer. Jour. Sci., **105**, 869; Nature, **8**, 77.

A theory of the constitution of the sun, founded upon spectroscopic obvations, original and other.
> Hastings (C. S.). Amer. Jour. Sci., (3) **21**, 33–44; Phil. Mag., (5) **11**, 91–103; Beiblätter, **5**, 588–592 (Abs.).

The Solar Spectrum.
> Herschel (J.). Nature, **6**, 454–455.

Action comparative des rayons solaires sous différentes latitudes.
> Herschel (J.). Comptes Rendus, **3**, 506.

Observations on the spectra of the Sun.
> Huggins (W.). Phil. Trans. (1868), 529.

Ueber die Längstreifen im Sonnenspectrum.
> Jahresber. d. Chemie, **1**, 198; **4**, 151; **5**, 125; **6**, 167.

Spectrum der Sonne.
> Jahresber. d. Chemie, **14**, 41, 43.

Fraunhofer Linien bei tiefem Stand der Sonne.
> Jahresber. d. Chemie, **15**, 26.

Constitution der Sonne.
> Jahresber. d. Chemie, **17**, 84.

Zusammenhang der Distanz der Spectrallinien mit den Dimensionem der Atome.
> Jahresber. d. Chemie, **19**, 78.

Sonnenspectrum.
> Jahresber. d. Chemie, **25**, 147.

Objective Darstellung des Sonnenspectrums.
>   Jahresber. d. Chemie, **29**, 158.

Lettre à M. Dumas sur les résultats des observations spectroscopiques
concernant la constitution du Soleil.
>   Janssen (J.).   Comptes Rendus, **68**, 312.

Constitution du Soleil.
>   Janssen (J.).   Comptes Rendus, **73**, 432–6.

Sur ce qu'ont jusqu'à ce jour d'incomplet les résultats fournis par l'analyse
spectrale pour nous faire connaître la constitution du Soleil.
>   Janssen (J.).   Comptes Rendus, **73**, 798.

Réponse à la note de M. Tacchini inserée au dernier "Comptes Rendus,"
séance du 14 Mai 1877.
>   Janssen (J.).   Comptes Rendus, **84**, 1182.

Notice sur les progrès récents de la physique solaire.
>   Janssen (J.).   Ann. du Bureau des Longitudes (1879), 623–685; Bei-
>   blätter, **4**, 277 (Abs.).

Die Chemie des Himmels.
>   Janssen (J.).   Archiv. f. Pharmacie (1875), 51.

Reply to Angström's observations on the solar lines.
>   Janssen (J.).   Phil. Mag., (4) **23**, 78.

Objective Darstellung des Sonnenspectrums.
>   Kessler (F.).   Ber. chem. Ges., **9**, 577.

Sur la loi de Stokes.
>   Lamansky (S.).   Comptes Rendus, **88**, 1192.

In feuchter Luft sind die Streifen des Sonnenspectrums breiter.
>   Lamansky (S.).   Ann. Phys. u. Chem., **146**, 208–221.

The solar atmosphere, an introduction to an account of researches made
at the Alleghany Observatory.
>   Langley (S. P.).   Amer. Jour. Sci., (3) **10**, 489–497.

A proposed new method in solar spectrum analysis.
>   Langley (S. P.).   Amer. Jour. Sci., (3) **14**, 140–146; Beiblätter, **1**, 621
>   (Abs.).

Solar spectrum at high altitudes.
>   Langley (S. P.).   Amer. Jour. Sci., (3) **24**, 393.

Observations du spectre solaire.

> Langley (S. P.). Comptes Rendus, **95**, 482–487; Jour. Chem. Soc., **44**, 137 (Abs.).

Procédé pour obtenir la récomposition de la lumière du spectre solaire.

> Lavaud de Lestrade. Comptes Rendus, **96**, 61.

On recent discoveries in solar physics made by means of the spectroscope.

> Lockyer (J. N.). Phil. Mag., (4) **38**, 142.

Spectroscopic Observations of the Sun.

> Lockyer (J. N.). Proc. Royal Soc., **15**, 256; **17**, 91, 128, 131, 350, 415, 506; **18**, 74; Ber. chem. Ges., **2**, 742; **3**, 578; Nature, **3**, 34.

Researches in spectrum analysis in connection with the spectrum of the sun, No. I.

> Lockyer (J. N.). Proc. Royal Soc., ·**21**, 88; Phil. Trans., **163**, 253–275; Amer. Jour. Sci., (3) **5**, 236–7 (Abs.).

Ditto, No. II.

> Lockyer (J. N.). Proc. Royal Soc., **21**, 285; Phil. Trans., **163**, 639–658; Jour. Chem. Soc., (2) **11**, 994–995 (Abs.); Phil. Mag., (4) **46**, 407–410 (Abs.); Ber. chem. Ges., **6**, 978 (Abs.).

Ditto, No. III.

> Lockyer (J. N.). Proc. Royal Soc., **21**, 508–514 (Abs.); Phil. Trans., **164**, 479–494; Phil. Mag., (4) **47**, 884–890.

Ditto, No. IV.

> Lockyer (J. N.). Proc. Royal Soc., **22**, 391; Phil. Trans., **164**, 805–818; Phil. Mag., (4) **49**, 326.

Ditto, No. V.

> Lockyer (J. N.). Proc. Royal Soc., **25**, 546.

Ditto, No. VI.

> Lockyer (J. N.). Proc. Royal Soc., **27**, 49, 279, 409.

Ditto, No. VII.

> Lockyer (J. N.). Proc. Royal Soc., **28**, 157–180; Amer. Jour. Sci., (3) **17**, 98–116; Beiblätter, **3**, 88–113; Nature, **19**, 197–201, 225–280; Ann. Chim. et Phys., (5) **16**, 107–144; Chem. News, **39**, 1–5, 11–16.

Note on a recent communication of Messrs. Liveing and Dewar.

> Lockyer (J. N.). Proc. Royal Soc., **29**, 45–7; Beiblätter, **3**, 710–711 (Abs.).

Recent researches in solar chemistry.

> Lockyer (J. N.). Proc. Physical Soc., **2**, 308–325; Phil. Mag., (5) **6**, 161–176; Beiblätter, **3**, 353–354 (Abs.).

Spectroscopic observations of the Sun.

> Lockyer (J. N.) and Seabroke (G. M.). Phil. Trans., 165, 577–586.

Lectures on solar physics; the chemistry of the Sun.

> Lockyer (J. N.). Nature, 24, 267–274, 296–301, 315–324, 365–370, 391–399.

Constitution physique du Soleil.

> Lockyer (J. N.). Comptes Rendus, 69, 121.

Réponse au Père Secchi.

> Lockyer (J. N.). Comptes Rendus, 69, 452.

Observations spectroscopiques du Soleil.

> Lockyer (J. N.). Comptes Rendus, 70, 1268.

Recherches expérimentales sur le spectre solaire.

> Lockyer (J. N.). Comptes Rendus, 75, 1816–19.

Recherches d'analyse spectrale au sujet du spectre solaire.

> Lockyer (J. N.). Comptes Rendus, 76, 1399.

Recherches sur les rapports d'analyse spectrale avec le spectre du Soleil.

> Lockyer (J. N.). Comptes Rendus, 88, 148–154; Jour. Chem. Soc., 36, 575–6 (Abs.).

Recherches sur l'analyse spectrale dans ses rapports avec le spectre solaire.

> Lockyer (J. N.). Ann. Chim. et Phys., (4) 29, 430.

On a new method of spectrum observation.

> Lockyer (J. N.). Amer. Jour. Sci., (3) 19, 303–311.

Solar spectroscopic observations.

> Maclear (J. P.). Nature, 6, 514.

Considérations sur le spectre solaire.

> Matthiessen. Comptes Rendus, 16, 917.

Spectrum of the Sun.

> Mellone (M.). Amer. Jour. Sci., 55, 1.

Spectrum analysis of the Sun.

> Miller (W. A.). Pop. Sci. Monthly, 8, 335.

Spectrum des durch Chlor gegangenen Sonnenlichtes.

> Morren. Ann. Phys. u. Chem., 137, 165.

On the physical constitution of the Sun.

> Norton (W. A.). Amer. Jour. Sci., (3) 1, 395–407; Phil. Mag., (4) 42, 55–67.

Spectrum of the Sun.
>   Olmstead (D.). Amer. Jour. Sci., (2) **48**, 187.

Les raies du spectre solaire.
>   Peslin. Comptes Rendus, **74**, 825.

Researches in circular solar spectra.
>   Pigott (G. West Royston). Proc. Royal Soc., **21**, 426.

Spectroscopic discoveries concerning the Sun.
>   Proctor (R. A.). Temple Bar, **25**, 281.

Réponse à une Note précédente du P. Secchi sur quelques particularités de la constitution du Soleil.
>   Respighi (L.). Comptes Rendus, **74**, 1887-90.

Réponse aux critiques présentées par le Père Secchi, à propos des observations faites sur quelques particularités de la constitution du Soleil.
>   Respighi (L.). Comptes Rendus, **75**, 134-138.

Sur la grandeur et les variations du diamètre solaire.
>   Respighi (L.). Comptes Rendus, **77**, 715-720, 774-778.

Sulla constituzione fisica del Sole.
>   Respighi (L.). R. Accad. dei Lincei, 10 April, 1871.

Osservazioni solari dirette et spettroscopiche esequite nel R. osservatorio di Palermo.
>   Riccò (A.). Mem. Spettr. ital., **9**, 25-86, 61-90, 161-189; **10**, 146-147.

Recherches sur les raies du spectre solaire et des différents spectres électriques.
>   Robiquet. Comptes Rendus, **49**, 606.

Solar spectrum in a hailstorm.
>   Romanes (C. H.). Nature, **25**, 507.

Italian spectroscopy.
>   Secchi (A.). Nature, **6**, 465-6.

Ueber den Einfluss der Atmosphäre auf die Linien des Spectrums.
>   Secchi (A.). Ann. Phys. u. Chem., **126**, 485.

Certain spectroscopic observations.
>   Secchi (A.). Chem. News, **27**, 244.

Notes sur les spectres solaires.
>   Secchi (A.). Comptes Rendus, **66**, 124, 898.

Existence d'une couche donnant un spectre continu entre la couche rose et le bord solaire.

> Secchi (A.). Comptes Rendus, **68**, 580.

Étude spectrale des taches solaires ; documents que peut fournir cette étude sur la constitution du Soleil.

> Secchi (A.). Comptes Rendus, **68**, 1082.

Remarques sur la lettre de M. Lockyer, du 2 Août.

> Secchi (A.). Comptes Rendus, **69**, 315.

Replique à la Note de M. Lockyer, du 16 Août.

> Secchi (A.). Comptes Rendus, **69**, 549.

Résultats de quelques observations spectrales du Soleil.

> Secchi (A.). Comptes Rendus, **70**, 908.

Note contenant une rectification numérique à sa dernière communication.

> Secchi (A.). Comptes Rendus, **70**, 1062.

Déplacement des raies observées dans le spectre solaire.

> Secchi (A.). Comptes Rendus, **70**, 1218.

Nouveaux observations concernant la constitution physique du Soleil.

> Secchi (A.). Comptes Rendus, **72**, 862.

Quelques nouveaux résultats d'analyse spectrale.

> Secchi (A.). Comptes Rendus, **74**, 598.

Sur quelques particularités de la constitution du Soleil.

> Secchi (A.). Comptes Rendus, **74**, 1087-91.

Réponse aux observations presentées par M. Respighi sur quelques particularités de la constitution du Soleil.

> Secchi (A.). Comptes Rendus, **74**, 1501-7.

Observations des variations des diamètres solaires.

> Secchi (A.). Comptes Rendus, **75**, 606-618.

Recherches spectroscopiques solaires.

> Secchi (A.). Comptes Rendus, **75**, 749.

Sur quelques observations spectroscopiques particulières.

> Secchi (A.). Comptes Rendus, **76**, 1052-56.

Nouvelles recherches sur la diamètre solaire.

> Secchi (A.). Comptes Rendus, **77**, 253-260.

Réponse à M. Respighi.

> Secchi (A.). Comptes Rendus, **77**, 904.

Note on a possible ultra-solar spectroscopic phenomenon.
>Smyth (C. Piazzi). Proc. Royal Soc., **20**, 136.

The visual, grating and glass-lens, solar spectrum, in 1884.
>Smyth (C. Piazzi). Trans. Roy. Soc. of Edinburgh, **32**, part III, 519–544, with plates; Monthly Notices Astronom. Soc., **47** (1887), 191-2.

On the Sun as a variable star.
>Stewart (B.). Lecture at the Royal Institution, April 12, 1867.

On the change of refrangibility of light; with a drawing of the fixed lines in the solar spectrum in the extreme violet, and in the invisible region beyond,
>Stokes (G. G.). Phil. Trans., 1852 II, 463.

Lecture on solar physics.
>Stokes (G. G.). Nature, **24**, 595–8, 618-18.

On the bearing of recent observations upon solar physics.
>Stoney. Phil. Mag., (4) **36**, 441.

Osservazioni solari dirette e spettroscopiche fatte a Palermo nel 1 trimestre del 1879, nel secondo trimestre del 1879, nel terzo e quarto trimestre del 1879, nel 1 trimestre del 1880, nel secondo trimestre del 1880, nel 3 trimestre del 1880, nel 4 trimestre del 1880, riassunto delle osservazioni, 1880,
>Tacchini (P.). Mem. Spettr. ital., **8**, 87–40, 52–54, 93–97, 102–104; **9**, 49–58, 105–110, 194-208; **10**, 5–11, 12; Comptes Rendus, **88**, 1131; **89**, 519.

Sull'andamento dell'attivitá solare del 1871 al 1878.
>Tacchini (P.). Mem. Spettr. ital., **8**, 65–72.

Nouvelles observations spectrales.
>Tacchini (P.). Comptes Rendus, **77**, 195-198.

Sur le magnésium dans le spectre solaire.
>Tacchini (P.). Comptes Rendus, **84**, 1450.

Résultats des observations solaires pendant le deuxième trimestre de 1878, et des observations pendant le troisième trimestre de 1878.
>Tacchini (P.). Comptes Rendus, **87**, 259, 1081.

Sur la cause des spectres fugitifs observés par M. Trouvelot sur la limbe solaire.
>Tacchini (P.). Comptes Rendus, **91**, 156-8.

7 T

Observations solaires faites à l'observatoire royal du Collège romain pendant le troisième, 1880.

> Tacchini (P.).   Comptes Rendus, **91**, 1053–4.

Observations solaires faites à l'Observatoire royal du Collège romain pendant le premier, le deuxième et le troisième trimestres de 1881.

> Tacchini (P.).   Comptes Rendus, **93**, 880; **94**, 880.

Comparaison entre le spectre normal du Soleil et celui de réfraction suivant l'échelle de Kirchhoff.

> Thalén (R.).   Ann. Chim. et Phys., (4) **18**, 211.

Déplacement des raies spectrales, dû au mouvement de rotation du Soleil.

> Thollon (L.).   Comptes Rendus, **88**, 169–171; Beiblätter, **3**, 855–6
> (Abs.); Jour. Chem. Soc., **36**, 574.

Observation faite sur un groupe de raies dans le spectre solaire.

> Thollon (L.).   Comptes Rendus, **91**, 868–70; Beiblätter, **4**, 790 (Abs.);
> Amer. Jour. Sci., (3) **20**, 430; Jour. Chem. Soc., **40**, 333.

Quelques phénomènes solaires observés à Nice.

> Thollon (L.).   Comptes Rendus, **91**, 487–92.

Études spectroscopiques faites sur le Soleil à l'Observatoire de Paris.

> Thollon (L.).   Comptes Rendus, **91**, 656–60.

Sur l'interprétation de quelques phénomènes de spectroscopie solaire.

> Thollon (L.).   Comptes Rendus, **97**, 747.

Études faites au sommet du Pic du Midi, en vue de l'établissement d'une station astronomique permanente.

> Thollon et Trépied.   Comptes Rendus, **97**, 884–886; Nature, **29**, 7–8;
> Beiblätter, **8**, 824 (Abs.).

Observations relatives à la réponse de M. Faye concernant divers phénomènes de spectroscopie solaire.

> Thollon (L.).   Comptes Rendus, **97**, 900.

Recherches sur la décomposition de l'acide carbonique dans le spectre solaire par les parties vertes des végétaux.

> Timiriasef (C.).   Ann. Chim. et Phys., (5) **12**, 355.

Spectres fugatifs observés près du limbe solaire.

> Trouvelot (L.).   Ann. Chim. et Phys., (5) **19**, 433–449; Beiblätter, **4**,
> 727 (Abs.).
> Note par M. Tacchini.   Comptes Rendus, **91**, 156–8.

Sur la constitution physique du Soleil; réponse aux critiques de M. Faye.

> Vicaire (E.).   Comptes Rendus, **75**, 527–31; **77**, 1491–95.

Vermehrung und Verdickung der Fraunhofer'schen Linien bei Sonnenuntergang.

Weiss (A.). Ann. Phys. u. Chem., 116, 191; Phil. Mag., (4) 24, 407.

Remarks on spectroscopic observations of the Sun, made at the Temple Observatory, Rugby School, in 1871-2-3.

Wilson (J. M.) and Seabroke (G. M.). Monthly Notices Astronom. Soc., 34, 26-29.

Application of the spectroscope to observations of the Sun.

Winlock (J.). Proc. Amer. Acad., 8, 330.

Note on the duplicity of the "1474" line in the solar spectrum.

Young (C. A.). Amer. Jour. Sci., (3) 11, 429-431.

Spectroscopic observations of the Sun.

Young (C. A.). Nature, 3, 84.

Spectroscopic Notes.

Young (C. A.). Amer. Jour. Sci , (3) 20, 353-8; (3) 26, 333; Nature, 23, 281; Chem. News, 20, 271; Beiblätter, 5, 287.

Anologia delle vibrazioni luminose e delle spettro solare, con 1 tav.

Zantedeschi (F.). Sitzungsber. Wiener Akad., 25, 145-165.

De mutationibus quae contingunt in spectro solari fixo elucabratio.

Zantedeschi (F.). Münchener Abhandlungen, 8, 99.

Ueber die Temperatur und die physische Beschaffenheit der Sonne.

Zöllner (F.). Der Naturforscher, 3, 93, 189, 233, 311; Ber. Sächs. Ges. Wiss., 25, 158-194; Phil. Mag., (4) 46, 290-304, 843-56.

2, Solar Absorption.

Sur la loi de répartition suivant l'altitude de la substance absorbant dans l'atmosphère.

Cornu (A.). Comptes Rendus, 90, 940-946; Beiblätter, 4, 727-8 (Abs.).

Sur l'intensité calorifiquo do la radiation solaire et son absorption par l'atmosphèro terrestre.

Crova (A.). Comptes Rendus, 81, 1205-7.

Sur la mesure de l'intensité des raies d'absorption et des raies obscures du spectre solaire.

Gouy. Comptes Rendus, 89, 1033-4; Beiblätter, 4, 369 (Abs.).

Absorption of solar rays by atmospheric ozone.

Hartley (W. N.). Jour. Chem. Soc., 39, 111-128; Ber. chem. Ges., 14, 1890 (Abs.).

The selective absorption of solar energy.

> Langley (S. P.). Amer. Jour. Sci., (3) **25**, 169–196; Ann. Phys. u.
> Chem., n. F. **19**, 226–244, 384–400; Phil. Mag., (5) **15**, 153–183;
> Ann. Chim. et Phys., (5) **29**, 497–542.

Observations of absorbing vapours upon the Sun.

> Trouvelot (E. L.). Monthly Notices Astronom. Soc., **39**, 374–379.

Spectral-photometrische Untersuchungen insbesondere zur Bestimmung
der Absorption der die Sonne umgebenden Gashülle.

> Vogel (H. C.). Monatsber. d. Berliner Akad. (1877), 104–142.

Ueber die Absorption der chemisch wirksamen Strahlen in der Atmosphäre der Sonne.

> Vogel (H. C.). Ber. Sächs. Ges. Wiss., **24**, 135–141; Ann. Phys. u.
> Chem., **148**, 161–168; Phil. Mag., (4) **45**, 345–350.
> Note by Schuster (A.). Phil. Mag., (4) **45**, 350.

3, *Solar Atmosphere.*

On hydrocarbons in the solar atmosphere.

> Abney (W. de W.). Rept. British Assoc. (1881), 524.

Mémoire sur l'atmosphère solaire.

> Angelot. Comptes Rendus, **68**, 245.

Atmospheric lines of the solar spectrum, with a map.

> Hennessey (J. B. N.). Phil. Trans., **165**, 157–160; Amer. Jour. Sci.,
> (3) **9**, 307.

Ursache der Spectren und Folgerungen über die Zustände der Sonnen-
atmosphäre.

> Jahresber. d. Chemie, **15**, 82.

Sur une atmosphère incandescente qui entoure la photosphère solaire.

> Janssen (J.). Comptes Rendus, **68**, 181.

Remarques à propos des résultats obtenus par M. Janssen et des connaissances précédemment acquises au sujet de l'atmosphère solaire.

> Leverrier. Comptes Rendus, **68**, 314.

Atmosphère du Soleil.

> Littrow. Comptes Rendus, **68**, 435.

Réfrangibilité de la raie jaune brilliante de l'atmosphère solaire.

> Rayet. Comptes Rendus, **68**, 820; Chem News, **19**, 158.

Spectre de l'atmosphère solaire.

> Rayet. Comptes Rendus, **68**, 1821; **71**, 301; **77**, 529; Ann. Chim. et
> Phys., (4) **24**, 5–80; Archiv. f. Pharmacie, **4**, 325–7.

Nouvelles observations sur l'atmosphère et les protubérances solaires.
Secchi (A.). Comptes Rendus, **68**, 1243.

Sur l'état actuel de l'atmosphère solaire.
Secchi (A.). Comptes Rendus, **84**, 1430-34.

Ueber den Einfluss der Atmosphäre auf die Linien des Spectrums.
Secchi (A.). Ann. Phys. u. Chem., **126**, 485.

Résultats des opérations faites en 1877 au bord du Soleil sur les raies b et 1474 k.
Tacchini. Comptes Rendus, **86**, 756.

Observation of absorbing vapours on the Sun.
Trouvelot. Monthly Notices Astronom. Soc., **39**, 874.

Spectral-photometrische Untersuchungen, insbesondere zur Bestimmung der Absorption der die Sonne umgebenden Gashülle.
Vogel (H. C.). Monatsber. d. Berliner Akad. (1877), 104-142.

Influence de la vapeur aqueuse visible dans l'atmosphère, et de la pluie sur le spectre solaire.
Zantedeschi. Comptes Rendus, **63**, 644.

4, *B lines in the solar spectrum.*

Measures of the Great B line in the spectrum of a high sun.
Smyth (C. Piazzi). Monthly Notices Astronom. Soc., **39**, 88-43.

Note on the Little b group of lines in the solar spectrum.
Smyth (C. Piazzi). Trans. Roy. Soc. Edinburgh, **32**, 87-44; Nature, **28**, 287 (Abs.); Amer. Jour. Sci., (3) **21**, 323.

Résultats des opérations faites en 1877, au bord du Soleil sur les raies b et 1474 k.
Tacchini. Comptes Rendus, **86**, 756.

Constitution et origine du groupe B du spectre solaire.
Thollon (L.). Jour. de Phys. **13**, 421; Nature, **30**, 520.

Mémoire sur la constitution et l'origine du groupe B du spectre solaire.
Thollon (L.). Bull. astronomique, 1883-4.
Note by Smyth (C. Piazzi). Nature, **30**, 585.

5, *Bright lines in the solar spectrum.*

On the existence of bright lines in the solar spectrum.
Christie (W. H. M.). Monthly Notices Astronom. Soc., **38**, 478-4.

On the coïncidence of the bright lines of the oxygen spectrum with bright lines in the solar spectrum.

> Draper (H.). Amer. Jour. Sci., (3) **18**, 262–76; Monthly Notices Astronom. Soc., **39**, 440–47; Beiblätter, **4**, 275 (Abs.).

Report to the Committee on Solar Physics on the basic lines common to Spots and Prominences.

> Lockyer (J. N.). Proc. Royal Soc., **29**, 247–65; Beiblätter, **4**, 45 (Abs.).

On a cause for the appearance of bright lines in the solar spectrum.

> Meldola (R.). Phil. Mag., (5) **6**, 50–61; Jour. Chem. Soc., **36**, 574; Amer. Jour. Sci., (3) **16**, 290–800; Beiblätter, **2**, 561–2 (Abs.).

Letter to the Superintendent of the U. S. Coast Survey, containing a catalogue of bright lines in the spectrum of the solar atmosphere.

> Young (C. A.). Amer. Jour. Sci., (3) **4**, 356–62; Nature, **7**, 17–20.

6, *Chemical effects of the solar spectrum.*

Sur l'action chimique des différents rayons du spectre solaire.

> Claudet. Comptes Rendus, **25**, 938.

On the chemical efficiency of sunlight.

> Dewar (J.). Phil. Mag., **44**, 807–811.

Wirkung der chemischen Strahlen verscniedener Theile der Sonnenscheibe.

> Jahresber. d. Chemie, **16**, 101.

Rayons violets qui renferment le maximum d'action chimique de toutes les couleurs du spectre solaire.

> Poey (A.). Comptes Rendus, **73**, 1288.

Expériences sur la transmission des rayons chimiques du spectre solaire à travers différents milieux.

> Somerville (Mrs.). Comptes Rendus, **3**, 478.

Beziehungen zwischen der chemischen Wirkung des Sonnenspectrums, der Absorption und anomalen Dispersion des Sonnenspectrums.

> Vogel (H.). Ber. chem. Ges., **7**, 976.

7, *Chromosphere and Corona.*

Spectre de la couronne.

> Blaserna (P.). Comptes Rendus, **74**, 879.

The comparative aggregate strength of the light from the red hydrogen stratum, and of that of the rest of the chromosphere.

Hammond (B. E.). Nature, 3, 487.

On the solar corona.

Harkness (W.). Bull. Philosoph. Soc. Washington, 3, 116–119; Be-.blätter, 5, 128.

Photographing the spectrum of the corona.

Huggins (W.). Nature, 27, 199.

The coronal atmosphere of the Sun.

Janssen (J.). Nature, 8, 127–9, 149–50.

Sur la photographie de la chromosphère.

Janssen (J.). Comptes Rendus, 91, 12; Beiblätter, 4, 615.

L'analyse spectrale de la lumière zodiacale et sur la couronne des éclipses.

Liais (E.). Comptes Rendus, 74, 262–4; Amer. Jour. Sci., (3) 3, 390–91.

Note on the unknown chromospheric substance of Young.

Liveing (G. D.) and Dewar (J.). Proc. Royal Soc., 28, 475–7; Beiblätter, 3, 709 (Abs.).

A new method of viewing the chromosphere.

Lockyer (J. N.) and Seabroke (G. M.). Proc. Royal Soc., 21, 105–107; Amer. Jour. Sci., (3) 5, 319 (Abs.); Comptes Rendus, 76, 863–5; Phil. Mag., (4) 45, 222–4.

Note on the existence of carbon in the coronal atmosphere of the Sun.

Lockyer (J. N.). Proc. Royal Soc., 27, 308; Jour. Chem. Soc., 38, 429 (Abs.).

Preliminary note on the substances which produce the chromospheric lines.

Lockyer (J. N.). Proc. Royal Soc., 28, 283–4; Nature, 19, 202; Amer. Jour. Sci., (3) 17, 250; (3) 18, 158; Beiblätter, 3, 420–422.

Discussion of " Young's List of Chromospheric Lines."

Lockyer (J. N.). Proc. Royal Soc., 28, 432–444; Beiblätter, 3, 420 (Abs.).

Photographie der Corona.

Lohse (O.). Astronom. Nachr., 104, 209–212; Beiblätter, 7, 291 (Abs.).

On the corona seen in total eclipses of the Sun.

Norton (W. A.). Amer. Jour. Sci., (3) 1, 5–15; Phil. Mag., (4) 41, 225–236.

Note on the chromosphere.

> Perry (S. J.). Monthly Notices Astronom. Soc., **43**, 426–7 ; Nature,
> **3**, 67.

Osservazioni spettroscopiche del Bordo e delle Protuberanze Solari.

> Respighi (L.). Roma, 1871.

La corona solare l'eclisse, 22 Dic. 1870.

> Ricca (V. S.). Palermo, 1871.

Osservazioni delle inversioni della coronale 1474 k, e delle b del magnesio
fatte nel Osservatorio di Palermo.

> Riccò (A.). Mem. Spettr. ital., **10**, 148–51.

Professor Young and the presence of ruthenium in the chromosphere.

> Roscoe (H. E.). Nature, **9**, 5.

On the spectrum of the corona.

> Sampson (W. T.). Amer. Jour. Sci., (8) **16**, 343–5 ; Beiblätter, **3**, 277
> (Abs.).

Résultats de quelques observations spectroscopiques des bords du Soleil.

> Secchi (A.). Comptes Rendus, **67**, 1018.

Note sur les spectres des trois étoiles de Wolf et sur l'analyse comparative
de la lumière du bord solaire et des taches.

> Secchi (A.). Comptes Rendus, **69**, 89.

Note sur la constitution de l'auréole solaire et sur quelques particularités
du tube de Geissler.

> Secchi (A.). Comptes Rendus, **70**, 27, 82.

Sur les relations qui existent, dans le Soleil, entre les facules, les protubérances et la couronne.

> Secchi (A.). Comptes Rendus, **72**, 829–832 ; **73**, 242–246, 593–599.

Hydrogène et la raie D₃ dans le spectre de la chromosphère solaire.

> Secchi (A.). Comptes Rendus, **73**, 1300.

Spectre de la chromosphère.

> Secchi (A.). Comptes Rendus, **74**, 305.

Observations de la chromosphère.

> Secchi (A.). Comptes Rendus, **75**, 606–618.

Magnésium dans la chromosphère du Soleil.

> Tacchini. Comptes Rendus, **75**, 23, 430 ; Phil. Mag., (4) **44**, 159–160,
> 479–80.

Présence du spectre du magnésium sur le bord entière du Soleil.

> Tacchini. Comptes Rendus, **76**, 1577; **77**, 606–9; **82**, 1385–7.

Observations on the Corona seen during the eclipse of Dec. 11 and 12, 1871.

> Winter (G. K.). Phil. Mag., (4) **43**, 191–4.

On the solar corona.

> Young (C. A.). Amer. Jour. Sci., (3) **1**, 311–373.

Note on the spectrum of the corona.

> Young (C. A.). Amer. Jour. Sci., (3) **2**, 53–55; Chem. News, **24**, 198–9.

Preliminary catalogue of the bright lines in the spectrum of the chromosphere.

> Young (C. A.). Amer. Jour. Sci., 8 **2**, 332–335; Phil. Mag., (4) **42**, 377–380; Nature, **5**, 312–313.

Spectrum of the corona of the Sun.

> Young (C. A.). Amer. Jour. Sci., (3) **2**, 53; Chem. News, **24**, 198.

Note on the chromosphere lines.

> Young (C. A.). Nature, **3**, 266–7.

Spectrum of the chromosphere.

> Young (C. A.). Nature, **5**, 312.

The corona line.

> Young (C. A.). Nature, **7**, 28.

Beobachtungen der Corona.

> Zöllner (F.). Der Naturforscher (Berlin), **2**, 167, 253, 379, 395; **3**, 91, 392; Les Mondes (Paris), **21**, 845, 602; **22**, 142; Nature, **1**, 15, 139, 146, 533, 543; **2**, 114, 164, 277; **3**, 163, 175, 262, 263, 278; Phil. Mag., (4) **38**, 281; **39**, 17; Monthly Notices Astronom. Soc., **30**, 193.

8, *The D group of lines in the solar spectrum.*

Monographie du groupe D dans le spectre solaire.

> Thollon. Jour. de Phys., (2) **3**, 5–11; Beiblätter, **8**, 647.

9, *Dark lines in the solar spectrum.*

Sur les raies sombres du spectre solaire et la constitution du Soleil.

> Cornu (A.). Comptes Rendus, **86**, 315.

Sur la distribution de la chaleur dans les régions obscures des spectres solaires.

> Desains (P.). Comptes Rendus, **95**, 433.

On the presence of dark lines in the solar spectrum which correspond closely to the lines of the spectrum of oxygen.

> Draper (J. C.). Amer. Jour. Sci., (3) 16, 256–65; Nature, 18, 654–7; Beiblätter, 3, 188 (Abs.); Jour. Chem. Soc., 36, 997.

Mesure de l'intensité de quelques raies obscures du spectre solaire.

> Gouy. Comptes Rendus, 91, 388; Jour. Chem. Soc., 40, 388 (Abs.); Beiblätter, 5, 46 (Abs.).

Dunkle Linien des Sonnenspectrums.

> Jahresber. d. Chemie, 16, 107, 110.

A method of examining refractive and dispersive powers by prismatic reflection.

> Wollaston (W. H.). Phil. Trans. (1802), 365.

Ursache der ungleichen Intensität der dunklen Linien im Spectrum der Sonne.

> Zöllner (F.). Ann. Phys. u. Chem., 141, 378.

10, *Displacement of the solar spectrum.*

Note on the displacement of the solar spectrum.

> Hennessey (J. H. N.). Proc. Royal Soc., 22, 219.

Observations on the displacement of lines in the solar spectrum caused by the Sun's rotation.

> Young (C. A.). Amer. Jour. Sci., (3) 12, 321–8.

11, *Eclipse Spectra.*

On the solar eclipse of Dec. 22, 1870, observed at Xeres, in Spain.

> Abbay (R.). Monthly Notices Astronom. Soc., 31, 60–62.

Observations on the total eclipse of the Sun of 1869.

> Abbe (C.). Amer. Jour. Sci., (3) 3, 264–267.

On the total solar eclipse of May 17, 1882.

> Abney (W. de W.) and Shuster (A.). Phil. Trans., 175, 253–271; Proc. Royal Soc., 35, 151 (Abs.); Beiblätter, 7, 896 (Abs.); Nature, 26, 465.

Eclisse totale del 22 Dic. 1870.

> Agnello (A.). Palermo, 1870.

On the results of the spectroscopic observations of the solar eclipse of July 29, 1878.

> Barker (G. F.). Amer. Jour. Sci., (3) 17, 121–5.

Observations sur un artifice semblable auquel ont songé en même temps M. Janssen dans l'Inde et M. Zantedeschi en Italie.

Beaumont (Élie de). Comptes Rendus, **68**, 314

The solar eclipse of July 29, 1878.

Draper (H.). Amer. Jour. Sci., (3) **16**, 227–30; Phil. Mag., (5) **6**, 818–320.

The Eclipse.

Draper (H.). Nature, **18**, 462–4.

Account of the expedition of the Jesuits from Manilla, eclipse of Aug. 18, 1868.

Faura (F.). Bull. meteorol. dell. Osservatorio del Collegio Romano, **7**, no. 12.

Suggestion relative à l'observation de l'éclipse de Soleil du 31 décembre 1861.

Faye. Comptes Rendus, **53**, 679.

Observations relatives à la coïncidence des méthodes employées séparément par M. Lockyer et par M. Janssen.

Faye. Comptes Rendus, **67**, 840.

Note sur une télégramme et sur une lettre de M. Janssen.

Faye. Comptes Rendus, **68**, 112.

Rapport au Bureau des Longitudes sur la prochaine éclipse du 6 mai 1883.

Fizeau, Cloué, Lewy et Janssen. Comptes Rendus, **95**, 881–885; Ann. du Bureau des Longitudes (1883), 818–820; Nature, **27**, 110–112.

Account of spectroscopic observations of the eclipse of the Sun, Aug. 18, 1868.

Haig (C. T.). Proc. Royal Soc., **17**, 74.

On the total eclipse of the Sun of Aug. 18, 1868.

Herschel (Alex.). Proc. Royal Institution, 1868–9.

The total eclipse of Aug. 7, 1869.

Hough (G. W.). Albany (J. Munsell), 1870.

Indication de quelques-uns des résultats obtenus à Cocanada pendant l'éclipse du mois d'août dernier, et à la suite de cette éclipse.

Janssen (J.). Comptes Rendus, **67**, 838.

Lettre sur l'éclipse du 18 août.

Janssen (J.). Comptes Rendus, **67**, 839.

Resumé des notions acquises sur la constitution du Soleil.
> Janssen (J.). Comptes Rendus, **68**, 312.

Observations spectrales prises pendant l'éclipse du 18 août 1868.
> Janssen (J.). Comptes Rendus, **68**, 867.

Sur l'éclipse totale du 22 décembre prochain, 1870.
> Janssen (J.). Comptes Rendus, **71**, 531.

Lettre sur les résultats du voyage pour observer en Algérie l'éclipse du Soleil du 22 Déc. 1870.
> Janssen (J.). Comptes Rendus, **72**, 220.

Remarques sur une dernière note de M. Cornu.
> Janssen (J.). Comptes Rendus, **73**, 793–794.

Télégrammes addressés à l'Académie sur les observations faites pendant l'éclipse du Soleil du 11 Déc. 1871, sur la côte de Malabar.
> Janssen (J.). Comptes Rendus, **73**, 1437.

Lettre sur l'éclipse du 12 Déc. 1871.
> Janssen (J.). Comptes Rendus, **74**, 111.

Les conséquences principales qu'il peut tirer de ses observations sur l'éclipse du 12 Déc. 1871.
> Janssen (J.). Comptes Rendus, **74**, 175, 514, 725; Monthly Notices Astronom. Soc., **32**, 69–70; Proc. Royal Soc., **20**, 138–9; Amer. Jour. Sci., (3) **3**, 226; Jour. Chem. Soc., (2) **10**, 590 (Abs.).

Sur l'éclipse solaire.
> Janssen (J.). Comptes Rendus, **96**, 1745; Nature, **28**, 216.

Rapport à l'Académie sur la mission en Océanie pour l'observation de l'éclipse totale de Soleil du 6 mai 1883.
> Janssen (J.). Comptes Rendus, **97**, 586–602; Mem. Spettr. ital., **12**, 201–216.

Rapport à l'Académie relatif à l'observation de l'éclipse du 12 Déc. 1871, observée à Schoolor (Indoustan).
> Janssen (J.). Ann. Chim. et Phys., (4) **28**, 474–99.

Applications utiles de la méthode graphique à la prédiction des éclipses de Soleil.
> Laussedat. Comptes Rendus, **70**, 240.

Report of observations, etc., of the total eclipse of the Sun taken at "Le Maria Louisa" Vineyard, Cadiz, Dec. 21–22, 1870.
> Lindsay (Lord). Monthly Notices Astronom. Soc., **31**, 49–60.

Remarks on the recent eclipse of the Sun as observed in the United States.

Lockyer (J. N.). Proc. Royal Soc., **18**, 179; Comptes Rendus, **70**, 1890; Nature, **1**, 14.

Note on the recent and coming total solar eclipses.

Lockyer (J. N.). Proc. Royal Soc., **34**, 291–300; Nature, **27**, 185–9; Beiblätter, **7**, 193 (Abs.).

The Mediterranean eclipse, 1870.

Lockyer (J. N.). Nature, **3**, 221–24, 321–2; Amer. Jour. Sci., (3) **3**, 226–30.

The solar eclipse.

Lockyer (J. N.). Nature, **5**, 217–19; Amer. Jour. Sci., (3) **3**, 226–30.

The Eclipse.

Lockyer (J. N.). Nature, **18**, 457–62.

Eclipse notes on the solar spectrum.

Lockyer (J. N.). Nature, **25**, 573–8; **26**, 100–101.

Spectrum of solar eclipses.

Lockyer (J. N.). Nature, **27**, 185.

Report on the total solar eclipse of April 6, 1875.

Lockyer (J. N.). Phil. Trans., **169**, 139–154.

The solar eclipse.

Lockyer (J. N.)., Maclear (J. P.). Nature, **5**, 219–21; Amer. Jour. Sci., (3) **3**, 310–12.

The total eclipse of the Sun of Aug. 7, 1869.

Morton (Henry). Jour. Franklin Inst , (3) **58**, 149, 150, 200.

The solar eclipse of Dec. 22, 1870, observed at San Antonio, near Puerto de Sta. Maria.

Perry (S. J.). Monthly Notices Astronom. Soc., **31**, 62–3, 149, 151.

Sur l'éclipse du 17 mai 1882.

Puiseux (A.). Comptes Rendus, **94**, 1643.

Analyse spectrale des protubérances observées à la presqu'île de Malacca pendant l'éclipse totale du Soleil du 18 août.

Rayet. Comptes Rendus, **67**, 757; Rept. Astronom. Soc., 1868–9, p. 152.

The solar eclipse.

Respighi (L.). Nature, **5**, 287–8; Amer. Jour. Sci., (3) **3**, 312–14.

Spectralbeobachtungen während der totalen Sonnenfinsterniss des Jahres 1868 zu Aden.

> Riha (J.). Sitzungsber. d. Wiener Akad., **58**, II, 655, 721-4.

Some remarks on the total solar eclipse of July 29, 1878.

> Schuster (A.). Monthly Notices Astronom. Soc., **39**, 44-7.

Essai, pendant une éclipse solaire, de la nouvelle méthode spectroscopique proposée pour le prochain passage de Vénus.

> Secchi (A.). Comptes Rendus, **76**, 1327-31; Chem. News, **27**, 320.

Observations de l'éclipse solaire du 10 octobre 1874, avec le spectroscope.

> Secchi (A.). Comptes Rendus, **79**, 885.

L'observation des protubérances solaires faites hors du moment d'une éclipse par M. Janssen et par M. Lockyer.

> Stewart (B.). Comptes Rendus, **67**, 904.

Sull'eclisse totale di sole del 17 maggio 1882, osservato à Sohage in Egitto.

> Tacchini (P.). Mem. Spettr. ital., **11**, Sept. 1-14; Comptes Rendus, **95**, 896.

The total solar eclipse of Dec. 12, 1871.

> Tennant (J. F.). Monthly Notices Astronom. Soc., **32**, 70-2; Nature, **6**, 492.

Report of the Indian Eclipse, Aug. 18, 1868.

> Tennant (J. F.). Royal Astronom. Soc. Memoirs, Vol. **7**; Nature, **1**, 536; Naturforscher (Berlin), **1**, 811, 819, 827, 851, 369, 393; **2**, 59; Les Mondes, **18**, 180, 168, 272, 296, 362, 418.

Eclipse totale de Soleil, observée à Souhage (haute Égypte) le 17 mai (temps civil) 1882.

> Thollon (L.). Comptes Rendus, **94**, 1630-35; Beiblätter, **6**, 878-80.

Observation de l'éclipse totale du 17 mai 1882.

> Trépied. Comptes Rendus, **94**, 1638.

Reports on the total eclipse of the Sun, Aug. 7, 1869.

> United States Naval Observatory (Commodore B. F. Sands and others), Washington, 1869.

On the results of the eclipse observations, Aug. 7, 1869.

> Young (C. A.). Amer. Jour. Sci., (3) **3**, 814; Nature, **1**, 14, 170, 203, 336, 552; Les Mondes, **21**, 288, 600; Naturforscher, **2**, 253, 879, 533; **3**, 16, 53, 142, 168, 175.

Spectroscopic observations of the American eclipse party in Spain.

> Young (C. A.). Nature, **3**, 261.

The Sherman astronomical expedition.

> Young (C. A.). Nature, **7**, 107–109.

Observations upon the solar eclipse of July 29, 1878, by the Princeton Eclipse Expedition.

> Young (C. A.). Amer. Jour. Sci., (3) **16**, 279–90.

Total solar eclipse of August 28–29, 1886.

> By various persons. Abstract in Monthly Notices Astronom. Soc., **47** (1887), 175.

12, *Spectra of the elements in the Sun.*

On sun-spots and terrestrial elements in the Sun.

> Liveing and Dewar. Phil. Mag., (5) **16**, 401–408; Beiblätter, **8**, 304–5 (Abs.); Jour. de Phys., **13**, 418.

Note préliminaire sur les éléments existant dans le Soleil.

> Lockyer (J. N.). Comptes Rendus, **77**, 1347–52; Ber. d. chem. Ges., **6**, 1554–5 (Abs.).

Les éléments présents dans la couche du Soleil qui produit le renversement des raies spectrales.

> Lockyer (J. N.) Comptes Rendus, **86**, 317.

Sur la composition élémentaire du spectre solaire.

> Matthiessen. Comptes Rendus, **19**, 112.

13, *Spectra of solar eruptions.*

Eruzione solare metallica dal 31 luglio, 1880, osservata a Palermo.

> Riccò (A.). Mem. Spettr. ital., **9**, 96–100.

Sur l'éruption solaire observée le 7 juilliet.

> Secchi (A.). Comptes Rendus, **75**, 314–322.

Sur les éruptions métalliques solaires observées à Palermo depuis 1871 jusqu'en avril 1877.

> Tacchini (P.). Comptes Rendus, **84**, 1448–50.

Disegni delle eruzioni etc. del Sole fatti à Roma dal giugno a dicembre 1879.

> Tacchini (P.). Mem. Spettr. ital., **4**, 5–7.

Sulle eruzioni solari metalliche osservate a Roma nel 1881.

> Tacchini (P.). Mem. Spettr. ital., **11**, 53–8; Comptes Rendus, **94**, 1031–8; **95**, 373–8; Beiblätter, **6**, 486 (Abs.).

An explosion on the Sun (Sept. 13, 1871).

> Young (C. A.) Boston Jour. Chemistry, 1871; Amer. Jour. Sci., (3) 2, 468-70; Nature, 4, 488-9; Phil. Mag., (4) 43, 76-79.

14, *Gas spectra in the Sun.*

Preliminary note of researches on gaseous spectra in relation to the physical constitution of the Sun.

> Franckland and Lockyer. Proc. Royal Soc., 17, 288; Comptes Rendus, 68, 420; 69, 264.

15, *Heat in the solar spectrum.*

Sur la distribution de la chaleur dans les régions obscures des spectres solaires.

> Desains (P.). Comptes Rendus, 95, 488.

Lage des Wärmemaximums im Sonnenspectrum.

> Knoblauch (H.). Ann. Phys. u. Chem., 120, 198.

Geschichtliches über das Wärmespectrum der Sonne.

> Lamansky (S.). Ann. Phys. u. Chem., 146, 200, 207, 209.

Observations on invisible heat-spectra and the recognition of hitherto unmeasured wave-lengths, made at the Allegheny Observatory, Pa.

> Langley (S. P.). Amer. Jour. Sci., (3) 31 (1886), 1-12; 32 (1886), 83-106; Phil. Mag., (5) 21 (1886), 394-409; 22 (1886), 149-173; Ann. Chim. et Phys., (6) 9 (1886), 433-506; Jour. de Phys., (2) 5, 377-380 (Abs.); Beiblätter, 11 (1877), 245 (Abs.).

Influence des différentes heures de la journée sur la position du maximum de température dans la partie obscure du spectre solaire.

> Melloni. Comptes Rendus, 11, 141.

Spectre calorifique normal du Soleil.

> Mouton. Comptes Rendus, 89, 295.
> Remarques par M. Thénard. Comptes Rendus, 89, 298.

Untersuchungen über die thermischen Wirkungen des Sonnenspectrums.

> Müller (J.). Ann. Phys. u. Chem., 105, 837.

Wellenlänge und Brechungsexponent der äussersten dunklen Wärmestrahlen des Sonnenspectrums.

> Müller (J.). Ann. Phys. u. Chem., 105, 543; Berichtigung dazu, do., 116, 644.

Sur les propriétés échauffantes des rayons solaires par de grandes et de faibles latitudes.

> Pentland. Comptes Rendus, 8, 310.

The solar spectrum in 1877–8, with some practical idea of its probable temperature of origination.

> Smyth (C. Piazzi). Trans. Royal Soc. Edinburgh, **29**, 285–342; Beiblätter, **4**, 276 (Abs.).

Sur la température du Soleil.

> Soret (J. L.). Archives de Genève, (2) **52**, 89–95; Phil. Mag., (4) **50**, 155–8.

16, *Hydrogen in the solar spectrum.*

La circulation de l'hydrogène solaire.

> Faye. Comptes Rendus, **76**, 597–601.

The comparative aggregate strength of the light from the red hydrogen-stratum, and of that from the rest of the Chromosphere.

> Hammond (B. E.). Nature, **3**, 487.

Dépêche télégraphique addressé de Simla au sujet des lignes de l'hydrogène dans le spectre des protubérances solaires.

> Janssen (J.). Comptes Rendus, **68**, 245.

17, *Intensity of light in the solar spectrum.*

On the variation in the intensity of the fixed lines of the solar spectrum.

> Draper (W.). Phil. Mag., (4) **25**, 342.

The comparative aggregate strength of the light from the red hydrogen-stratum, and of that from the rest of the Chromosphere.

> Hammond (B. E.). Nature, **3**, 487.

Distribution de l'énergie dans le spectre solaire normal.

> Langley (S. P.). Comptes Rendus, **92**, 701.

Confronto fra la radiazione e l'intensità chimica della luce dèl sole.

> Macagno (J.). Mem. Spettr. ital., **8**, App. 13–18.

Étude de la distribution de la lumière dans le spectre solaire.

> Macé (J.) et Nicati (W.). Comptes Rendus, **91**, 623, 1078; Beiblätter, **5**, 301 (Abs.).

Ueber die Vertheilung der chemischen Lichtintensität im Sonnenspectrum.

> Monckhoven. Photographische Mittheilungen, **16**, 145–6; Beiblätter, **4**, 49 (Abs.).

Untersuchungen über die Helligkeitsänderungen in verschiedenen Theilen des Sonnenspectrums bei abnehmender Höhe der Sonne über dem Horizont.

> Müller (G.). Astronom. Nachr., **103**, 241–252; Beiblätter, **7**, 111 (Abs.).

18, *Iron lines in the solar spectrum.*

On the iron lines widened in solar spots.

> Lockyer (J. N.). Proc. Royal Soc., **31**, 348–9 ; Beiblätter, **5**, 288 (Abs.);
> Comptes Rendus, **92**, 904–910 ; Jour. Chem. Soc., **40**, 669 (Abs.).

19, *Magnesium in the solar spectrum.*

Spectre du magnésium en rapport avec la constitution du Soleil.

> Fievez (Ch.). Ann. Chim. et Phys., (5) **23**, 866.

20, *Maps of the solar spectrum.*

On the photographic method of mapping the least refrangible end of the
solar spectrum (with a map of the spectrum from 7600 to 10750).
Bakerian Lecture.

> Abney (W. de W.). Phil. Trans., **171**, 637–667 ; Comptes Rendus,
> **90**, 182–3 ; Beiblätter, **4**, 875 (Abs.).

Sur le spectre normal du Soleil, partie ultra-violette.

> Cornu (A.). Paris, Gauthier-Villars, 1881, 4°. Extrait des Annales de
> l'École normale supérieur, (2) **9**, (1880). Avec deux planches. (Maps
> drawn by wave-lengths.)

Étude du spectre solaire.

> Fievez (Ch.). Bruxelles, F. Hayez, 1882, 4°. Extrait des Annales de
> l'Observatoire royal de Bruxelles, n. sér., tome IV. Avec une planche.
> (Wave-lengths, lines 6899 to 4522.)

Étude de la région rouge (A–C) du spectre solaire.

> Fievez (Ch.). F. Hayez, Bruxelles, 1883, 4°. Extrait des Annales de
> l'Observatoire royal de Bruxelles, n. sér., tome V. Avec deux planches.
> (Wave-lengths, lines 7500 to 6500.)

Untersuchungen über das Sonnenspectrum und die Spectren der chemischen Elemente.

> Kirchhoff (G.). Berlin, Dümmber, 1866–1875, 2 Theile, 4°. Mit vier
> Tafeln. Besondere Abdrück aus den Abhandlungen der Berliner
> Akademie der Wissenschaften, 1861 und 1862. (He used an arbi-
> trary scale.)

Recherches sur le spectre solaire ultra-violet, et sur la détermination des
longueurs d'onde, suivies d'une note sur les formules de dispersion.

> Mascart (E.). Extrait des Annales scientifiques de l'École normale
> supérieure, tome I (1864). Paris, Gauthier-Villars, 1864, 4°. Avec
> un planche.

[A photographic map of the solar spectrum is being made by Prof.
Rowland, and some thirty parts of it have been distributed privately.
At the end of the year 1887 it extended from wave-length 0.0003675 to
wave-length 0.0005796.]

Large Maps of the Solar Spectrum,

> [by Thollon, in the Annals of the Academy of Nice, Tome I. Not yet .
> published, but about to be so; and Tome II. is to contain another,
> smaller, map.]

21, Oscillation-frequencies.

Catalogue of the oscillation-frequencies of solar rays.

> Rept. British Assoc. for 1878.

22, Oxygen in the solar spectrum.

Discovery of oxygen in the Sun by photography, and a new theory of the solar spectrum.

> Draper (H.). Amer. Jour. Sci., (3) 14, 89–96; Nature, 16, 364; 17,
> 339; Comptes Rendus, 85, 613: Beiblätter, 2, 86–90.

On a photograph of the solar spectrum showing the dark lines of oxygen.

> Draper (J. C.). Monthly Notices Astronom. Soc., 40, 14–17; Amer.
> Jour. Sci., (3) 17, 448–452; Jour. Chem. Soc., 38, 201 (Abs.); Bei-
> blätter, 3, 872.

Telluric oxygen lines in the solar spectrum.

> Egoroff. Amer. Jour. Sci., 126, 477; Comptes Rendus, Aug. 27, 1883.

On the presence of oxygen in the Sun.

> Schuster (A.). Nature, 17, 148–9; Beiblätter, 2, 90–91.

23, Photography of the solar spectrum.

Preliminary note on photographing the least refracted portion of the solar spectrum.

> Abney (W. de W.). Monthly Notices Astronom. Soc., 36, 276–7;
> Phil. Mag., (5) 1, 414–415.

Photography at the least refrangible end of the solar spectrum.

> Abney (W. de W.). Monthly Notices Astronom. Soc., 38, 348–51;
> Phil. Mag., (5) 6, 154–7.

On the photographic method of mapping the least refrangible end of the solar spectrum (with a map of the spectrum from 7600 to 10750). Bakerian Lecture.

> Abney (W. de W.). Phil. Trans., 171, 653–67; Proc. Royal Soc., 30,
> 67 (Abs.); Beiblätter, 4, 875 (Abs.); 5, 507–9; Comptes Rendus, 90,
> 182–3; Jour. Chem. Soc., 38, 429.

Use of the spectroscopic camera during the total solar eclipse of May 17, 1882.

> Abney and Schuster. Proc. Royal Soc., 35, 152.

Photography of the ultra-red portions of the solar spectrum.

> Abney (W. de W.). Chem. News, 40, 811.

Photographs of the solar spectrum.
>   Amory (R.).   Proc. Amer. Acad., 11, 70, 279, with plates.

Image photographique colorée du spectre solaire.
>   Becquerel (Éd.).   Comptes Rendus, 26, 181.

De l'image photochromatique du spectre solaire, et des images obtenus dans la chambre obscure.
>   Becquerel (Éd.).   Comptes Rendus, 27, 483.
>   Rapport sur ce mémoire, par M. Regnault, do., 28, 200.

Sur les phosphorographies du spectre solaire.
>   Becquerel (Éd.).   Jour. de Phys., (2) 1, 189.

Observations sur un mémoire de M. E. Marchand relatif à la mesure de la force chimique contenu dans la lumière du Soleil.
>   Becquerel (Éd.).   Ann. Chim. et Phys., (4) 30, 572–3 ; Jour. Chem. Soc., (2) 12, 942 (Abs.).

Janssen's new method of solar photography.
>   Blanford (H. F.).   Nature, 18, 643–645.

Ueber directe Photographirung der Sonnenprotuberanzen.
>   Braun (C.).   Astronom. Nachr., 80, 34–42 ; Ann. Phys. u. Chem., 148, 475–488.

The solar spectrum.
>   Capron (J. R.).   Nature, 6, 492.

Sur la photographie du spectre solaire.
>   Conche (E.).   Comptes Rendus, 90, 689–90.

On the phosphorograph of a solar spectrum, and on the lines of its infra-red region.
>   Draper (J. W.).   Amer. Jour. Sci., (3) 21, 171–182; Phil. Mag., (5) 11, 157–169 ; Beiblätter, 5, 509–510.

On a method of photographing the solar corona without an eclipse.
>   Huggins (W.).   Proc. Royal Soc., 34, 409–414 ; Nature, 27, 199–201 ; Amer. Jour. Sci., (3) 25, 126–130 ; 27, 27–32 ; Ann. Chim. et Phy.. (6) 3, 540–550 ; Beiblätter, 7, 194 (Abs.); Astronom. Nachr.. 104. 113–118 ; Jour. de Phys., (2) 2, 178 (Abs.); Comptes Rendus, 96, 51–53.

Photographische Darstellung des Sonnenspectrums.
>   Jahresber. d. Chemie, 16, 101 ; 17, 116.

Objective Darstellung des Sonnenspectrums; Vorlesungsversuch.
>   Kessler (F.).   Ber. chem. Ges., 9, 577–8; Jour. Chem. Soc., 2, 266.

On the use of the reflecting grating in eclipse photography.
> Lockyer (J. N.). Proc. Royal Soc., **27**, 107-8.

Rutherfurd's Photographie des Sonnenspectrums.
> Müller (J.). Ann. Phys. u. Chem., **126**, 485.

Photographie de l'image du spectre solaire.
> Niepce de Saint Victor. Comptes Rendus, **45**, 814; **46**, 451, 490.

Photography of the infra-red region of the solar spectrum.
> Pickering (H. W.). Proc. Amer. Acad., **20**, 473.

On recent progress in photographing the solar spectrum.
> Rowland (H. A.). Rept. British Assoc. (1884), 635.

On photographs of the solar spectrum.
> Rowland (H. A.). Amer. Jour. Sci., (3) **31**, 319.

Étude photographique du Soleil à l'observatoire impérial de Paris.
> Sourel. Comptes Rendus, **71**, 225.

Le fotografie del Sole fatte all'osservatorio di Meudon dal Professor Janssen.
> Tacchini (P.). Mem. Spettr. ital., **9**, 1-5.

Photographie der weniger brechbaren Theile des Sonnenspectrums.
> Vogel (H. C.) und Lohse (O.). Ann. Phys. u. Chem., **159**, 297; **160**, 292.

On reversed photographs of the solar spectrum beyond the red, obtained on a collodion plate.
> Waterhouse (Capt. J.). Proc. Royal Soc., **24**, 186-9.

Ueber den Einfluss des Eosins auf die photographische Wirkung des Sonnenspectrums auf das Silberbromid und Silberbromjodid.
> Waterhouse (Capt. J.). Ann. Phys. u. Chem., **159**, 616-622; Proc. Royal Soc. Bengal for 1876.

Photographie directe des protubérances solaires sans l'emploi du spectroscope.
> Zenger (C. W.). Comptes Rendus, **88**, 374.

24, *Pressure on the Sun.*

On a method of determining the pressure on the solar surface.
> Wiedemann (E.). Monthly Notices Astronom. Soc., **40**, 627-8.

On a means to determine the pressure at the surface of the Sun and stars, and some spectroscopic remarks.

> Wiedemann (E.). Proc. Physical Soc., **4**, 31–34; Phil. Mag., (5) **10**, 123–5; Beiblätter, **4**, 613 (Abs.).

25, *Spectra of solar protuberances.*

Quadri statistici delle protuberanze e macchie solari osservati all' Collegio Romano nel 1 semestre, 1879.

> Barbieri (E.). Mem. Spettr. ital., **8**, 75–80.

Constitution des protubérances solaires.

> Bianchi. Comptes Rendus, **68**, 276.

La découverte du moyen qui permet d'observer en tout temps les protubérances solaires.

> Delaunay. Comptes Rendus, **67**, 867.

Travaux de M. Respighi pour l'observatiou spectrale des protubérances solaires.

> Faye. Comptes Rendus, **70**, 886.

Sur les taches et protubérances solaires observées à l'équatorial du Collège romain.

> Ferrari. Comptes Rendus, **87**, 971–3.

Spectroscopic observations of the solar prominences.

> Herschel (Capt.). Proc. Royal Soc., **18**, 62, 119, 355.

Note on a method of viewing the solar prominences without an eclipse.

> Huggins (W.). Proc. Royal Soc., **17**, 302.

Note on the wide-slit method of viewing the solar prominences.

> Huggins (W.). Proc. Royal Soc., **21**, 127.

Étude spectrale des protubérances solaires.

> Janssen (J.). Comptes Rendus, **68**, 93.

Méthode qui permet de constater la matière protubérantielle sur tout le contour du disque solaire.

> Janssen (J.). Comptes Rendus, **68**, 713.

On the solar protuberances.

> Janssen (J.). Proc. Royal Soc., **17**, 276.

Notice of an observation of the spectrum of a solar prominence.

> Lockyer (J. N.). Proc. Royal Soc., **17**, 91, 104, 128.

Report to the Committee on Solar Physics on the Basic Lines common to Spots and Prominences.

> Lockyer (J. N.). Proc. Royal Soc., **29**, 247-265; Beiblätter, **4**, 45 (Abs.).

Protubérances solaires.

> Lockyer (J. N.). Comptes Rendus, **67**, 949.

Analyse spectrale des protubérances observées à la presqu'île de malacca pendant l'éclipse totale du Soleil du 18 août 1868.

> Rayet. Comptes Rendus, **67**, 757.

Sur le spectre des protubérances solaires.

> Rayet. Comptes Rendus, **68**, 62; Ann. Chim. et Phys., (4) **24**, 56.

Renversement de deux lignes du sodium dans le spectre de la lumière d'une protubérance.

> Rayet. Comptes Rendus, **70**, 1388.

Osservazioni spettroscopiche del Bordo e delle Protuberanze Solari [with lithographic plate of the prominences].

> Respighi (L.), Roma, 1871.

Sulle protuberanze solari.

> Respighi (L.). Bull. meteorol. dell'osservat. del Coll. Rom., **9**, 89-91; Amer. Jour. Sci., (3) **1**, 283-287.

Spectre des protubérances solaires.

> Respighi (L.). Comptes Rendus, **77**, 716, 774.

Noch einmal meine Bedenken gegen die Zöllner'sche Erklärung der Sonnenflecke und Protuberanzen.

> Reye (T.). Ann. Phys. u. Chem., **151**, 166-178.

Quelques particularités du spectre des protubérances solaires.

> Secchi (A.). Comptes Rendus, **67**, 1123.

Remarques sur la rélation entre les protubérances et les taches solaires.

> Secchi (A.). Comptes Rendus, **68**, 237-8.

Sur les relations qui existent, dans le Soleil, entre les facules, les protubérances et la couronne.

> Secchi (A.). Comptes Rendus, **72**, 829-32; **73**, 242-6, 593-9.

Sur les divers aspects des protubérances.

> Secchi (A.). Comptes Rendus, **73**, 826-36, 979-83.

Sur un nouveau moyen de mesurer les hauteurs des protubérances solaires.

    Secchi (A.). Comptes Rendus, 74, 218–224.

Spectre des protubérances solaires.

    Secchi (A.). Comptes Rendus, 74, 218–24.

Resumé des observations des protubérances solaires du 1 janvier au 29 avril.

    Secchi (A.). Comptes Rendus, 74, 1815–20; Monthly Notices Astronom. Soc., 32, 318–20 (Abs.).

Sur les protubérances et les taches solaires.

    Secchi (A.). Comptes Rendus, 76, 251.

Quelques observations spectroscopiques particulières.

    Secchi (A.). Comptes Rendus, 76, 1052.

Nouvelle série d'observations sur les protubérances solaires; spectre du sodium, de l'hydrogène, du fer, du magnésium, peutêtre des oxydes.

    Secchi (A.). Comptes Rendus, 76, 1522–26.

Protubérances solaires.

    Secchi (A.). Comptes Rendus, 77, 977.

Observations spectrales des protubérances solaires pendant le dernier trimestre de l'année 1873.

    Secchi (A.). Comptes Rendus, 78, 606.

Tableaux des observations des protubérances solaires, du 26 décembre 1873 au 2 août 1874.

    Secchi (A.). Comptes Rendus, 79, 885–9.

Études des taches et des protubérances solaires de 1871 à 1875.

    Secchi (A.). Comptes Rendus, 80, 1278–8.

Résultats des observations des protubérances et des taches solaires du 23 avril au 28 juin 1875.

    Secchi (A.). Comptes Rendus, 81, 563, 605.

Suite des observations spectroscopiques des protubérances solaires, 1875.

    Secchi (A.). Comptes Rendus, 82, 717.

Nouvelle série d'observations sur les protubérances et les taches solaires.

    Secchi (A.). Comptes Rendus, 83, 26–7.

Observations des protubérances solaires pendant le second trimestre de 1876.

    Secchi (A.). Comptes Rendus, 84, 423.

Observations des protubérances solaires, pendant le premier semestre de l'année 1877.

Secchi (A.). Comptes Rendus, 86, 98.

Ueber eine ausgezeichnete Protuberanz.

Spörer. Ann. Phys. u. Chem., 148, 171-2.

L'observation des protubérances solaires faites du moment une éclipse par M. Janssen et M. Lockyer.

Stewart (Balfour). Comptes Rendus, 67, 904.

Observations des taches et des protubérances solaires, pendant le 1 trimestre de 1878.

Tacchini (P.). Comptes Rendus, 86, 1008.

Observations des taches et protubérances solaires pendant les troisième et quatrième trimestres de 1879.

Tacchini (P.). Comptes Rendus, 90, 358-60.

Observations des protubérances, des facules et des taches solaires pendant le premier semestre de l'année 1880.

Tacchini (P.). Comptes Rendus, 91, 466-7.

Observations des taches, des facules et des protubérances solaires, faites à l'observatoire du Collège romain pendant le dernier trimestre, 1880.

Tacchini (P.). Comptes Rendus, 92, 502-4.

Protuberanze solari osservate a Palermo nel quarto trimestre del 1878.

Tacchini (P.). Mem. Spettr. ital., 8, 10-11.

Riassunto delle protuberanze e delle macchie solari osservate alla specola del Collegio Romano nel mese di Settembre, Ottobre e Dicembre.

Tacchini (P.). Mem. Spettr. ital., 8, 13-16.

Sulla distribuzione delle macchie, facole e protuberanze solari sulla superficie del Sole, durante l'anno 1880.

Tacchini (P.). Mem. Spettr. ital., 10, 122-3.

Observations des protubérances, des facules et des taches solaires faites à l'observatoire royal du Collège romain pendant le premier semestre 1882.

Tacchini (P.). Comptes Rendus, 95, 276-8.

Observations des protubérances, facules et taches solaires faites à l'Observatoire royal du Collège romain pendant le troisième et le quatrième trimestre de 1882.

Tacchini (P.). Comptes Rendus, 96, 1290-1; Nature, 28, 48 (Abs.).

Forms of solar protuberances.
> Tacchini (P.).   Nature, **6**, 293.

Taches et protubérances solaires observées avec un spectroscope à grande dispersion.
> Thollon (L.).   Comptes Rendus, **89**, 855.

Observation spectroscopique d'une protubérance solaire le 30 août 1880.
> Thollon (L.).   Comptes Rendus, **91**, 432.

Perturbations solaires nouvellement observées.
> Thollon (L.).   Comptes Rendus, **97**, 144.

Taches et protubérances solaires observées avec un spectroscope à très grande dispersion.
> Thollon (L.).   Jour. de Phys., **9**, 118.

Sudden extinction of the light of a solar protuberance.
> Trouvelot (E.).   Amer. Jour. Sci., (3) **15**, 85–8.

Observations of the solar prominences.
> Tupman (Capt.).   Monthly Notices Astronom. Soc., **33**, 105–115; Amer. Jour. Sci., (3) **5**, 319.

Sur une méthode employée par M. Lockyer pour observer en temps ordinaire les spectres des protubérances signalées dans les éclipses de Soleil.
> Warren de la Rue.   Comptes Rendus, **67**, 836.

Beobachtung der Sonnenprotuberanzen in monochromatischem Lichte.
> Zenker (W.).   Ann. Phys. u. Chem., **142**, 172–176.

Einrichtung des Spectroskops zur Wahrnehmung der Protuberanzen.
> Zöllner (F.).   Ann. Phys. u. Chem., **138**, 42.

Beobachtungen von Protuberanzen der Sonne.
> Zöllner (F.).   Der Naturforscher, **1**, 417; **2**, 9, 83, 51, 74, 91, 116, 133, 213, 245, 338; **3**, 89, 175, 189, 205, 262, 263, 278; Les Mondes, **18**, 362, 413; **19**, 213, 215, 232, 498; Nature, **1**, 172, 195, 607; **2**, 131.

26, *Radiation and the solar spectrum.*

Recherches sur les effets de la radiation chimique de la lumière solaire, au moyen des courants électriques.
> Becquerel (Éd.).   Comptes Rendus, 9, 145.
> Remarques sur cette note, par M. Biot, do., 169.
> Réponse, do., 172–3.

Sur de nouveaux procédés pour étudier la radiation solaire, tant directe que diffuse, dans ses rapports avec la phosphorescence.

Biot. Comptes Rendus, 8, 259, 315.

Sur la répartition de la radiation solaire à Montpellier pendant l'année 1875.

Crova (A.). Comptes Rendus, 82, 375-7.

On the present state of our knowledge of solar radiations.

Hunt (R.). Rep'ts British Assoc. for 1850, 1852, 1853.

Étude des radiations superficielles du Soleil.

Langley (S. P.). Comptes Rendus, 81, 486-9.

27, *Red end of the solar spectrum.*

Photography of the ultra-red portions of the solar spectrum.

Abney (W. de W.). Chem. News, 40, 811.

Work in the infra-red of the spectrum.

Abney (W. de W.). Nature, 27, 15-18; Jour. de Phys., (2) 3, 48; Beiblätter, 7, 695 (Abs.).

Atmospheric absorption in the infra-red of the solar spectrum.

Abney (W. de W.) and Festing (Lieut. Col.). Nature, 26, 45; Proc. Royal Soc., 35, 80.

On the fixed lines in the ultra-red region of the spectrum.

Abney (W. de W.). Phil. Mag., (5) 3, 222; Beiblätter, 1, 289.

On lines in the infra-red region of the solar spectrum.

Abney (W. de W.). Phil. Mag., (5) 11, 800; Beiblätter, 5, 509.

Sur l'observation de la partie infra-rouge du spectre solaire au moyen des effets de phosphorescence.

Becquerel (Éd.). Comptes Rendus, 83, 249-255; Archives de Genève, (2) 57, 306-318; Amer. Jour Sci., (3) 13, 379-80 (Abs.); Ann. Chim. et Phys., (5) 10, 5-18.

La détermination des longueurs d'onde des rayons de la partie infra-rouge du spectre au moyen des effets de phosphorescence.

Becquerel (Édm.). Comptes Rendus, 77, 302; Amer. Jour. Sci., (3) 28, 391, 459.

On the fixed lines in the ultra-red invisible region of the spectrum.

Draper (J. W.). Phil. Mag., (5) 3, 86-89; Beiblätter, 1, 289-40 (Abs.).

Optical spectroscopy of the red end of the solar spectrum.

Hennessey (J B. N). Nature, 17, 28.

Der infra-rothe Theile des Sonnenspectrums.

    Lang (V. von).   Carl's Repert, **19**, 107–9; Beiblätter, **7**, 374 (Abs.).

On certain remarkable groups in the lower spectrum.

    Langley (S. P.).   Proc. Amer. Acad., **14**, 92–105; Beiblätter, **4**, 208.

Photography of the infra-red region of the solar spectrum.

    Pickering (W. H.).   Proc. Amer. Acad., **20**, 473.

Eine Wellenlängenmessung im ultrarothen Sonnenspectrum.

    Pringsheim (E.).   Ann. Phys. u. Chem., n. F. **18**, 32; Amer. Jour. Sci., (3) **25**, 230.

Optical spectroscopy of the red end of the solar spectrum.

    Smyth (C. Piazzi).   Nature, **16**, 264.

28, *Spectroscopic effect of rotation.*

Sur la loi de rotation du Soleil; réponse à une réclamation du P. Secchi et à un mémoire du Dr. Zöllner.

    Faye.   Comptes Rendus, **73**, 1122–31.

Ueber die spectroscopische Beobachtung der Rotation der Sonne, und ein neues Reversionspectroscop.

    Zöllner (F.).   Ann. Phys. u. Chem., **144**, 449.

29, *Storms and cyclones on the Sun.*

Sur la nouvelle hypothèse du P. Secchi.

    Faye.   Comptes Rendus, **76**, 593–7.

Note sur quelques points de la théorie des cyclones solaires, en réponse à une critique par M. Vicaire.

    Faye.   Comptes Rendus, **76**, 733–41.

Réponse au P. Secchi et à M. Vicaire.

    Faye.   Comptes Rendus, **76**, 919–923, 977–982.

Note sur les cyclones solaires, avec une réponse de M. Respighi à M. M Vicaire et Secchi.

    Faye.   Comptes Rendus, **76**, 1229–82.

Sur les cyclones du Soleil comparés à ceux de notre atmosphère.

    Tarry (H.).   Comptes Rendus, **77**, 44–8.

Spectre d'une cyclone solaire.

    Thollon (L.).   Comptes Rendus, **90**, 87–9.

Observations sur la théorie des cyclones solaires.

    Vicaire (E.).   Comptes Rendus, **76**, 703–6, 948–52.

30, *Sun-spots.*

On the spectrum of a solar spot observed at the Royal Observatory, Greenwich.

    Airy (G. B.). Monthly Notices Astronom. Soc., **38**, 82–3.

On the spectrum of a sun-spot observed at the Royal Observatory, Greenwich, 1880.

    Airy (G. B.). Monthly Notices Astronom. Soc., **41**, 68–4.

Dessin des taches solaires observées le 23 mai à 7 heures du soir.

    Baudin. Comptes Rendus, **70**, 1193.

On a periodicity of cyclones and rainfalls in connection with sun-spot periodicity.

    British Assoc. Rep'ts for 1878–8.

Bands observed in the spectra of sun-spots at Stonyhurst Observatory.

    Cortie (A.). Monthly Notices Astronom. Soc., **47** (1886), 19.

Complément de la théorie physique du Soleil; explication des taches.

    Faye. Comptes Rendus, **75**, 1664–72, 1798–6; **76**, 301–10, 389–97 (réponse aux critiques de M. M. Secchi et Tacchini).

Réponse à de nouvelle objections de M. Tacchini.

    Faye. Comptes Rendus, **77**, 881–8, 621–7.

Théorie des scories solaires selon M. Zöllner.

    Faye. Comptes Rendus, **77**, 501–9.

Sur l'explication des taches solaires proposée par M. le Dr. Raye.

    Faye. Comptes Rendus, **77**, 855–61.

Réponse aux remarques de M. Tarry sur la théorie des taches solaires.

    Faye. Comptes Rendus, **77**, 1122–30.

Théories solaires; réponse à quelques critiques récentes.

    Faye. Comptes Rendus, **78**, 1668–70.

Observations au sujet de la dernière note M. Tacchini, et du récent mémoire de M. Langley.

    Faye. Comptes Rendus, **79**, 74–82.

Double série de dessins réprésentant les trombes terrestres et les taches solaires executée par M. Faye.

    Faye. Comptes Rendus, **79**, 265–73.

Sur le dernier numéro des " Memorie dei Spettroscopisti italiani."

    Faye. Comptes Rendus, **80**, 935–6.

Spectrum of the great sun-spot of 1882, Nov. 12–25.
    Greenwich Observatory, Monthly Notices Astronom. Soc., **43**, 77.

On sun-spots and terrestrial elements in the Sun.
    Liveing (G. D.) and Dewar (J.). Phil. Mag., (5) **16**, 401–8; Beiblätter, **8**, 804 (Abs.); Jour. de Phys., **13**, 418.

Temperature of sun-spots.
    Liveing (G. D.) and Dewar (J.). Phil. Mag., (5) **17**, 302–4; Beiblätter, **8**, 768 (Abs.).

On a sun-spot observed Aug. 31, 1880.
    Lockyer (J. N.). Proc. Royal Soc., **31**, 72; Beiblätter, **5**, 129 (Abs.).

Note on the reduction of the observations of the Spectra of 100 sun-spots observed at Kensington.
    Lockyer (J. N.). Proc. Royal Soc., **32**, 208–6.

Preliminary Report to the Solar Physics Committee on the Sun-spot Observations made at Kensington.
    Lockyer (J. N.). Proc. Royal Soc., **33**, 154; Chem. News, **44**, 297–8; Beiblätter, **6**, 281–2 (Abs.).

On the most widened lines in sun-spot spectra; first and second series, from November 12, 1879, to October 15, 1881.
    Lockyer (J. N.). Proc. Royal Soc., **36**, 443–6; **42** (1887), 37–46.

Observations of sun-spot spectra in 1883.
    Perry (S. J.). Monthly Notices Astronom. Soc., **44**, 244–8.

On the sun-spot spectrum from D to B.
    Perry (S. J.). Rept. British Assoc. (1884), 635.

Analyse spectrale d'une tache solaire.
    Rayet. Comptes Rendus, **70**, 846.

Réponse à M. Faye concernant les taches solaires.
    Reye (T.). Comptes Rendus, **77**, 1178–81.

Les minima des taches du Soleil en 1881.
    Riccò (A.). Comptes Rendus, **94**, 1169–71.

Sulla diversa attività dei due emisferi solari nel 1881.
    Riccò (A.). Astronom. Nachr., **103**, 155–6.

Remarques sur la relation entre les protubérances et les taches solaires.
    Secchi (A.). Comptes Rendus, **68**, 237.

Présence de la vapeur d'eau dans le voisinage des taches solaires.
    Secchi (A.). Comptes Rendus, **68**, 858.

L'analyse comparative de la lumière du bord solaire et des taches.

> Secchi (A.). Comptes Rendus, **69**, 39.

Note sur les taches solaires.

> Secchi (A.). Comptes Rendus, **69**, 163, 589, 652.

Sur les taches et le diamètre solaires.

> Secchi (A.). Comptes Rendus, **75**, 1581–4.

Taches solaires.

> Secchi (A.). Comptes Rendus, **76**, 519–27.

La théorie des taches solaires, réponse à M. Faye.

> Secchi (A.). Comptes Rendus, **76**, 911–19.

Études des taches et des protubérances solaires.

> Secchi (A.). Comptes Rendus, **80**, 1273–78; **83**, 26–7.

Note sur les taches du Soleil.

> Sonrel. Comptes Rendus, **70**, 1033.

Report to the Solar Physics Committee on a Comparison between apparent Inequalities of Short-period in Sun-spot Areas, and in Diurnal Temperature-ranges at Toronto and at Keno.

> Stewart (B.) and Carpenter (W. L.). Proc. Royal Soc., **37**, 22, 290.

Macchie solari e facole osservate a Palermo nei mesi di gennaio, febbraio, e marzo 1879 (e durante l'anni 1879 e 1880).

> Tacchini (P.). Mem. Spettr. ital., **8**, 35–6, 50–1, 55–6, 90–2, 97–101; **9**, 45–8, 91–2, 190–2; **10**, 1–4, 122–128.

Sur la théorie des taches solaires; réponse à deux notes précédentes de M. Faye.

> Tacchini (P.). Comptes Rendus, **76**, 633–5.

Sur la théorie émise par M. Faye des taches solaires.

> Tacchini (P.). Comptes Rendus, **76**, 826–30.

Nouvelles observations spectrales, en désaccord avec quelques-unes des théories émises sur le taches solaires.

> Tacchini (P.). Comptes Rendus, **77**, 195–8.

Observations spectroscopiques sur les taches solaires; réponse à M. Faye.

> Tacchini (P.). Comptes Rendus, **79**, 39.

Sur les taches solaires.

> Tacchini (P.). Comptes Rendus, **84**, 1079–81.

Spectre d'une tache solaire observée pendant le mois de juin 1877.

> Tacchini (P.). Comptes Rendus, **84**, 1500.

Observations des taches et des protubérances solaires pendant le 1 trimestre de 1878.

> Tacchini (P.).   Comptes Rendus, 86, 1006.

Observations des taches et des protubérances solaires (pendant les années 1879, 1880, 1881, et 1882).

> Tacchini (P.).   Comptes Rendus, 90, 358-60; 91, 316-7, 466-7; 93, 382; 95, 276-8; 96, 1290.

Sur la grande tache solaire de novembre 1882, et sur les perturbations magnétiques qui en ont accompagné l'apparition.

> Tacchini (P.).   Comptes Rendus, 95, 1212-14.

Macchie solari e facole osservate in Roma all'equatoriale di Cauchoix nel terzo trimestre, e nel ultimo trimestre 1879.

> Tacchini (P.) e Millosevich (E.).   Mem. Spettr. ital., 8, 73-4, 83-9.

Macchie solari e facole osservate a Roma nel mese di gennaio, 1880.

> Tacchini (P.) e Millosevich (E.).   Mem. Spettr. ital., 9, 8.

Observations des taches du Soleil, faites à l'Observatoire de Toulouse en 1874 et 1875.

> Tisserand (F.).   Comptes Rendus, 82, 765-7.

Sur deux taches solaires actuellement visibles à l'œil nu.

> Tremeschini.   Comptes Rendus, 70, 340.

On the veiled solar spots.

> Trouvelot (L.).   Proc. Amer. Acad., 11, 62-69; Amer. Jour. Sci., (3) 11, 169-176.

Sur la théorie des taches et sur le noyau obscur du Soleil.

> Vicaire (E.).   Comptes Rendus, 76, 1896-9.

Sur la constitution du Soleil, et la théorie des taches.

> Vicaire (E.).   Comptes Rendus, 76, 1540-4; 77, 40-4.

Note on the temperature of sun-spots.

> Wiedemann (E.).   Phil. Mag., (5) 17, 247-8; Beiblätter, 8, 768 (Abs.).

Études sur la fréquence des taches du Soleil et sa relation avec la variation de la déclinaison magnétique.

> Wolf.   Comptes Rendus, 70, 741.

Spectroscopic Notes; Spot-spectra.

> Young (C. A.).   Jour. Franklin Inst., 60, 831-40; Nature, 3, 110-118.

Ueber die Periodicität und heliographische Verbreitung der Sonnenflecken.

> Zöllner (F.). Ber. Sächs. Ges. d. Wiss., **22**, 838–850; Ann. Phys. u. Chem., **142**, 524–589.

Ueber den Aggregatzustand der Sonnenflecken.

> Zöllner (F.). Ann. Phys. u. Chem., **152**, 291–310.

31, *Telluric (terrestrial) rays of the solar spectrum.*

Étude spectrale du groupe de raies telluriques nommé *α* (Alpha) par Angström.

> Cornu (A.). Comptes Rendus, **95**, 801; **98**, 169–76; Nature, **29**, 351; Beiblätter, **8**, 305–7 (Abs.); Jour. de Phys., (2) **3**, 109–117.

Les bandes telluriques du spectre solaire.

> Crova (A.). Comptes Rendus, **87**, 107.

Sur les raies telluriques du spectre solaire.

> Egoroff (N.). Comptes Rendus, **93**, 885, 788; Chem. News, **44**, 256 (Abs.); Beiblätter, **5**, 871–2 (Abs.); **6**, 100–101 (Abs.).

Sur la production des groupes telluriques fondamentaux A et B du spectre solaire par une couche absorbante d'oxygène.

> Egoroff (N.). Comptes Rendus, **97**, 555–7; Beiblätter, **7**, 859–60 (Abs.); Amer. Jour. Sci., (8) **26**, 477 (Abs.).

Tellurische Linien der Sonne und der Gestirne.

> Jahresber. d. Chemie, **18**, 92; **19**, 77.

Sur les raies telluriques du spectre solaire.

> Janssen (J.). Comptes Rendus, **54**, 1280; **56**, 189, 538; **57**, 1008; **60**, 213; **95**, 885; Ann. Chim. et Phys., (4) **23**, 274–299; Ann. Phys. u. Chem., **126**, 480; Phil. Mag., (4) **30**, 78.

In feuchter Luft sind die Wärmestreifen des Sonnenspectrums breiter.

> Lamansky (S.). Ann. Phys. u. Chem., **146**, 217.

Étude sur les raies telluriques du spectre solaire.

> Thollon (L.). Comptes Rendus, **91**, 520–522; Beiblätter, **4**, 891 (Abs.).

32, *Ultra-violet part of the solar spectrum.*

Étude du spectre solaire ultra-violet.

> Cornu (A.). Comptes Rendus, **86**, 101; Jour. de Phys., **7**, 285.

Deux planches relatives au spectre solaire.

> Cornu (A.). Comptes Rendus, **86**, 988.

Sur l'absorption atmosphériques des radiations ultra-violettes.

     Cornu (A.).   Jour. de Phys., **10**, 5.

Sur la limite ultra-violette du spectre solaire.

     Cornu (A.).   Comptes Rendus, **88**, 1101–8; Proc. Royal Soc., **29**, 47–
55; Jour. Chem. Soc., **36**, 861 (Abs.); Beiblätter, **4**, 39–40 (Abs.).

Observation de la limite ultra-violette du spectre solaire à diverses altitudes.

     Cornu (A.).   Comptes Rendus, **89**, 808–814; Jour. Chem. Soc., **38**,
201 (Abs.); Amer. Jour. Sci., (3) **19**, 406.

Loi de repartition, suivant l'altitude, de la substance absorbant dans
l'atmosphère des radiations solaires ultra-violettes.

     Cornu (A.).   Comptes Rendus, **90**, 940.

Sur le spectre normal du Soleil; partie ultra-violette.

     Cornu (A.).   Ann. de l'École Nòrmale, (2) **9**, 21–106; Beiblätter, **4**,
871–4 (Abs.).

Sur les longueurs d'onde et les caractères des raies violettes et ultra-
violettes du Soleil, données par une photographie faite au moyen
d'un réseau.

     Draper (H.).   Comptes Rendus, **78**, 682–6.

Influence des rayons ultra-violets du spectre solaire sur la matière verte
des végétaux et sur la flexion des tiges.

     Guillemin.   Comptes Rendus, **45**, 62, 543.

Ultra-violette Strahlen des Sonnenspectrums.

     Jahresber. d. Chemie (1872), 134.

Sur les raies du spectre solaire ultra-violet.

     Mascart.   Comptes Rendus, **57**, 789; Phil. Mag., (4) **27**, 159.

Sur l'absorption du nouveau violet extrême par diverses matières.

     Matthiessen.   Comptes Rendus, **19**, 112.

Rayons violets qui renferment le maximum d'action chimique de toutes
les couleurs du spectre solaire.

     Poey (A.).   Comptes Rendus, **73**, 1288.

Nouvelles expériences tendant à démontrer qu'il existe une force magné-
tisante dans l'extrémité violette du spectre solaire.

     Ridolfi (C.).   Ann. Chim. et Phys., (5) **3**, 823–4.

33, *Water in the solar spectrum.*

The influence of water in the atmosphere on the solar spectrum and solar temperature.

> Abney (W. de W.) and Festing (R.). Proc. Royal Soc., **35**, 328–41; Jour. Chem. Soc., **46**, 241; Beiblätter, **8**, 507 (Abs.).

Aqueous lines in the spectrum of the Sun.

> Cooke (J. P., Jr.). Amer. Jour. Sci., **91**, 178; Phil. Mag., (4) **31**, 337.

Influence de la vapeur aqueuse visible dans l'atmosphère, et de la pluie sur le spectre solaire.

> Zantedeschi. Comptes Rendus, **63**, 644.

34, *Wave-lengths of the solar spectrum.*

Wave-lengths of A, a, and of prominent lines in the infra-red of the solar spectrum.

> Abney (W. de W.). Proc. Royal Soc., **36**, 187.

Détermination des longueurs d'onde des raies et bandes principales du spectre solaire infra-rouge.

> Becquerel (H.). Comptes Rendus, **99**, 417; Amer. Jour. Sci., **128**, 391, 459.

Détermination des longueurs d'onde des raies du spectre solaire au moyen des bandes d'interférence.

> Bernard (F.). Comptes Rendus, **58**, 1158; **59**, 82.

Sur la photométrie solaire.

> Crova (A.). Comptes Rendus, **94**, 1271; **95**, 1271–3; **96**, 126; Beiblätter, **7**, 113 (Abs.).

Bestimmung der Wellenlängen der Fraunhofer'schen Linien des Sonnenspectrums, mit 2 Tafeln.

> Ditscheiner (L.). Sitzungsber. d. Wiener Akad., **50** II, 286, 296–341.

Sur les longueurs d'onde et les caractères des raies violettes et ultra-violettes du Soleil, données par une photographie faite au moyen d'un réseau.

> Draper (H.). Comptes Rendus, **78**, 682–6.

On the normal solar spectrum (giving wave-lengths of the principal lines of the solar spectrum).

> Gibbs (Wolcott). Amer. Jour. Sci., **93**, 1.

Mesures spectrophotométriques en divers points du disque solaire.

> Gouy et Thollon. Comptes Rendus, **95**, 834–6; Beiblätter, **7**, 113–114 (Abs.).

Wellenlänge und Brechungsexponent der äussersten dunklen Wärmestrahlen des Sonnenspectrums.

> Müller (J.).   Ann. Phys. u. Chem., **115**, 543.
> Berichtigung dazu, **116**, 644.

Eine Wellenlängenmessung im ultrarothen Sonnenspectrum.

> Pringsheim (E.).   Ann. Phys. u. Chem., n. F. **18**, 32; Nature, **28**, 72.

Relative wave-length of the lines of the solar spectrum.

> Rowland (H. A.).   Amer. Jour. Sci., (3) **38** (1887), 182–190; Phil. Mag., (5) **23** (1887), 257–65.

Note on Sir David Brewster's Line Y in the infra-red of the solar spectrum.

> Smyth (C. Piazzi).   Edinburgh Transactions, **32** II, 223–238.

Spectralphotometrische Untersuchungen.

> Vogel (H. C.).   Monatsber. d. Berliner Akad., (1877) 104–142.

35, *White lines in the solar spectrum.*

White lines in the solar spectrum.

> Hennessey (J. H. N.).   Proc. Royal Soc., **22**, 221; Phil. Mag., (4) **48**, 303–6; **53**, 259 (appendix to the preceding note).

k, TWINKLING OF STARS.

Ueber das Funkeln der Sterne und die Scintillation überhaupt.

> Exner (K.).   Sitzungsber. d. Wiener Akad., **84** II, 1038–81; Ann. Phys. u. Chem., n. F. **17**, 305–22; Jour. de Phys., (2) **1**, 373 (Abs.).

Analyse prismatique de la lumière des étoiles scintillantes.

> Montigny (Ch.).   Bull. de l'Acad. de Belgique, (2) **37**, 165–90; Comptes Rendus, **66**, 910; Ann. Phys. u. Chem., **153**, 277–98.

Nouvelles recherches sur la fréquence de la scintillation des étoiles dans ses rapports avec la constitution de leur lumière d'après l'analyse spectrale.

> Montigny (Ch.).   Bull. de l'Acad. roy. de Belgique, (2) **38**, 300–320; Ann. Phys. u. Chem., Ergänzungsband, **7**, 605–624.

ATMOSPHERIC SPECTRA.

Atmospheric transmission of visual and photographically active light.

Abney (W. de W.). Monthly Notices Astronom. Soc., 47 (1887), 260–5.

Spectre de l'air atmosphérique.

Becquerel (H.). Comptes Rendus, 90, 1407.

La radiation atmosphérique comme agent chimique.

Biot. Comptes Rendus, 8, 598.

Observations of the lines of the solar spectrum, and on those produced by the Earth's atmosphere.

Brewster (Sir D.). Phil. Mag., (3) 8, 384.

On the aqueous lines of the solar spectrum.

Cooke (J. P.). Amer. Jour. Sci., (2) 41, 178; Phil. Mag., (4) 31, 337.

Sur l'absorption par l'atmosphère des radiations ultra-violettes.

Cornu (A.). Comptes Rendus, 88, 1285; Jour. de Phys., 10, 5.

Sur l'observation comparative des raies telluriques et métalliques comme moyen d'observer les pouvoirs absorbants de l'atmosphère.

Cornu (A.). Comptes Rendus, 95, 801–6; Jour. de Phys., (2) 2, 58; Beiblätter, 7, 110 (Abs.); Amer. Jour. Sci., (3) 25, 78; Bull. Soc. franç. de Phys. (1882), 241–7.

Étude spectrale du groupe de raies telluriques nommé α (alpha) par Angström.

Cornu (A.). Comptes Rendus, 98. 169; Ann. Chim. et Phys., (6) 7 (1886), 5–102; Phil. Mag., (5) 22 (1886), 458–63; Amer. Jour. Sci., (3) 33 (1887), 70 (Abs.); Beiblätter, 11 (1887), 87 (Abs.).

s bandes telluriques du spectre solaire.

Crova (A.). Comptes Rendus, 87, 107.

Recherches sur les raies telluriques du spectre solaire.

Egoroff (N.). Comptes Rendus, 93, 385, 788.

Recherches sur le spectre d'absorption de l'atmosphère terrestre.

Egoroff (N.). Comptes Rendus, 95, 447; Beiblätter, 6, 987; Jour. Chem. Soc., 44, 187.

Sur la production des groupes telluriques fondamentaux A et B du spectre solaire, par une couche d'oxygène.

Egoroff (N.). Comptes Rendus, 97, 555.

Note on the atmospheric lines of the solar spectrum and on certain spectra of gases.

> Gladstone (J. H.). Proc. Royal Soc., **11**, 805.

Bandenspectrum der Luft.

> Goldstein. Sitzungsber. d. Wiener Akad., **84** II, 698; Ann. Phys. u. Chem., n. F. **15**, 280.

On the absorption of solar rays by atmospheric ozone.

> Hartley (W. N.). Jour. Chem. Soc., **39**, 111–28; Ber. chem. Ges., **14**, 1890 (Abs.).

Atmospheric lines of the solar spectrum.

> Hennessey (J. H.). Proc. Royal Soc., **19**, 1; **23**, 201.

Zustand der Atmosphäre.

> Jahresber. d. Chemie, **13**, 607; **14**, 45; **16**, 108; **19**, 77.

Spectres telluriques.

> Janssen (J.). Comptes Rendus, **101** (1885), 111.

Analyse spectrale des éléments de l'atmosphère terrestre.

> Janssen (J.). Comptes Rendus, **101** (1885), 649.

In feuchter Luft sind die Wärmestreifen des Sonnenspectrums breiter.

> Lamansky (S.). Ann. Phys. u. Chem., **146**, 217.

Abhängigkeit des Brechungsquotienten der Luft von der Temperatur.

> Lang (V. von). Ann. Phys. u. Chem., **153**, 448–65; Sitzungsber. Wiener Akad., **69** II, 451-68.

Amount of atmospheric absorption.

> Langley (S. P.). Phil. Mag., (5) **18**, 289–307; Jour. Chem. Soc., **23**, 819; Amer. Jour. Sci., (8) **28** (1885), 163, 242.

Ueber die Absorption der Sonnenstrahlung durch die Kohlensäure unserer Atmosphäre.

> Lecher (E.). Sitzungsber. Wiener Akad., **82** II, 851–868.

On the spectrum of the atmosphere.

> Maclear (J. P.). Nature, **5**, 341.

Sur la théorie de l'absorption atmosphérique.

> Maurer (J.). Archives de Genève, (3) **9**, 874–91.

Opalescence of the atmosphere for the chemically active rays.

> Roscoe (H. E.). Chem. News, **14**, 28.

On the atmospheric lines between the D lines.

> Russell (H. C.). Monthly Notices Astronom. Soc., **38**, 30–32.

Spectrum des electrischen Glimmlichts in atmosphärischer Luft.

Schimkow (A.).   Ann. Phys. u. Chem., **129**, 518.

Sur l'influence de l'atmosphère sur les raies du spectre.

Secchi (A.).   Comptes Rendus, **60**, 879.

Spectrum von atmosphärischer Luft.

Vogel (H. C.).   Ann. Phys. u. Chem., **146**, 580.

AURORA AND ZODIACAL LIGHT.

The aurora and its spectrum.

> Abercromby (R.). Nature, 27, 173; Beiblätter, 7, 193.

Magnetic disturbances, auroras and earth-currents.

> Adams (W. G.). Nature, 25, 66–71.

Spectrum of aurora borealis.

> Angström (A. J.). Nature, 10, 210; Ann. Phys. u. Chem., Jubel-
> band, 424–9; Arch. de Genève, (2) 50, 204 (Abs.); Jour. de Phys.,
> 3, 210.

Observations of the zodiacal light at Cadiz.

> Arcimis (A. T.). Monthly Notices Astronom. Soc., 36, 48–51.

Spectrum of the Aurora.

> Backhouse (T. W.). Nature, 4, 66; 7, 182, 463; 28, 209.

A line in the green between b and F; a line in the yellow-green between
D and E (principal auroral line); a line in the green-blue at or
near F, assumed to be 485 of Alvan Clarke, Jr.; a line in the red
between C and D, almost equidistant between C and D; a line in
the green at or near b, at 517.

> Barker (G. F.). Nature, 7, 182.

Spectrum of the Aurora.

> Barker (G. F.). Amer. Jour. Sci., (3) 2, 465–8; 5, 81–84; Jour.
> Chem. Soc., (2) 10, 119 (Abs.); Chem. News, 24, 270.

On the spectrum of the aurora borealis.

> Browning (J.). Monthly Notices Astronom. Soc., 31, 17; Phil. Mag.,
> (4) 41, 79; Amer. Jour. Sci., (3) 1, 215.

Comparison of some tube and other spectra with the spectrum of the
aurora.

> Capron (J. R.). Phil. Mag., (4) 49, 249–66.

Spectrum of aurora.

> Capron (J. R.). Nature, 3, 28; Phil. Mag., (4) 49, 481.

The aurora borealis of Feb. 4, 1872.

> Capron (J. R.). Nature, 5, 284–5. (See below under Cornu, Key,
> Maclear, Murphy, Perry, Prazmowski, Respighi, Secchi, Smyth,
> Stone, Tacchini, Twining, and Watts.)

Spectrum of the aurora and of the zodiacal light (with a list of authorities on the subject, included here).
  Capron (J. R.). Nature, 7, 182–186.

The aurora spectrum.
  Capron (J. R.). Nature, 7, 201.

The aurora and its spectrum.
  Capron (J. R.). Nature, 25, 58; Jour. de Phys., (2) 2, 97 (Abs.).

The aurora.
  Capron (J. R.). Nature, 27, 83–4, 189, 198.

Magnetic storm, aurora and sun-spot.
  Christie (W. H. M.). Nature, 27, 83.

Spectrum of the Aurora.
  Church (A. H.). Chem. News, 22, 225.

A line in the green-blue at or near F; at 485; assumed to be 486 F hydrogen.
  Clark (Alvan, Jr.). Nature, 7, 182.

A line in the green near E (corona line?); at 532; assumed to be 531·6 (corona line).
  Clark (Alvan, Jr.). Nature, 7, 182.

A line in the yellow-green between D and E (principal auroral line).
  Clark (Alvan, Jr.). Nature, 7, 182.

Line in the indigo at or near G; at 435; supposed to be G hydrogen.
  Clark (Alvan, Jr.). Nature, 7, 183.

Observations of the aurora on Aug. 12 and 13, 1880
  Copeland (R.). Nature, 22, 510.

Spectre de l'aurore boréale du 4 février.
  Cornu (A.). Comptes Rendus, 74, 890.

Sur l'intensité calorifique de la radiation solaire et son absorption par l'atmosphère terrestre.
  Crova (A.). Comptes Rendus, 81, 1205–7.

The aurora.
  Eiger (T. G.). Nature, 3, 6–7; 7, 182; 27, 85–6.

Spectrum of the aurora.
  Ellery (R. J.). Nature, 4, 280.

Spectrum of the aurora.
> F. (T.). Nature, **3**, 6.

Sur les aurores boréales.
> * Faye. Comptes Rendus, **77**, 546.

The continuous spectrum; faint green reaching from the aurora line to F.
> Flögel. Nature, **7**, 183.

Spectroscopic examination of the aurora, April 10, 1872.
> Frazer (P.). Proc. Amer. Philosoph. Soc., **12**, 579.

On the spectrum of the aurora.
> Herschel (A. S.). Phil. Mag., (4) **49**, 65–71; Nature, **3**, 486.

Line in the yellow-green between D and E (principal auroral line).
> Herschel (A. S.). Nature, **7**, 182.

Spectrum of the aurora.
> Holden (E. S.). Amer. Jour. Sci., (3) **4**, 423; Phil. Mag., (4) **44**, 478.

Spectrum of the aurora.
> Hyatt. Nature, **3**, 105.

Das Nordlichtspectrum.
> Jahresber. d. Chemie, (1868) 128, (1869) 180, (1872) 148, (1873) 151, (1875) 128.

Spectrum des Zodiacal-Lichtes.
> Jahresber. d. Chemie, (1872) 148.

The aurora borealis of Feb. 4, 1872.
> Key (H. Cooper). Nature, **5**, 302.

Spectrum of the aurora.
> Kirk (E. B.). Observatory, (1882) 271, (1886) 311.

Spectrum of the aurora.
> Kirkwood (D.). Nature, **3**, 126.

Sur la décharge électrique dans l'aurore boréale, et le spectre du même phénomène.
> Lemström (S.). Archives de Genève, (2) **50**, 225–42, 355–86; Nature, **28**, 60–3, 107–9, 128–30; Jour. de Phys., (2) **2**, 315–17 (Abs.). (See Tresca in Comptes Rendus, **96**, 1885.)

L'analyse spectrale de la lumière zodiacale et sur la couronne des éclipses.
> Liais (É.). Comptes Rendus, **74**, 262.

Spectrum of the aurora.
> Lindsay (Lord). Nature, **4**, 347, 866; **7**, 182.

The aurora borealis of Feb. 4, 1872.
> Maclear (J. P.). Nature, **5**, 283.

Spectrum of aurora.
> Maclear (J. P.). Nature, **6**, 329

Spectrum of aurora australis.
> Maclear (J. P.). Nature, **17**, 11.

Swan lamp spectrum and the aurora.
> Munro (J.). Nature, **27**, 173; Beiblätter, **7**, 198.

The aurora borealis of Feb. 4, 1872.
> Murphy (J. J.). Nature, **5**, 283.

Spectrum of the aurora.
> Newlands (J. A. R.). Chem. News, **23**, 218.

Das Nordlichtspectrum.
> Oettigen (A. J.). Ann. Phys. u. Chem., **146**, 284-7; Ann. Chim. et Phys., (4) **26**, 269-78.

The aurora borealis of Feb. 4, 1872.
> Perry (S. J.). Nature, **5**, 808.

Spectrum of the aurora.
> Pickering (E. C.). Nature, **3**, 104.

Étude spectrale de la lumière de l'aurore boréale du 4 février.
> Prazmowski. Comptes Rendus, **74**, 891.

Spectrum of the aurora.
> Pringle (G. H.). Nature, **6**, 260.

Spectra of the aurora and corona.
> Proctor (H. R.). Nature, **3**, 6, 68, 346, 369, 468; **6**, 161, 220; **7**, 242.

Spectrum of the aurora.
> Proctor (H. R.). Nature, **7**, 132.

Sur le spectre de l'aurore boréale.
> Rayet (G.). Jour. de Phys., **1**, 368.

L'analyse spectrale de la lumière zodiacale.
> Respighi (L.). Comptes Rendus, **74**, 514.

Le spectre de la lumière zodiacale et le spectre de l'aurore boréale sont identicales.

    Respighi (L.).   Comptes Rendus, **74**, 743.

Observations of the aurora borealis of Feb. 4 and 5, 1872.

    Respighi (L.).   Nature, **5**, 511; Gazz. Ufficiale d. Regno d'Italia, Feb. 5, 1872.

The aurora.

    Robinson (H.).   Nature, **27**, 85.

The aurora.

    Romanes (C. H.).   Nature, **27**, 86.

On the auroral spectrum.

    Rowland (H. A.).   Amer. Jour. Sci., **5**, 320.

Spectre de l'aurore boréale.

    Salet (G.).   Bull. Soc. chim. Paris, 1 Mars 1872; Ber. chem. Ges., **5**, 222.

Spectrum of the aurora.

    Schmidt.   Nature, **7**, 182–3.

The aurora borealis of Feb. 4, 1872.

    Seabroke (G. M.).   Nature, **5**, 288.

Sur l'aurore boréale du 4 février observée à Rome, et sur quelques nouveaux résultats d'analyse spectrale.

    Secchi (A.).   Comptes Rendus, **74**, 583–8.

Aurore boréale observée à Rome le 10 août à 10 heures du matin.

    Secchi (A.).   Comptes Rendus, **75**, 606–613.

La luce zodiacale confronto tra le osservazioni del P. Dechevrens e quelle di G. Jones.

    Serpieri (A.).   Mem. Spettr. ital., **9**, 133–42.

Mémoire sur des faits dont on peut déduire: 1. une théorie des aurores boréales et australes, fondée sur l'existence de marées atmosphériques; 2. l'indication, à l'aide des aurores, de l'existence d'essaims d'étoiles filantes à proximité du globe terrestre.

    Silbermann (J.).   Comptes Rendus, **74**, 553–7, 638–42.

Spectra of aurora, corona and zodiacal light.

    Smyth (C. Piazzi).   Nature, **3**, 509–10.

Spectroscopic observations of the zodiacal light in April, 1872, at the Royal Observatory, Palermo.

    Smyth (C. Piazzi).   Monthly Notices Astronom. Soc., **32**, 277–288; Amer. Jour. Sci., (8) **4**, 245 (Abs.).

The aurora borealis of Feb. 4, 1872.
>   Smyth (C. Piazzi). Nature, **5**, 282–3.

Spectrum of the aurora.
>   Smyth (C. Piazzi). Nature, **7**, 182.

The aurora of Feb. 4, 1872.
>   Stone (E. J.). Nature, **5**, 448; Amer. Jour. Sci., (8) **3**, 391–2.

Beobachtung eines Nordlichtspectrum (Aurora Borealis).
>   Struve (Otto von). Bull. de l'Acad. de St. Pétersbourg, **3**, 49.

Observations of the aurora.
>   Sueur (A. Le). Proc. Royal Soc., **19**, 19.

Spectrum of the aurora.
>   T. (F.). Nature, **7**, 182–3.

Sur l'aurore boréale du 4 février 1872.
>   Tacchini (P.). Comptes Rendus, **74**, 540–2.

Sur l'origine des aurores polaires.
>   Tarry (H.). Comptes Rendus, **74**, 549–53.

Sur les observations de M. Lemström en Laponie.
>   Tresca. Comptes Rendus, **96**, 1635–6.

The aurora of Feb. 4, 1872.
>   Twining (A. C.). Amer. Jour. Sci., (8) **3**, 278–81.

Untersuchungen über das Spectrum des Nordlichtes.
>   Vogel (H. C.). Ber. Sächs. Ges. d. Wiss., **23**, 285–99; Ann. Phys. u.
>   Chem., **146**, 569–85; Jour. Chem. Soc., (2) **10**, 1061 (Abs.); Amer.
>   Jour. Sci., (3) **4**, 487 (Abs.).

Spectrum des Nordlichtes.
>   Vogel (H. C.). Astronom. Nachr., **78**, 247–8.

Spectrum of the aurora.
>   Watts (W. M.). Phil. Mag., (4) **49**, 410–11.

The aurora borealis of Feb. 4, 1872.
>   Watts (W. M.). Nature, **5**, 308.

Observations sur le spectre de l'aurore boréale.
>   Wijkander (A.). Arch. de Genève, (2) **51**, 25–30.

Line in the green near E (corona line).
>   Winlock. Nature, **7**, 182.

On the spectrum of the zodiacal light.

> Wright (A. W.). Amer. Jour. Sci., (3) **8**, 39–46; Ann. Phys. u. Chem., **154**, 619–29.

Ueber das Spectrum des Nordlichtes.

> Zöllner (F.). Ber. Sächs. Ges. Wiss., **22**, 254–260; Ann. Phys. u. Chem., **141**, 574–581; Phil. Mag., (4) **41**, 122–127; Amer. Jour. Sci., (3) **1**, 372–3 (Abs.).

Spectrum of the aurora.

> Zöllner (F.).   Nature, **7**, 182–3.

AUSTRIUM.

Spectrum of austrium.

Linnemann (E.). Monatschr., 7, 121–8; Jour. Chem. Soc., 50 (1886), 778 (Abs.).

BARIUM.

Ueber den Einfluss der Temperatur auf die Brechungsexponenten der natürlichen Sulfate des Baryum.

Arzruni (A.). Zeitschr. Krystallogr. u. Mineralog., 1, 165–192; Jahrb. f. Mineral. (1877), 526 (Abs.); Jour. Chem. Soc., 34, 189 (Abs.).

Barium spark spectrum.

Capron (J. R.). Photographed Spectra, London, 1877, p. 21.

Spectre de chlorure de baryum.

Gouy. Comptes Rendus, 84, 231.

Sur les caractères des flammes chargées du chlorure de baryum.

Gouy. Comptes Rendus, 85, 439.

Spectre continu du baryum.

Gouy. Comptes Rendus, 86, 878.

Spectrum von Baryum.

Jahresber. d. Chemie (1870), 174.

Chemische Analyse durch Spectralbeobachtungen, Baryum.

Kirchhoff und Bunsen. Ann. Phys. u. Chem., 110, 182

Chlorure de Baryum (ou Ba O) dans le gaz.

>Lecoq de Boisbaudran (F.). Spectres Lumineux, Paris, 1874, p. 57, 62, planche VII.

Bromure de baryum dans le gaz chargé de brome; iodure de baryum dans le gaz chargé d'iode.

>Lecoq de Boisbaudran (F.). Spectres Lumineux, Paris, 1874, p. 63, 65, planche VIII.

BERYLLIUM OR GLUCINUM.

Beryllium arc spectrum.

>Capron (J. R.). Photographed Spectra, London, 1877, p. 22.

Spectrum of beryllium.

>Hartley (W. N.). Chem. News, **47**, 201; Jour. Chem. Soc , **43**, 316-19; Ber. chem. Ges., **16**, 1859 (Abs.); Amer. Jour. Sci., (3) **26**, 316–17.

Remarks on the atomic weight of beryllium.

>Hartley (W. N.). Proc. Royal Soc., **36**, 462–4; Chem. News, **49**, 171–2; Beiblätter, **8**, 820 (Abs.).

Spectrum of beryllium.

>Nature, **29**, 90.

Propriétés principales du glucinum.

>Nilson (L. F.) et Petterson (O.). Comptes Rendus, **91**, 169.

Note on the atomic weight of beryllium.

>Reynolds (J. E.). Proc. Royal Soc., **35**, 248-50; Beiblätter, **8**, 3-4 (Abs.).
>Reply by Humpidge (T. S.). Proc. Royal Soc., **35**, 358–9.

BISMUTH.

Le bismuth n'a donné aucune apparence de renversement.
> Cornu (A.). Comptes Rendus, **73**, 882.

Fluorescence des composés de bismuth.
> Lecoq de Boisbaudran (F.). Comptes Rendus, **103** (1887), 629–31,
> 1064–8; Jour. Chem. Soc., **52**, 4 (Abs.), 189 (Abs.).

BLUE GROTTO.

Spectroscopische Untersuchung der blauen Grotte auf Capri.
> Vogel (H. W.). Ann. Phys. u. Chem., **156**, 825.

BORAX.

Boron arc spectrum.
> Capron (J. R.). Photographed Spectra, London, 1877, p. 22.

L'acide borique.
> Dieulafait (L.). Ann. Chim. et Phys., (5) **12**, 318–54; Jour. Chem.
> Soc., **34**, 11 (Abs.).

10 T

Existence de l'acide borique dans les eaux de la Mer Morte.

> Dieulafait (L.).   Comptes Rendus, **94**, 1852–4; Jour. Chem. Soc., **42**, 1087 (Abs.); Ann. Chim. et Phys., (5) **25**, 145–167.

L'acide borique dans les eaux minérales de Contrexeville et Schinznach (Suisse).

> Dieulafait (L.).   Comptes Rendus, **95**, 999–1001; Jour. Chem. Soc., **44**, 801 (Abs.).

Les salpêtres naturels du Chili et du Pérou au point de vue de l'acide borique.

> Dieulafait (L.).   Comptes Rendus, **98**, 1545–8; Chem. News, **50**, 45 (Abs.).

On line spectra of boron.

> Hartley (W. N.).   Proc. Royal Soc., **35**, 301–4; Chem. News, **48**, 1–2; Jour. Chem. Soc., **46**, 242 (Abs.); Beiblätter, **8**, 120 (Abs.).

Spectra of boric acid and blowpipe beads.

> Horner (Charles).   Chem. News, **29**, 66.

Spectre de l'acide borique dans le gaz.

> Lecoq de Boisbaudran (F.).   Spectres Lumineux, Paris, 1874, p. 191. planche XXVIII.

Spectre de l'acide borique.

> Lecoq de Boisbaudran (F.).   Comptes Rendus, **76**, 833.

Spectrum von Fluorborgas.

> Plücker.   Ann. Phys. u. Chem., **104**, 125.

Propriétés optiques de borax.

> Senarmont (H. de).   Ann. Chim. et Phys., (8) **41**, 886.

Spectra der verschiedenen grünen Flammen, Borax.

> Simmler (R. Th.).   Ann. Phys. u. Chem., **115**, 249.

Spectre du bore.

> Troost et Hautefeuille.   Comptes Rendus, **63**, 620; Bull. Soc. chim. Paris, n. s. **16**, 229.

BROMINE.

Action des rayons différemment réfrangible sur l'iodure et le bromure
  d'argent.
  > Becquerel (E.). Comptes Rendus, **79**, 185-90; Jour. Chem. Soc., (2)
  > **13**, 80 (Abs.).

Spectre du brome dans les tubes de Geissler.
  > Chautard (J.). Comptes Rendus, **82**, 278.

De l'action des différentes lumières colorées sur une couche de bromure
  d'argent impregnée de diverses matières colorantes organiques.
  > Cros (Ch.). Comptes Rendus, **88**, 879-81; Jour. Chem. Soc., **36**,
  > 504-5.

Spectre de bromure de cuivre.
  > Diacon (E.). Ann. Chim. et Phys., (4) **6**, 1.

Spectre d'absorption de protobromure de tellure et de protobromure
  d'iode.
  > Gernez (D.). Bull. Soc. chim. Paris, n. s. **18**, 172.

Spectre du brome.
  > Gouy. Comptes Rendus, **85**, 70.

Absorptionsspectrum des Bromtellurs, des Bromselens, und des Bromjods.
  > Jahresber. d. Chemie (1872), 140.

On the action of the less refrangible rays of light on silver iodide and
  bromide.
  > Lea (M. Carey). Amer. Jour. Sci., (3) **9**, 269-78; Jour. Chem. Soc.,
  > **1** (1876), 28 (Abs.).

Notes on the sensitiveness of silver bromide to the green rays as modified
  by the presence of other substances.
  > Lea (M. Carey). Amer. Jour. Sci., (3) **11**, 459-64.

Réaction spectrale du Brome.
  > Lecoq de Boisbaudran (F.). Comptes Rendus, **91**, 902-3; Phil. Mag.,
  > (5) **11**, 77-8; Beiblätter, **5**, 118 (Abs.).

Bromure de baryum dans le gaz chargé de brome.
  > Lecoq de Boisbaudran. Spectres Lumineux, Paris, 1874, p. **68**, 65,
  > planche VIII.

Verbindungsspectrum zur Entdeckung von Brom.

> Mitscherlich. Jour. prackt. Chem., **97**, 218.

Entdeckung sehr geringer Mengen von Brom in Verbindungen.

> Mitscherlich. Ann. Phys. u. Chem., **125**, 629.

Absorption spectra of bromine.

> Roscoe (H. E.) and Thorpe (T. E.). Proc. Royal Soc., **25**, 4.

Ueber die Lichtempfindlichkeit des Bromsilbers.

> Vogel (H.). Ber. chem. Ges., **6**, 1302–6; Ann. Phys. u. Chem., **150**,
> 458–9; Jour. Chem. Soc., (2) **12**, 217 (Abs.); Amer. Jour. Sci., (3)
> **7**, 140–1; Phil. Mag., (4) **47**, 273–7.

Ueber die chemische Wirkung des Lichtes auf reines und gefärbtes
Bromsilber.

> Vogel (H. W.). Ber. chem. Ges., **8**, 1685–6; Jour. Chem. Soc., **1**
> (1876), 510 (Abs.); Amer. Jour. Sci., (3) **11**, 215–16 (Abs.).

Neue Betrachtungen über die Lichtempfindlichkeit des Bromsilbers.

> Vogel (H. W.). Ber. chem. Ges., **9**, 667–70; Jour. Chem. Soc., **2**
> (1876), 265 (Abs.).

Ueber die Empfindlichkeit trockner Bromsilberplatten gegen das Sonnenspectrum.

> Vogel (H. W.). Ber. chem. Ges., **14**, 1024–8; Beiblätter, **5**, 521
> (Abs.); Jour. Chem. Soc., **40**, 778 (Abs.).

Ueber die verschiedenen Modificationen des Bromsilbers.

> Vogel (H. W.). Ber. chem. Ges., **16**, 1170–79; Beiblätter, **7**, 536
> (Abs.).

Sur la sensibilité du bromure d'argent à l'égard des radiations considérées
comme chimiquement inactives.

> Vogel (H. W.). Bull. Soc. chim. Paris, n. s. **21**, 233.

Ueber die Brechung und Dispersion des Lichtes im Bromsilber.

> Wernicke (W.). Ann. Phys. u. Chem., **142**, 560–73; Jour. Chem.
> Soc., (2) **9**, 653 (Abs.); Ann. Chim. et Phys., (4) **26**, 287.

Uebereinstimmung des Absorptionsspectrums von Brom mit dem Spectrum dessen Dampfes.

> Wüllner (A.). Ann. Phys. u. Chem., **120**, 150.

CADMIUM.

Ultra-violet spectrum of cadmium.

Bell (L.). Amer. Jour. Sci., **31** (1886), 426–31; Jour. Chem. Soc., **50**, 957 (Abs.).

Cadmium arc spectrum.

Capron (J. R.). Photographed Spectra, London, 1877, 23.

Spectrum of chloride of cadmium.

Chem. News, **35**, 107.

Déterminations des longueurs d'onde des radiations très réfrangibles du cadmium.

Cornu (A.). Arch. de Genève, (3) **2**, 119–126; Beiblätter, **4**, 34 (Abs.); Jour. de Phys., **10**, 425–31.

Renversement des raies spectrales du cadmium.

Cornu (A.). Comptes Rendus, **73**, 332.

Spectre de chlorure de cadmium.

Gouy. Comptes Rendus, **84**, 231.

Spectrum von Cadmium.

Jahresber. d. Chemie (1872), 145.

Chlorure de cadmium en solution, étincelle.

Lecoq de Boisbaudran (F.). Spectres Lumineux, Paris, 1874, 189.

Spectrum of cadmium at elevated temperatures.

Lockyer (J. N.). Chem. News, **30**, 98.

Indice du quartz pour les raies du cadmium.

Sarasin·(Ed.). Comptes Rendus, **85**, 1230.

CÆSIUM.

Observations on cæsium.

> Allen (O. D.). Phil. Mag., **25**, 189; Amer. Jour. Sci., (2) **34** (1862), 367.

On the equivalent and spectrum of cæsium.

> Allen (O. D.) and Johnson (S. W.). Phil. Mag., **25**, 196; Amer. Jour. Sci., (2) **35** (1868), 94.

On cæsium.

> Bunsen (R.). Phil. Mag., **26**, 241.

Les salpêtres naturels du Chili et du Pérou au point de vue du cæsium.

> Dieulafait. Comptes Rendus, **98**, 1545–8; Chem. News, **50**, 45 (Abs.).

Recherches sur la présence du cæsium dans les eaux naturelles.

> Grandeau (L.). Ann. Chim. et Phys., (3) **67**, 155.

Spectrum von Cæsium.

> Kirchhoff (G.) und Bunsen (R.). Ann. Phys. u. Chem., **113**, 337, 379; Phil. Mag., (4) **22**, 498.

Chlorure de cæsium.

> Lecoq de Boisbaudran (F.). Spectres Lumineux, Paris, 1874, p. 44, planche III.

On pollux, a silicate of cæsium.

> Pisani. Comptes Rendus, **58**, 714.

CALCIUM.

Sur la phosphorescence du sulfure de calcium.

Becquerel (Edm.). Comptes Rendus, **103** (1887), 551–3 ; Chem. News, **55** (1887), 128.

Action du manganèse sur le pouvoir de phosphorescence du carbonate de chaux.

Becquerel (Edm.). Comptes Rendus, **103** (1886), 1098–1101.

Ueber das Calciumspectrum.

Blochmann (R.). Jour. prackt. Chem., (2) **4**, 282–6 ; Jour. Chem. Soc., (2) **9**, 1149–1150 (Abs.).

Calcium (Zinc) spark spectrum.

Capron (J. R.). Photographed Spectra, London, 1877, p. 23.

Spectre de chlorure de calcium.

Gouy. Comptes Rendus, **84**, 231.

Recherches photométriques, spectre du calcium.

Gouy. Comptes Rendus, **85**, 70.

Sur les flammes chargées du chlorure de calcium.

Gouy. Comptes Rendus, **85**, 489.

Spectre continu du calcium.

Gouy. Comptes Rendus, **85**, 878, 1078.

Spectrum von Kalk.

Jahresber. d. Chemie (1870), 174.

Linien von Calcium.

Kirchhoff (G.) und Bunsen (R.). Ann. Phys. u. Chem., **110**, 177.

Das Wärmespectrum des Kalklichtes.

Lamansky (S.). Monatsber. d. Berliner Akad. (1871), 632–41 ; Phil. Mag., (4) **43**, 282–9 ; Ann. Phys. u. Chem., **146**, 200–32.

Ueber die Dispersion des Aragonits nach arbiträrer Richtung.

Lang (V. von). Sitzungsber. d. Wiener Akad., **83** II, 671–6.

Note on the spectra of calcium fluoride.

Liveing (G. D.). Proc. Philosoph. Soc. Cambridge, **3**, 96–8 ; Beiblätter, **4**, 611–12 (Abs.).

Sur de nouvelles raies de calcium.

Lockyer (J. N.). Comptes Rendus, **82**, 660–2; Ann. Chim. et Phys., (5) **7**, 569–72; Chem. News, **33**, 166–7; Jour. Chem. Soc., **2** (1876), **35** (Abs.); Ber. chem. Ges., **9**, 505 (Abs.); Ann. Phys. u. Chem.. **158**, 327–9 (Abs.); Bull. Soc. chim. Paris, n. s. **26**, 267.

Remarques à propos de la dernière communication de M. Lockyer sur de nouvelles raies de calcium, par M. C. Sainte-Claire Deville. Comptes Rendus, **82**, 709–10.

Calcium comme corps composé d'après le spectroscope.

Lockyer (J. N.). Comptes Rendus, **87**, 673.

Fluorescenz von Kalkspar.

Lommel (E.). Ann. Phys. u. Chem., n. F. **21**, 422–7; Jour. Chem. Soc., **46**, 649 (Abs.).

Sur l'origine de l'arsénic et de la lithine dans eaux sulfatées calciques.

Schlagdenhauffen. Jour. de Pharm., (5) **6**, 457–68; Jour. Chem. Soc., **44**, 302 (Abs.).

Sur les causes déterminantes de la phosphorescence du sulfure de calcium.

Verneuil (A.). Comptes Rendus, **103** (1887), 601–4; Beiblätter, **11** (1887), 253; Jour. Chem. Soc., **52**, 2.

Ueber die neuen Wasserstofflinien und die Dissociation des Calciums.

Vogel (H. W.). Ber. chem. Ges., **13**, 274–6; Jour. Chem. Soc., **33**, 597 (Abs.); Beiblätter, **4**, 274, 786; Monatsber. d. Berliner Akad. (1880), 192–8; Nature, **21**, 410.

Expériences sur divers échantillons de chaux.

Volpicelli (M.). Comptes Rendus, **56**, 493; **57**, 571.

Coïncidence of the spectrum lines of iron, calcium, and titanium.

Williams (W. Mattieu). Nature, **8**, 46.

CARBON.

1, CARBON IN GENERAL.

Note on the spectrum of carbon.

> Attfield (J.). Phil. Mag., (4) **49**, 106–8; Phil. Trans. (1862), 221.

Carbon points ruled out.

> Capron (J. R.). Photographed Spectra, London, 1877, 23.

Spectroscopic researches in carbon and cyanogen.

> Ciamician (G. L.). Chem. News, **44**, 216.

On the refraction equivalents of the diamond and the carbon compounds.

> Gladstone (J. H.). Chem. News, **42**, 175; Jour. Chem. Soc., **40**, 333
> (Abs.); Beiblätter, **5**, 43 (Abs.); Proc. Royal Soc., **31**, 327–30; Ber.
> chem. Ges., **14**, 1553 (Abs.).

Carbon and carbon compounds.

> Herschel (A. S.). Nature, **22**, 320; Beiblätter, **5**, 118–122.

Spectrum von Kohlenstoff.

> Jahresber. d. Chemie, (1862).33, (1863) 113, (1864) 109, (1865) 89, (1869)
> 176, 178, (1875) 122.

Refractionsäquivalente der Elemente C, etc.

> Landolt (R.). Versammlung deutscher Aertzte und Naturforscher,
> Aug. 12–18, 1872; Ber. chem. Ges., **5**, 808; Chem. Centralblatt, (3)
> **3**, 705; Jour. Chem. Soc., (2) **11**, 460 (Abs.).

Note on the history of the carbon spectrum.

> Liveing (G. D.) and Dewar (J.). Proc. Royal Soc., **30**, 490–4; Bei-
> blätter, **5**, 118–22; Nature, **23**, 265–6, 338.

Spectrum of Carbon.

> Liveing (G. D.) and Dewar (J.). Proc. Royal Soc., **33**, 403–410;
> Chem. News, **45**, 155 (Abs.); Nature, **25**, 545; Jour. Chem. Soc.,
> **44**, 1–2 (Abs.); Beiblatter, **6**, 675 (Abs.).

General observations on the spectra of carbon and its compounds.

> Liveing (G. D.) and Dewar (J.). Proc. Royal Soc., **34**, 123–30.

Spectrum of carbon at elevated temperatures.

> Lockyer (J. N.). Chem. News, **30**, 98.

Note on the spectrum of carbon.

> Lockyer (J. N.). Proc. Royal Soc., **30**, 385–43, 461–3; Beiblätter, **5**, 118–22 (Abs.).

Sulla questione dei doppi legami tra carbonio e carbonio dal punto di vista della chimica ottica.

> Nasini (R.). Gazz. chim. ital., **14**, 150–6; Ber. chem. Ges., **17**, Referate, 559–61 (Abs.); Atti R. Ac. dei Lincei, **8**, 169–73; Beiblätter, **8**, 577.

On the spectrum of carbon.

> Roscoe (H. E.). Nature, **23**, 313–14.

Spectre du carbone.

> Salet (G.). Bull. Soc. chim. Paris, 1 Mars 1872; Ber. chem. Ges., **5**, 222 (Abs.).

Ueber das Dispersionsäquivalent von Diamant.

> Schrauf (A.). Ann. Phys. u. Chem., n. F. **22**, 424–9; Jour. Chem. Soc., **48**, 14 (Abs.).

Note on the identity of the spectra obtained from the different allotropic forms of carbon.

> Schuster (A.) and Roscoe (H. E.). Proc. Manchester Philosoph. Soc., **19**, 46–49; Beiblätter, **4**, 208 (Abs.).

Carbon and hydrocarbon in the modern spectroscope.

> Smyth (C. Piazzi). Phil. Mag., (4) **49**, 24–33.

Carbon and carbo-hydrogen, spectroscoped and spectrometed.

> Smyth (C. Piazzi). Phil. Mag., (5) **8**, 107–19; Beiblätter, **4**, 36 (Abs.).

Spectre du carbone.

> Troost et Hautefeuille. Comptes Rendus, **73**, 620; Bull. Soc. chim. Paris, n. s. **16**, 229.

Spectra of carbon.

> Watts (W. M.). Phil. Mag., (4) **38**, 249; **41**, 12; **48**, 369, 456; **49**, 104; Nature, **23**, 197, 266; Beiblätter, **5**, 118; Chem. News, **22**, 172; Jour. prackt. Chemie, **104**, 422.

2, CARBON COMPOUNDS.

a, In general.

Influence of the molecular grouping in organic bodies on their absorption in the infra-red region of the spectrum.

> Abney (W. de W.) and Festing (Lieut. Col.). Proc. Royal Soc., **31**, 416; Chem. News, **43**, 92, 126; Beiblätter, **5**, 506.

Action des rayons différemment réfrangible sur l'iodure et le bromure d'argent; influence des matières colorantes.

> Becquerel (E.). Comptes Rendus, **79**, 185–90; Jour. Chem. Soc., (2) **13**, 80 (Abs.).

Sulla relazioni esistenti tra il potere rifrangente e la constituzione chimic ι della combinazioni organiche.

> Bernheimer e Nasini. Atti della R. Accad. dei Lincei, Transunti, (3) **7**, 227–80; Gazz. chim. ital., **13**, 817–20; Beiblätter, **7**, 523 (Abs.).

Influence des diverses couleurs sur la végétation.

> Bert (P.). Comptes Rendus, **73**, 1444.

Sur la région du spectre solaire indispensable à la vie végétale.

> Bert (P.). Comptes Rendus, **87**, 695–7; Jour. Chem. Soc., **36**, 836 (Abs.).

Vergleichung von Pigmentfarben mit Spectralfarben.

> Bezold (W. von). Ann. Phys. u. Chem., **158**, 165, 606.

On the action of various colored bodies on the spectrum.

> Brewster (Sir D.). Phil. Mag., (4) **24**, 441.

Die Beziehungen zwischen den physikalischen Eigenshaften organischer Körper und ihrer chemischen Constitution.

> Brühl (J. W.). Ber. chem. Ges., **12**, 2185–48; **13**, 1119–80, 1520–85; **14**, 2533–39; Jour. Chem. Soc., **38**, 293–5 (Abs.); Beiblätter, **4**, 776–86; Amer. Jour. Sci., (3) **23**, 284–6 (Abs.).

Die chemische Constitution organischer Körper in Beziehung zu deren Dichte und ihren Vermögen das Licht fortzupflanzen. Drei Theile und Nachtrag.

> Brühl (J. W.). Ann. Chem. u. Pharm., **200**, 139–231; **203**, 1–63, 255–285, 363–868; Jour. Chem. Soc., **38**, 295–7 (Abs.); **38**, 781–3 (Abs.); Beiblätter, **4**, 776–86.

Ueber den Zusammenhang zwischen den optischen und den thermischen Eigenschaften flüssiger organischer Körper.

> Brühl (J. W.). Sitzungsber. d. Wiener Akad., **84** II, 817–75; Monatschr. f. Chemie, **2**, 716–74; Ann. Phys. u. Chem., **211**, 121–178; Jour. Chem. Soc., **42**, 263 (Abs.); Beiblätter, **6**, 377 (Abs.).
> Berichtigung, Ann. Phys. u. Chem., **211**, 871–2.

Untersuchungen über die Molecularrefraction organischer flüssiger Körper von grossen Farbenzerstreuungsvermögen.

> Bruhl (J. W.). Ber. chem. Ges , **19** (1886), 2746.

De l'action des différentes lumières colorées sur une couche de bromure d'argent impregnée de diverses matières colorantes organiques.

 Cros (Ch.). Comptes Rendus, **88**, 879–81 , Jour. Chem. Soc., **36**, 504 (Abs.).

Relation between the chemical constitution of certain organic compounds and their action upon the ultra-violet rays.

Dunstan (W. R.). Pharmaceutical Trans., (3) **11**, 54–6.

Note concernant le mémoire de M. Kanonikoff sur le pouvoir réfringent des substances organiques.

Flavitsky (F.). Jour. Soc. phys. chim. russe, **16**, 260–7.

On the refraction equivalents of the diamond and the carbon compounds.

Gladstone (J. H.). Chem. News, **42**, 175 ; Jour. Chem. Soc., **40**, 333 (Abs.); Beiblätter, **5**, 43 (Abs.).

Refraction equivalents of organic compounds.

Gladstone (J. H.). Jour. Chem. Soc., **45**, 241–59 ; Chem. News, **49**, 283 (Abs.); Nature, **30**, 119 (Abs.); Ber. chem. Ges., **17**, Referate, 556 (Abs.).

Spectres des carbonates.

Gouy. Comptes Rendus, **85**, 70.

Influence of certain rays of the spectrum on plants growing in an iron manure.

Griffiths (A. B.). Jour. Chem. Soc., **45**, 74.

Ueber das Verhalten einiger Farbstoffe im Sonnenspectrum.

Haerlin (J.). Ann. Phys. u. Chem., **118**, 70.

Researches on the absorption of the ultra-violet rays of the spectrum by organic substances.

Hartley (W. N.) and Huntington (A. K.). Proc. Royal Soc., **28**, 233 ; **31**, 1 ; Chem. News, **40**, 269 ; Phil. Trans., **170**, 257–74 ; Beiblätter, **4**, 370.

Researches on the relation between the molecular structure of carbon compounds and their absorption spectra.

Hartley (W. N.). Jour. Chem. Soc., **39**, 153–68; **41**, 45–49 ; Beiblätter, **6**, 375 (Abs.); Amer. Chem. Jour., **3**, 373.

Das Auge empfindet alle Strahlen die brechbarer sind als die rothen.

Helmholtz (H.). Ann. Phys. u. Chem., **94**, 205.

Absorptionsstreifen färbiger Lösungen.

Jahresber. d. Chemie, (1864) 108, (1865) 85, (1867) 825, (1868) 129, (1873) 147.

On the chemical circulation in the body.

Jones (H. Bence). Proc. Royal Institution, May 26, 1865.

Zur Frage über den Einfluss der Structur auf das Lichtbrechungsvermögen organischer Verbindungen.

Kanonnikoff (J.). Jour. russ. phys. chem. Ges. (1881), 268; Ber. chem. Ges., 14, 1697–1700.

Sur le pouvoir réfringent des substances organiques dans les dissolutions.

Kanonnikoff (J.). Jour. Soc. phys. chim. russe, 15, 112–13; Ber. chem. Ges., 16, 950 (Abs.); Jour. prackt. Chemie, n. F. 27, 362–4; Beiblätter, 7, 598 (Abs.); Jour. Chem. Soc., 44, 1041 (Abs.).

Sur la relation du pouvoir réfringent et la composition des composés organiques.

Kanonnikoff (J.). Jour. Soc. phys. chim. russe, 15, 434–79; Ber. chem. Ges., 16, 3047–3051 (Abs.); Bull. Soc. chim. Paris, 41, 318 (Abs.); Beiblätter, 8, 875 (Abs.).

Sur les relations entre la composition et le pouvoir réfringent des composés chimiques.

Kanonnikoff (J.). Jour. Soc. phys. chim. russe, 16, 119–181; Ber. chem. Ges., 17, Referate, 157 (Abs.); Nature, 30, 84 (Abs.); Bull. Soc. chim. Paris, 12, 549.

Réponse à la note de M. Flavitsky.

Kanonnikoff (J.). Jour. Soc. phys. chim. russe, 16, 448–50; Jour. prackt. Chemie, (2) 31, 321–3 (Abs.).

Spectrum of colour-blind.

König (Dr.). Nature, 29, 168.

Beziehungen zwischen der Zusammensetzung und den Absorptionsspektren organischer Verbindungen.

Krüss (G.) und Oeconomides (S.). Ber. chem. Ges., 16, 2051–6; Jour. Chem. Soc., 44, 1041–2 (Abs.); Beiblätter, 7, 897 (Abs.).

Ueber die Gränzen der Empfindlichkeit des Auges für Spectralfarben.

Lamansky (S.). Ann. Phys. u. Chem., 143, 688–48.

Zur Kenntniss der Absorptionsspectra von Verbindungen.

Landauer (J.). Ber. chem. Ges., 14, 891–4; Jour. chem. Soc., 40, 591 (Abs.); Beiblätter, 5, 441.

Ueber die Molecularrefraction flüssiger organischer Verbindungen.

Landolt (H.). Sitzungsber. d. Berliner Akad. (1882), 64–91; Ann. Phys. u. Chem., 213, 75–112; Jour. Chem. Soc., 42, 909 (Abs.).

On the theory of the action of certain organic substances in increasing
the sensitiveness of silver haloids.

> Lea (M. Carey). Amer. Jour. Sci., (8) 14, 96–9; Beiblätter, 1, 5C3
> (Abs.).

Ueber die Aenderung der Absorptionsspectra einiger Farbstoffe in verschiedenen Lösungsmitteln.

> Lepel (F. von). Ber. chem. Ges., 11, 1146–51; Jour. Chem. Soc.,
> 34, 925 (Abs.).

Planzenfarbstoffe als Reagentien auf Magnesiumsalze.

> Lepel (F. von). Ber. chem. Ges., 13, 766–8; Jour. Chem. Soc., 40,
> 63 (Abs.).

Contributions to our knowledge of the spectra of the flames of gases containing carbon.

> Lielegg (A.). Phil. Mag., (4) 37, 208.

General observations on the spectra of carbon and its compounds.

> Liveing (G. D.) and Dewar (J.). Proc. Royal Soc., 34, 123–30; Jour.
> Chem. Soc., 44, 261 (Abs.).

New organic spectra.

> MacMunn (Dr. C. A.). Proc. Roy. Physiolog. Soc. (1884), No. 4;
> Nature, 31 (1885), 826–7.

De la flamme de quelques gaz carburés (avec une planche du spectre du
carbone).

> Morren (A.). Ann. Chim. et Phys., (4) 4, 305.

Sur les effets de coloration.

> Nickles. Comptes Rendus, 62, 93.

Les rapports entre les propriétés spectrales des corps simples avec leurs
propriétés physiologiques.

> Papillon. Comptes Rendus, 73, 791.

Quantitative Bestimmung von Farbstoffen durch den Spectralapparat.

> Preyer (W.). Ber. chem. Ges., 4, 404.

Du spectre musculaire.

> Ranvier (L.). Comptes Rendus, 78, 1572–5.

Absorptionsspectren verschiedener Farbenlösungen.

> Reynolds. Jour. prackt. Chemie, 105, 358.

Versuche über Farbenmischung.

> Schelske (R.). Ann. Phys. u. Chem., n. F. 16, 849–58.

Quantitative Bestimmung von Farbstoffen durch den Spectralapparat.

> Schiff (H.). Ber. chem. Ges., **4**, 474; Bull. Soc. chim. Paris, n. s. **16**, 97.

On a definite method of qualitative analysis of animal and vegetable colouring matters by means of the spectrum-microscope.

> Sorby (H. C.). Proc. Royal Soc., **15**, 433.

Comparative vegetable chromatology.

> Sorby (H. C.). Proc. Royal Soc., **21**, 442.

On the colouring matters derived from the decomposition of some minute organisms.

> Sorby (H. C.). Monthly Microscop. Jour., **3**, 229-31.

On the examination of mixed colouring matters with the spectrum-microscope.

> Sorby (H. C.). Monthly Microscop. Jour., **6**, 124-34.

Zur Spectralanalyse gefärbter Flüssigkeiten und Gläser.

> Stein. Jour. prackt. Chemie, n. F. **9**, 833; **10**, 368; Jour. Chemical Soc., (2) **13**, 412-14 (Abs.).

On the discrimination of organic bodies by their optical properties.

> Stokes (G. G.). Phil. Mag., (4) **27**, 388.

Prismatic spectra of the flames of compounds of carbon and hydrogen.

> Swan (W.). Edinburgh Philosoph. Trans., **21**, 411; Ann. Phys. u. Chem., **100**, 306.

Longueur d'ondes des bandes spectrales données par les composés du carbone.

> Thollon (L.). Comptes Rendus, **93**, 260; Ann. Chim. et Phys., (5) **25**, 287-8.

Absorptionsspectren verschiedener Farbenlösungen.

> Thudichum. Jour. prackt. Chemie, **106**, 414-15.

Der Gebrauch des Spectroscops zu physiologischen und ärtztlichen Zwecken.

> Valentin (G.). Leipzig, Winter'sche Buchhandlung, 1863.

Quantitative Bestimmung von Farbstoffen durch den Spectralapparat.

> Vierordt (K.). Ber. chem. Ges., **4**, 327, 457, 519; Phil. Mag., (4) **41**, 482-4; Amer. Jour. Sci., (3) **2**, 138 (Abs.); Bull. Soc. chim. Paris, n. s. **16**, 96.

Ueber die abnorme Wirkung mancher Farbstoffe auf die Lichtempfind-lichkeit photographischer Platten.

     Vogel (H. W.). Ber. chem. Ges., **8**, 95–6.

Ueber das Spectrum der Sell'schen Schwefelkohlenstofflampe.

     Vogel (H. W.). Ber. chem. Ges., **8**, 96–8; Jour. Chem. Soc., (2) **13**, 604 (Abs.).

Ueber die Absorptionsspectren verschiedener Farbenstoffe und ihre An-wendung zur Entdeckung von Verfälschungen.

     Vogel (H. W.). Ber. chem. Ges., **8**, 1246–54; Dingler's Journal, **219**, 78–81; Bull. Soc. chim. Paris, n. s. **26**, 475.

Ueber die Wandlung der Spectren verschiedener Farbstoffe.

     Vogel (H. W.). Ber. chem. Ges., **11**, 622–4; Jour. Chem. Soc., **34**, 545 (Abs.).

Ueber den Zusammenhang zwischen Absorption der Farbstoffen und deren sensibilisirender Wirkung auf Bromsilber.

     Vogel (H. V.). Ann. Phys. u. Chem., (2) **26** (1885), 527–30.

Untersuchungen über die Spectra der Kohlenverbindungen.

     Wesendonck (K.). Ann. Phys. u. Chem., n. F. **17**, 427–67; Jour. Chem. Soc., **44**, 761 (Abs.); Monatsber. d. Berliner Akad. (1880), 791–4.

     Bemerkungen, Wüllner (A.). 'Ann. Phys. u.' Chem., n. F. **14**, 863.

b, Carbon compounds in particular.

ACETIC ACID.

Indices de réfraction des dissolutions aqueuses d'acide acétique et d'hypo-sulfite de soude.

     Damien. Comptes Rendus, **91**, 828–5; Beiblätter, **5**, 41–42 (Abs.).

ACETYLENE.

Bemerkung zu Herrn Wüllner's Aufsatz; Ueber die Spectra des Wasser-stoffs und des Acetylens.

     Hasselberg (B.). Ann. Phys. u. Chem., n. F. **15**, 45–49.

Spectrum des Acetylens.

     Jahresber. d. Chemie (1869), 182.

De la flamme de quelques gaz carburés, et en particulier de celle de l'acetylène.

     Morren (A.). Ann. Chim. et Phys., (4) **4**, 305; Jour. prackt. Chem., **37**, 50.

Spectrum des Acetylens.

> Wüllner (A.). Ann. Phys. u. Chem., n. F. **14**, 855.
> Bemerkung, Hasselberg (B.), do., **15**, 45–9.

Spectrum of acid brown.

> Hartley (W. N.). Jour. Chem. Soc., **51** (1887), 198.

Spectrum of agarythrine, an alcaloid contained in agaricus ruber.

> Phipson (T. L.). Chem. News, **46**, 199–200; Ber. chem. Ges., **16**, 244 (Abs.).

Farbenreactionen des Albumin.

> Adamkiewicz (A.). Pflüger's Arch. f. Physiol., **9**, 156–162; Jour. Chem. Soc., (2) **13**, 172 (Abs.).

Spectroscopic notes on the carbohydrates and albumenoids from grain.

> Hartley (W. N.). Jour. Chem. Soc., **51** (1887), 58–61.

Misura dell'indice di rifrazione dell'alcool anisico e dell'alcool metil-salicilico.

> Blaserna (P.). Gazz. chim. ital., **2**, 69–75.

Brechungscoefficienten einiger Gemische von Anilin und Alkohol.

> Johst (W.). Ann. Phys. u. Chem., n. F. **20**, 47–62.

Spectre de l'alcohol.

> Masson (A.). Comptes Rendus, **32**, 129

Ueber die Absorption des Lichtes durch Alcohol, etc.

> Schönn (J. L.). Ann. Phys. u. Chem., Ergänzungsband, **8**, 670–675; Jour. Chem. Soc., **34**, 698 (Abs.).

Notiz über künstliches Alizarin.

> Boettger (R.) und Petersen (T.). Ber. chem. Ges., **4**, 778–9.

Spectre d'absorption d'alizarine.

> Gernez (D.). Bull. Soc. chim. Paris, n. s. **18**, 172.

Absorptionsspectrum des Alizarins.

> Jahresber. d. Chemie (1872), 140.

11 T

On artificial alizarine.

> Perkin (W. H.). Jour. Chem. Soc., (2) **8**, 133-43; Ann. Chem. u. Pharm., **158**, 315-19 (Abs.); Ann. Chim. et Phys., (4) **26**, 186 (Abs.).

Absorptionsspectrum des Alizarins.

> Reynolds. Jour. prackt. Chem., **105**, 358.

L'alizarine nitrée.

> Rosenstiehl (A.). Ann. Chim. et Phys., (5) **12**, 519-529; Jour. Chem. Soc., **34**, 281-2.

Sur les spectres d'alizarine et de quelques matières colorantes qui en derivent.

> Rosenstiehl (A.). Comptes Rendus, **88**, 1194-6; Jour. Chem. Soc., **36**, 807 (Abs.); Beiblätter, **3**, 798.

Zur Kenntniss der Alizarin-Farbstoffe.

> Vogel (H. W.). Ber. chem. Ges., **11**, 1371-4; Jour. Chem. Soc., **36**, 88-5 (Abs.).

ALKANNA.

Der Alkannafarbstoff, ein neues Reagens auf Magnesiumsalze.

> Lepel (F. von). Ber. chem. Ges., **13**, 763-6.

ALLYLDIPROPYLCARBINOL.

Untersuchungen über einen aus Allyldipropylcarbinol erhaltenen Kohlenwasserstoff.

> Reformatsky (S.). Jour. prackt. Chemie, n. F. **27**, 389-407; Beiblätter, **7**, 689 (Abs.).

ALUM.

Sur les aluns crystallisés.

> Soret (C.). Arch. d. Genève, (3) **10**, 300; Beiblätter, **8**, 374.

AMIDO-AZO-α-NAPHTHALENE.

Spectrum of amido-azo-α-naphthalene, $C_{10} H_7 \cdot N : N \cdot C_{10} H_6 \cdot N H_2$.

> Hartley (W. N.). Jour. Chem. Soc., **51** (1887), 190.

AMIDO-AZO-β-NAPHTHALENE.

Spectrum of amido-azo-β-naphthalene.

> Hartley (W. N.). Jour. Chem. Soc., **51** (1887), 191.

ANILINE.

Die Brechungscoefficienten einiger Gemische von Anilin.

> Johst (W.). Ann. Phys. u. Chem.. n. F. **20**, 47-62.

Lo Spettroscopio applicato alla ricerca dei colori di anilina introdati nei vini rossi per sofisticazione.
> Macagno (J.). Mem. Spettr. ital. (1881), 35–40; Ber. chem. Ges., **14**, 1584 (Abs.).

Aniline colours in the spectroscope.
> Reimann (M.). Chem. News, **33**, 260.

Absorptionslinien der Anilinfarbstoffe im Spectralapparat.
> Schiff. Jour. prackt. Chemie, **89**, 229.

Application of the spectroscope in the manufacture of aniline colours.
> Schoop (P.). Chemische Industrie, **9** (1886), No. 8; Chem. News, **53** (1886), 287 (Abs.).

Zur Kenntniss der grünen Anilinfarben.
> Vogel (H. W.). Ber. chem. Ges., **11**, 1871–4; Jour. Chem. Soc., **36**, 83–5 (Abs.).

ANTHRACEN.

Ueber Anthracen-disulfosäure und deren Umwandlung in Antrarufin.
> Liebermann (C.) und Boeck (K.). Ber. chem. Ges., **11**, 1613–18; Jour. Chem. Soc., **36**, 257–9.

Ueber die der Chrysazinreihe augehörigen Anthracenverbindungen.
> Liebermann (C.). Ber. chem. Ges., **12**, 182–8.

Use of the spectroscope in discriminating anthracens.
> Nickels (B.). Chem. News, **41**, 52, 95, 117; Jour. Chem. Soc., **38**, 757 (Abs.); Ber. chem. Ges., **13**, 829 (Abs.).

ANTHRAPURPURIN.

Absorptionsspectrum des Anthrapurpurins.
> Jahresber. d. Chemie (1873), 451.

Absorptionspectra of anthrapurpurin.
> Perkin (W. H.). Jour. Chem. Soc., (2) **11**, 433.

ANTHRARUFIN.

Ueber Anthracen-disulfosäure und deren Umwandlung in Anthrarufin.
> Liebermann (C.) und Boeck (K.). Ber. chem. Ges., **11**, 1613–18; Jour. Chem. Soc., **36**, 257–9 (Abs.).

APHIDES.

On the colouring matter of some aphides.
> Sorby (H. C.). Quar. Jour. Microscop. Sci., **11**, 352–61.

Spectrum of aurin.

> Hartley (W. N.). Jour. Chem. Soc., **51** (1887), 167-8.

Spectrum of a poisonous Australian lake.

> Francis (G.). Pharmaceutical Trans., (3) **8**, 1047-8; Jour. Chem. Soc., **34**, 907 (Abs.).

Spectrum of azobenzene.

> Hartley (W. N.). Jour. Chem. Soc., **51** (1887), 176-8.

Spectrum of amido-azo-α-naphthalene, and of amido-azo-β-naphthalene.

> Hartley (W. N.). Jour. Chem. Soc., **51** (1887), 190-1.·

On the spectra of the azo-colours.

> Stebbins (J. H.). Jour. Amer. Chem. Soc., **6** (1884), 117-20, 149-50.

Spectralanalytische Notiz; rothe Rüben in Weinverfälschungen.

> Lepel (F. von). Ber. chem. Ges., **10**, 1875-7; Jour. Chem. Soc., **34**, 168 (Abs.); Bull. Soc. chim. Paris, n. s. **30**, 573.

Description and measurements of the spectrum of benzene.

> Hartley (W. N.). Jour. Chem. Soc., **47** (1885), 694-6.

Spectrum of benzene-azo-β-naphtholsulphonic acid.

> Hartley (W. N.). Jour. Chem. Soc., **51** (1887), 196.

Misura dell'indice di rifrazione del cimene, della benzina e di alcuni derivati del timol naturale e del timol sintetico.

> Pisati (G.) e Paterno (E.). Gazz. chim. ital., **4**, 557-64; Ber. chem. Ges., **8**, 71 (Abs.).

Spectrum of biebrich scarlet.

> Hartley (W. N.). Jour. Chem. Soc., **51** (1887), 194.

Le reazioni dei pigmenti biliari.

> Capranica (S.). Gazz. chim. ital., **11**, 430-1; Ber. chem. Ges., **15**, 262-3 (Abs.); Jour. Chem. Soc., **42**, 282.

Researches into the colouring matters of human urine, with an account of their artificial production from bilirubin and from hæmatin.

> MacMunn (C. A.). Proc. Royal Soc., **31**, 206–37; Jour. Chem. Soc., **40**, 1056–8 (Abs.); Beiblätter, **5**, 281.

Observations on the so-called bile of invertebrates.

> . MacMunn (C. A.). Proc. Royal Soc., **35**, 370–408.

Künstliche Umwandlung von Bilirubin in Harnfarbstoff.

> Maly (R.). Ann. Chem. u. Pharm., **161**, 868–70; **163**, 77–95; Jour. Chem. Soc., (2) **10**, 514 (Abs.), 835 (Abs.).

A reducible by-product of the oxidation of bile-pigment.

> Stockvis (B. J.). Neues Repertorium f. Pharm., **21**, 123, 732–7; Jour. Chem. Soc., (2) **10**, 808 (Abs.); **11**, 288; Bull. Soc. chim. Paris, n. s. **18**, 265.

Researches on bilirubin and its compounds.

> . Thudichum (J. L. W.). Jour. Chem. Soc., (2) **13**, 389–403.

BIRDS.

Spectres observés au travers d'une plume.

> Hugo (L.). Comptes Rendus, **83**, 602.

Ueber die Färbungen der Vogeleierschalen.

> Liebermann (C.). Ber. chem. Ges., **11**, 606–610; Amer. Jour. Sci., (3) **16**, 66 (Abs.).

BISMARCK BROWN.

Spectrum of bismarck brown.

> Hartley (W. N.). Jour. Chem. Soc., **51** (1887), 180–1.

BLOOD.

Ueber das Verhalten von Blut und Ozon zu einander.

> Binz (C.). Medicinalisches Centralblatt, **20**, 721–5; Chemisches Centralblatt (1882), 810–11; Jour. Chem. Soc., **44**, 486 (Abs.).

Dosage de l'hemoglobine dans le sang par les procédés optiques.

> Branly (E.). Ann. Chim. et Phys., (5) **27**, 238–78; Jour. Chem. Soc., **44**, 394 (Abs.); Z. analyt. Chem., **22**, 629–32 (Abs.); Jour. de Phys., (2) **2**, 480 (Abs.).

Absorptionsspectrum des durch Wasserstoffsuperoxyd gebräunten blausäurehaltigen Blutes.

> Buchner. Jour. prackt. Chem., **104**, 845.

On the action of nitrates on the blood.

Gamge (A.). Phil. Trans. (1868), 589; Ber. chem. Ges., 9, 838; Jour. prackt. Chemie, 105, 287.

Absorptionslinien in Blutspectrum.

Hoppe-Seyler (F.). Jahrb. d. gesammt. Medicin, 114, 8.

Ueber das Verhalten des Blutfarbestoffs in Spectrum des Sonnenlichtes.

Hoppe-Seyler (F.). Virchow's Annalen, 22, 446; 29, 238; Chem. Centralblatt, 1862, 170.

Untersuchungen zur physicalischen Chemie des Blutes.

Hüfner (G.). Jour. prackt. Chemie, (2) 22, 362–88; Jour. Chem. Soc., 40, 111–13 (Abs.).

Untersuchungen über den Blutfarbestoff und seine Derivate.

Jäderholm (A.). Zeitschr. f. Biologie, 13, 198–255; Jour. Chem. Soc., 34, 236–7 (Abs.).

Spectren des Blutfarbstoffs.

Jahresber. d. Chemie, 15, 585 (Abs. See Hoppe-Seyler, above.)

Photometrie des Absorptionsspectrums der Blutkörperchen.

Jessen (E.). Zeitschr. f. Biologie, 17, 251–72; Ber. chem. Ges., 15, 952 (Abs.).

Spectrum der Sanguinarlösung.

Naschold. Jour. prackt. Chemie, 106, 407.

Beträge zur Kentniss der Blutfarbstoffe.

Otto (J. G.). Pflüger's Archiv. f. Physiol., 31, 240–44; Ber. chem. Ges., 16, 2688–9.

On some improvements in the spectrum method of detecting blood.

Sorby (H. C.). Monthly Microscop. Jour., 6, 9–17.

On some compounds derived from the colouring matter of blood.

Sorby (H. C.). Quar. Jour. Microscop. Sci., 10, 400–2.

Application of spectrum analysis to microscopical investigations, and especially to the detection of blood stains.

Sorby (H. C.). Chem. News, 11, 186, 194, 232, 256.

On the blood spectrum.

Sorby (H. C.). Nature, 4, 505; 5, 7.

Spectre d'absorption du sang dans la partie violette et ultra-violette.

Soret (J. L.). Comptes Rendus, 97, 1269.

Reduction and oxidation of the colouring matter of the blood.

> Stokes (G. G.). Proc. Royal Soc., **13**, 853.

Ueber das Vorkommen eines neuen, das Absorptionsspectrum des Blutes zeigenden, Körper's im thierischen Organismus.

> Struve (H.). Ber. chem. Ges., **9**, 623; Bull. Soc. chim. Paris, n. s. **18**, 471.

Ueber die spectralanalytische Reaction auf Blut.

> Vogel (H. W.). Ber. chem. Ges., **9**, 587, 1472; Bull. Soc. chim. Paris, n. s. **27**, 83.

BONELLIA VIRIDIS.

Der grüne Farbstoff von Bonellia Viridis.

> Schenck (L. S.). Sitzungsber. Wiener Akad., **72** II, 581–5.

On the colouring matter of bonellia viridis.

> Sorby (H. C.). Quar. Jour. Microscop. Soc., **15**, 166.

BRUCINE.

Absorption spectrum of brucine, etc.

> Meyer (A.). Archives of the Pharmaceutical Soc., (3) **13**, 418–16; Jour. Chem. Soc., **36**, 269.

BUTTER.

Ueber einige Methylester aus der Propionsäure-und Buttersäuregruppe.

> Kahlbaum (G. W. A.). Ber. chem. Ges., **12**, 843–4; Jour. Chem. Soc., **36**, 521 (Abs.).

CARBOHYDRATES.

Spectroscopic notes on the carbohydrates and albuminoids from grain.

> Hartley (W. N.). Jour. chem. Soc., **51** (1887), 58–61.

CARMINE.

Spectrum von ammoniakalischer Carminlösung und von Blut.

> Campani. Ber. chem. Ges., **5**, 287.

Spectre du carmin d'indigo.

> Vogel (H. W.). Bull. Soc. chim. Paris, n. s. **27**, 88

CARYOPHYLLACEÆ.

Colouring matter of the caryophyllaceæ.

> Hilger (A.) and Bischoff (H.). Landwirthschaftl. Versuch-Statistik, **23**, 456–61; Jour. Chem. Soc., **36**, 780 (Abs.).

CHINIZARIN.

Ueber Chinizarin.

Grimm (F.). Ber. chem. Ges., **6**, 506–12.

Absorptionsspectrum des Chinizarins.

Jahresber. d. Chemie (1873), 455 (Abs.). See Grimm.

CHINOLIN.

Ueber einige im Pyridinkern substituirte Chinolinderivate.

Friedländer (P.) und Weinberg (A.). Ber. chem. Ges., **15**, 2679–2685.

CHINON.

Ueber den im Ag. atrotomentosus vorkommenden chinonartigen Körper.

Thörner (W.). Ber. chem. Ges., **12**, 1630–5.

CHOTELIN.

Ueber Chotelin.

Liebermann (L.). Pflüger's Archiv. f. Physiol., **11**, 181–90; Jour. Chem. Soc. (1876), **1**, 407–8 (Abs.).

CHROMOGENE.

Ueber einige Chromogene des Harns und deren Derivate.

Plósz (P.). Zeitschr. f. physiolog. Chemie, **8**, 85–94; Ber. chem. Ges., **16**, 2933 (Abs.).

CHRYSOIDINE.

Das Chrysoidin, eine antiphotogenische Farbe.

Bardy (C.). Chemisches Centralblatt, (8), **9**, 109; Jour. Chem. Soc., **34**, 618 (Abs.).

Spectrum of chrysoidint.

Hartley (W. N.). Jour. Chem. Soc., **51** (1887), 178.

CITRACON.

Ueber die Molecularrefraction der Citracon und Mesaconsäureather.

Brühl (J. W.). Ber. chem. Ges., **14**, 2736–44; Jour. Chem. Soc., **42**, 829–30; Beiblätter, **6**, 876.

COAL.

Soda flames in coal fires.

Herschel (J.). Nature, **27**, 78, 106.

COLEÏN.

Spectrum of coleïn.

Church (J. H.). Jour. Chem. Soc., 1877, **1**, 260.

CROCEÏNE SCARLET.

Spectrum of croceïne scarlet.

Hartley (W. N.). Jour. Chem. Soc., 51 (1887), 195.

CROTON ACID.

Ueber die Molecularrefraction der Crotonsäure.

Brühl (J. W.). Ber. chem. Ges., 14, 2797-2801; Jour. Chem. Soc., 42, 827 (Abs.); Beiblätter, 6, 477 (Abs.).

CRYSTALLOIDS.

On the rate of passage of crystalloids in and out of the body.

Jones (H. Bence). Proc. Royal Soc., 14, 400.

CUMENE.

Spectrum of cumene-azo-β-naphtholdisulphonic acid.

Hartley (W. N.). Jour. Chem. Soc., 51 (1887), 187.

CURCUMIN.

Ueber Curcumin, den Farbstoff der Curcumawurzel.

Daube (F. U.). Neues Repert. d. Pharm., 20, 86; Ber. chem. Ges., 3, 609-13; Jour Chem. Soc., (2) 9, 152 (Abs.).

CYANOGEN.

Photographed spectrum of cyanogen.

Capron (J. R.). Photographed Spectra, London, 1877, 71.

Spectroscopic researches in carbon and cyanogen.

Ciamician. Chem. News, 44, 216.

Spectrum von Cyanogen.

Dibbits (H. C.). Ann. Phys. u. Chem., 122, 507.

Constitution of cyanuric acid.

Hartley (W. N.). Jour. Chem. Soc., 41, 45-9; Beiblätter, 6, 375 (Abs.).

Note on the reversal of the spectrum of cyanogen.

Liveing (G. D.) and Dewar (J.). Chem. News, 44, 258; Proc. Royal Soc., 33, 8; Ann. Chim. et Phys., (5) 23, 571.

Sur le chromocyanure de potassium.

Moissan (H.). Comptes Rendus, 93, 1079-81; Chem. News, 45, 22 (Abs.); Ber. chem. Ges., 15, 248 (Abs.).

De la flamme du cyanogen.

Morren (M. A.).   Ann. Chim. et Phys., (4) 4, 305.

Bestimmung der Brechungsquotienten einer Cyaninlösung.

. Pulfrich (C.).   Ann. Phys. u. Chem., n. F. 16, 385.

Cyanogen in small induction sparks in free air.

Smyth (C. Piazzi).   Nature, 28, 340.

CYMENE.

An examination of terpenes for cymene by means of the ultra-violet spectrum.

Hartley (W. N.).   Jour. Chem. Soc., 37, 676–8.

(Look above under Cumene.)

DECAY.

Zur Lehre von den Fäulnissalkaloïden.

Poehl (A.).   Ber. chem. Ges., 16, 1975–88.

DIAMOND.

On the refraction equivalents of the diamond and the carbon compounds.

Gladstone (J. H.).   Chem. News, 42, 175; Jour. Chem. Soc., 40, 333 (Abs.); Beiblätter, 5, 43 (Abs.).

DIAZO.

Spectrum of diazo.

Hartley (W. N.).   Jour. Chem. Soc., 51 (1887), 196.

DIPHENYL.

Ueber Diphenyldüsoindolazofarbstoffe.

Möhlau (R.).   Ber. chem. Ges., 15, 2490–7; Jour. Chem. Soc., 44, 342 (Abs.).

DIPYRIDENE.

Description and measurement of the spectrum of dipyridene (Dr. Ramsay).

Hartley (W. N.).   Jour. Chem. Soc., 47 (1885), 717.

DROSERA WHITTAKERI.

Absorption spectra of the colouring matter of Drosera Whittakeri.

Rennie (E. H.).   Jour. Chem. Soc., 51 (1887), 377.

EBONITE.

On the transmission of radiation of low refrangibility through ebonite.

> Abney (W. de W.) and Festing (R.).   Proc. Physical Soc., **4**, 256–9; Phil. Mag., (5) **11**, 466–9; Chem. News, **43**, 175 (Abs.); Beiblätter, **5**, 506 (Abs.).

Note on the index of refraction of ebonite.

> Ayrton (W. E.) and Perry (J.).   Proc. Physical Soc., **4**, 845–8; Phil. Mag., (5) **12**, 196–9; Nature, **23**, 519; Beiblätter, **5**, 741 (Abs.).

EOSIN.

Photographic action of eosin.

> Waterhouse (J.).   Photographic Journal, **16**, 185–6; Jour. Chem. Soc., 1876, **2**, 282 (Abs.).

ETHER VAPOUR.

Spectrum or ether vapour.

> Capron (J. R.).   Photographed Spectra, London, 1877, p. 74.

EXCREMENTS.

Swei pathologische Harnfarbstoffe.

> Baumstark (F.).   Pflüger's Arch. f. Physiol., **9**, 568–84; Jour. Chem. Soc., (2) **13**, 480 (Abs.).

Ueber das Urorosein, einen neuen Harnfarstoff.

> Nencki (M.) und Sieber (N).   Jour. prackt. Chemie, **26**, 888–6; Chem. News, **42**, 12 (Abs.); Jour. Chem. Soc., **44**, 101 (Abs.); Ber. chem. Ges., **15**, 8067.

Ueber einen neuen krystallinischen farbigen Harnbestantheil.

> Plósz (P.).   Zeitschr. physiol. Chemie, **6**, 504–7; Ber. chem. Ges., **15**, 2626–7 (Abs.).

Ueber einige Chromogene des Harns und deren Derivate.

> Plósz (P.).   Zeitschr. physiol. Chemie, **8**, 85–94; Ber. chem. Ges., **16**, 2988–4 (Abs.).

FAST RED

Spectrum of fast red.

> Hartley (W. N.).   Jour. Chem. Soc., **51** (1887), 197.

FISH.

Spectrum of fish pigment.

> Francis (G.).   Nature, **13**, 167.

Spectroscopic notes on the carbohydrates and albuminoids from grain.

Hartley (W. N.). Jour. Chem. Soc., **51** (1887), 58–61.

Matière colorante se forment dans la colle de farine.

Lecoq de Boisbaudran (F.). Comptes Rendus, **94**, 562–3; Jour. Chem. Soc., **42**, 789 (Abs.).

Ueber den Nachweis von Mutterkorn im Mehle auf spectroscopischem Wege.

Petri (J.). Zeitschr. analyt. Chemie, **18** 211-20; Jour. Chem. Soc., **36**, 977–9 (Abs.).

Ueber Blumenblau.

Schönn (L.). Zeitschr. analyt. Chemie, **9**, 827–8.

The colouring matter of the petals of Rosa Gallica.

Senier (H.). Pharmaceutical Trans., (3), **7**, 650–652; Jour. Chem. Soc., 1877, **2**, 502 (Abs.).

Ueber die Brechungsverhältnisse des Fuchsins.

Christiansen (C.). Oversight k. Danske Vidensk. Selskabs, 1871, 5–17; Ann. Phys. u. Chem., **143**, 250–9; Ann. Chim. et Phys., (4) **25**, 400 (Abs.).

Zur Farbenzerstreuung des Fuchsins.

Christiansen (C.). Ann. Phys. u. Chem., **146**, 154–155; Jour. Chem. Soc., (2) **11**, 236.

Nachweis von Fuchsin im Weine.

Liebermann (L.). Ber. chem. Ges., **10**, 866; Jour. Chem. Soc., 1877, **2**, 989 (Abs.).

Ueber die optischen Eigenschaften des festen Fuchsins.

Voigt (W.). Göttinger gelehrten Nachr. (1884), 262.

Ueber den Nachweis von Fuchsin in damit gefärbten Weinen durch Stearin.

Wolff (C. H.). Repert. analyt. Chem., **2**, 193–4; Chemisches Central-blatt, (3) **13**, 670, (Abs.); Jour. Chem. Soc., **44**, 384 (Abs.).

Fluorescence of the pigments of fungi.

Weiss (A.). Chem. Centralblatt, 1886, 670–1; Jour. Chem. Soc., **44**, 384–5 (Abs.).

GALL.

Die Oxydationsproducte der Gallenfarbstoffe und ihre Absorptionsstreifen.
Heynsius (A.) und Campbell (J. F. F.).   Pflüger's Archiv. f. Physiol.,
4, 497–547; Jour. Chem. Soc., (2) 10, 807–8 (Abs.).

Absorptionsspectren der Gallenfarbstoffe.
Jaffe.   Jour. prackt. Chemie, 104, 401.

Untersuchungen über die Gallenfarbstoffe.
Maly (R.).   Wiener Anzeigen, 9, 39–41; Chem. Centralblatt, (8) 8,
180–1; Jour. Chem. Soc., (2) 10, 638 (Abs.); Jour. prackt. Chem.,
103, 255; 104, 88.

Untersuchungen über die Gallenfarbstoffe und ihre Erkennung mittelst
des Spectroscops.
Stockvis (B. J.).   Ber. chem. Ges., 5, 583–5; Jour. Chem. Soc., (2)
11, 78 (Abs.).

GELATINE.

Emploi de la gélatine pour montrer l'absorption dans le spectre.
Lommel (E.).   Ann. Chim. et Phys., (4) 26, 279.

GUN-COTTON.

Spectrum explodirender Schiessbaumwolle.
Jahresber. d. Chemie (1873), 151.

Spectrum des Lichtes explodirender Schiessbaumwolle.
Lohse (O.).   Ann. Phys. u. Chem., 150, 641.

Spectrum des Lichtes explodirender Schiessbaumwolle.
Vogel (H. W.).   Ann. Phys., u. Chem., n. F. 3, 615.

Spectrum of $H\,S\,O_3 \cdot C_6\,H_4 \cdot N : N \cdot C_{10}\,H_4\,(H\,S\,O_3)_2 \cdot O\,H\,\beta$ (Na Salt).
Hartley (W. N.).   Jour. Chem. Soc., 51 (1887), 188–9.

HELIANTHIN.

Spectrum of helianthin.
Hartley (W. N.).   Jour. Chem. Soc., 51 (1887), 192–3.

HEMATINE.

Action de l'hydrosulfite de soude sur l'hématine du sang (hématine
reduite).
Cazeneuve (P.).   Bull. Soc. chim. Paris, (2) 27, 258–60; Jour. Chem.
Soc., 1877, 2, 346 (Abs.).

Ueber Assimilation von Hæmatococcus.

> Englemann (T. W.). Onderzoekingen physiol. Lab. Utrecht, (3) **7**, 200–8; Proc. Verb. K. Akad. Wetenschappen, Amsterdam, March 25, 1882, 3–6 (Abs.); Beiblätter, **7**, 877–8 (Abs.).

Researches into the colouring matters of human urine, with an account of their artificial production from bilirubin and from hematine.

> MacMunn (C. A.). Proc. Royal Soc., **31**, 206–337; Jour. Chem. Soc., **40**, 1056–8 (Abs.); Beiblätter, **5**, 281.

On hemine, hematine and a phosphorized substance contained in blood corpuscules.

> Thudichum (J. L. W.) and Kingzett (C. T.). Jour. Chem. Soc., 1876, **2**, 255–64.

HÆMOGLOBIN.

Dosage de l'hémoglobine dans le sang par les procédés optiques.

> Branly (E.). Ann. Chim. et Phys., (5) **27**, 288–278; Jour. Chem. Soc., **44**, 894 (Abs.); Zeitschr. analyt. Chem., **22**, 629–32 (Abs.); Jour. de Phys., (2), **2**, 430 (Abs.).

Ueber die Bestimmung des Hæmoglobin-und Sauerstoff-gehaltnes im Blute.

> Hüfner (G.). Zeitschr. physiol. Chem., **3**, 1–18; Ber. chem. Ges., **12**, 702 (Abs.); Jour. Chem. Soc., **36**, 885.

On the evolution of hemoglobine.

> Sorby (H. C.). Quar. Jour. Microscop. Sci., **16**, 76–85.

Spectralanalytische Bestimmung des Hæmoglobingehaltes des menschlichen Blutes.

> Wiskemann (M.). Zeitschr. f. Biologie, **12**, 434–47; Jour. Chem. Soc., 1877, **2**, 808–9.

HOFFMANN'S VIOLET.

Spectrum of Hoffmann's violet.

> Hartley (W. N.). Jour. Chem. Soc., **51** (1887), 171–4.

HYDROCARBONS.

Hydrocarbons in the solar atmosphere.

> Abney (W. de W.). Rept. British Assoc., 1881, 524.

Sur le pouvoir réfringent de l'hydrocarbure $C_{12} H_{20}$.

> Albitsky (A.). Jour. Soc. phys. chim. russe, **15**, 524–6.

Spectrum von Kohlenwasserstoff.

> Angström (A. J.). Ann. Phys. u. Chem., **94**, 157.

On the spectra of the compounds of carbon with hydrogen and nitrogen.

> Liveing (G. D.) and Dewar (J.). Proc. Royal Soc., **30**, 494–509; Nature, **22**, 620–3.

On the origin of the hydrocarbon flame spectrum.

> Liveing (G. D.) and Dewar (J.). Proc. Royal Soc., **34**, 418–20; Nature, **27**, 257–9; Chem. News, **46**, 293–7; Beiblätter, **7**, 288–9 (Abs.).

Nuovo metodo spettroscopico per discoprire nei miscugli gassosi e nelle acque le puì piccole quantità d'un idrocarburo gassoso od almeno molto volatile.

> Negri (A. e G. de). Gazz. chim. ital., **5**, 488; Jour. Chem. Soc., 1876, **2**, 659 (Abs.); Chem. News, **33**, 76.

Untersuchungen über einen aus Allildipropylcarbinol erhaltenen Kohlenwasserstoff, $C_{10}H_{18}$.

> Reformatsky (S.). Jour. prackt. Chem., n. F. **27**, 389–407; Beiblätter, **7**, 689 (Abs.).

Carbon and hydrocarbon in the modern spectroscope.

> Smyth (C. Piazzi). Phil. Mag., (4) **49**, 24–33.

Carbon and carbohydrogen, spectroscoped and spectrometed in 1879.

> Smyth (C. Piazzi). Phil. Mag., (5) **8**, 107–119; Beiblätter, **4**, 86 (Abs.).

Hydrocarbons of the formula $(C_5H_8)_n$.

> Tilden (W. A.). Chem. News, **46**, 120–1; Jour. Chem. Soc., **44**, 75–6 (Abs.).

Carbon and hydrocarbon in the modern spectroscope.

> Watts (W. M.). Phil. Mag., (4) **49**, 104–6.

HYDROBILIRUBIN.

Ueber Choletelin und Hydrobilirubin.

> Liebermann (L.). Pflüger's Arch. Physiol., **11**, 181–90; Jour. Chem. Soc., 1876, **1**, 407–8 (Abs.).

HYDROCHINON.

Ueber das Phthaleïn des Hydrochinons.

> Grimm (F.). Ber. chem. Ges., **6**, 506–12.

HYDROXYANTHRAQUINONE.

Spectra of the methyl derivatives of hydroxyanthraquinone.

> Liebermann (C.) und Kostanecki (S. von). Ber. chem. Ges., **19**, 2827–32; Jour. Chem. Soc., **52** (1887), 1 (Abs.).

INDIGO.

Spectre de l'indigo.

> Lallemand (A.). Comptes Rendus, **78**, 1272.

Sur la diffusion de l'indigo, etc.

> Lallamand (A.). Comptes Rendus, **79**, 693.

Spectre du carmin de l'indigo.

> Vogel (H. W.). Bull. Soc. chim. Paris, n. s. **27**, 88.

Spectralanalytische Werthbestimmung verschiedener reiner Indigosorten.

> Wolff (C. H.). Zeitschr. analyt. Chem., **23**, 29–82.

IODINE GREEN.

Spectrum of iodine green.

> Hartley (W. N.). Jour. Chem. Soc., **51** (1887), 174–6.

LAMP-BLACK.

Spectre du noir de fumée.

> Lallemand (A.). Comptes Rendus, **78**, 1272.

LEAVES.

Das Grün der Blätter.

> Müller (J.). Ann. Phys. u. Chem., **142**, 615–16; Jour. Chem. Soc.,
> (2) **9**, 654.

Ueber Blattgrün.

> Schönn (L.). Zeitschr. analyt. Chemie, **9**, 827–8; Ann. Phys. u.
> Chem., **145**, 166–7; Arch. de Genève, (2) **43**, 282–3.

On the various tints of autumnal foliage.

> Sorby (H. C.). Chem. News, **23**, 137–9, 148–50; Jour. Chem. Soc.,
> (2) **9**, 184 (Abs.).

On the colour of leaves at different seasons of the year.

> Sorby (H. C.). Quar. Jour. Microscop. Sci., **11**, 215–234.

Ueber die Lichtwirkung verschieden gefärbter Blätter.

> Vogel (H. W.). Sitzungsber. d. Münchener Akad., 1872, 183–7.

LUTEÏNE.

Results of researches on luteïne and the spectra of yellow organic substances contained in animals and plants. Researches conducted
for the medical department of the Privy Council.

> Thudichum (J. L. W.). Proc. Royal Soc., **17**, 253; Jour. prackt.
> Chem., **106**, 414.

MESACON.

Ueber die Molecularrefraction der Citracon-und Mesacon-säureather.
> Brühl (J. W.). Ber. chem. Ges., 14, 2786–44; Jour. chem. Soc., 42, 829–80; Beiblätter, 6, 876.

METAXYLENE.

Description and measurement of the spectrum of metaxylene (Kahlbaum).
> Hartley (W. N.). Jour. Chem. Soc., 47 (1885), 700–7.

METHYLENE BLUE.

On the spectroscopic examination of methylene blue and of South's violet.
> Stebbins (J. H., Jr.). Jour. Amer. Chem. Soc., 6 (1884), 804–5.

METHACRYL.

Ueber die Molecularrefraction der Methacrylsäure.
> Brühl (J. W.). Ber. chem. Ges., 14, 2797–2801; Jour. Chem. Soc., 42, 827 (Abs.); Beiblätter, 6, 477 (Abs.).

METHÄMOGLOBIN.

Studien über das Methämoglobin.
> Otto (J. G.). Pflüger's Arch. f. Physiol., 31, 245–67; Ber. chem. Ges., 16, 2689 (Abs.).

Ueber das Methämoglobin.
> Saarbach (H.). Pflüger's Arch. f. Physiol., 28, 882–8; Ber. chem. Ges., 15, 2752 (Abs.).

MORINDON.

Spectrum der Morindonlösungen.
> Stein. Jour. prackt. Chemie, 97, 241.

Spectrum der Morindonlösungen.
> Stenhouse. Jour. prackt. Chemie, 98, 127.

MORPHINE.

Absorption spectrum of morphine.
> Meyer (A.). Archives of the Pharmaceutical Soc., (3) 13, 413–16; Jour. Chem. Soc., 36, 269.

NAPHTHALENE.

Description and measurement of the spectrum of naphthalene.
> Hartley (W. N.). Jour. Chem. Soc., 47 (1885), 691–701.

Spectrum of amido-azo-α-naphthalene.

Hartley (W. N.).   Jour. Chem. Soc., **51** (1887), 190.

Spectrum of amido-azo-β-naphthalene.

Hartley (W. N.).   Jour. Chem. Soc., **51** (1887), 191.

Absorptionsspectrum von Naphthalin.

Jahresber. d. Chemie (1873), 157.

Spectre de naphthaline pure.

Lallemand (A.).   Comptes Rendus, **77**, 1218.

Ueber die Fluorescenz des Naphthalinrothes.

Wesendonck (K.).   Ann. Phys. u. Chem., (2) **26** (1885), 521-7; Jour. Chem. Soc., **50** (1886), 585; Jour. de Phys., (2) **5** (1886), 517 (Abs.).

OILS.

Olefiant spectrum.

Capron (J. R.).   Photographed Spectra, London, 1877, p. 73.

Spectrum analysis of oils.

Doumer and Thibaut.   Chem. News, **51** (1885), 229.

The spectroscope applied to the detection of adulterations of fixed oils.

Gilmour (W.).   Pharmaceutical Jour. Trans., (3) **6**, 931-2; **7**, 22-3.

On essential oils.

Gladstone (J. H.).   Jour. Chem. Soc., (2) **10**, 1-12; Ber. chem. Ges., **5**, 60 (Abs.).

Examination of essential oils.

Hartley (W. N.) and Huntington (A. K.).   Proc. Royal Soc., **29**, 290.

Ueber gefärbte ætherische Oele.

Hock (K.).   Archiv. f. Pharm., (3) **21**, 17-18, 437-8; Zeitschr. analyt. Chemie, **23**, 241 (Abs.).

Spectrum fetter Oele.

Jahresber. d. Chemie (1870), 175.

Objective Darstellung des Spectrums der Oele.

Jahresber. d. Chemie (1876), 963.

Reports of the committee for investigating the constitution and optical properties of essential oils.

Reports of the British Assoc., 1872, 1873, and 1874.

ORTHO—TOLUIDINE.

Description and measurement of the spectrum of ortho-toluidine.
  Hartley (W. N.). Jour. Chem. Soc., **47** (1885), 789.

Ueber einige Derivate der Orthotoluysäure.
  Jacobsen (O.) und Weiss (F.). Ber. chem. Ges., **16**, 1956–02; Jour.
    Chem. Soc., **44**, 1121 (Abs.).

ORTHO—XYLENE.

Description and measurement of the spectrum of ortho-xylene (Kahl-
    baum).
  Hartley (W. N.). Jour. Chem. Soc., **47** (1885), 702–4.

CARBONIC ACID (CARBON AND OXYGEN).

Spectrum von Kohlensäure.
  Angström (A J.). Ann. Phys. u. Chem., **94**, 155.

Spectre de l'acide carbonique.
  Becquerel (H.). Comptes Rendus, **90**, 1407.

Spectrum of carbonic acid.
  Capron (J. R.). Photographed Spectra, London, 1877, p. 68.

Action of the spectral rays on the decomposition of carbonic acid in
    plants.
  Crookes (W.). Chem. News, **27**, 133.

Spectrum der Flamme von Kohlenoxyd.
  Dibbits (H. C.). Ann. Phys. u. Chem., **122**, 508.

Combustion of carbonic oxide under pressure.
  Franckland (E.). Proc. Royal Soc., **16**, 419, 421; Jour. prackt.
    Chemie, **105**, 190.

Erkennung der Vergiftung mit Kohlenoxyd.
  Hoppe-Seyler (F.). Zeitschr. f. analyt. Chem., **3**, 489; Phil. Mag.,
    (4) **30**, 456.

Funkenspectrum von kohlensäurem Lithium.
  Jahresber. d. Chemie (1873), 152.

Absorption of radiant heat by carbon dioxide.
  Keeler (J. E.). Amer. Jour. Sci., (3) **28**, 190–198; Nature, **31**, 46
    (Abs.).

Die Wirkung der Spectralfarben auf die Kohlensäurezersetzung in Pflanzen.

> Pfeffer (W.). Versuchs-Stationen Organ, 15, 356–67; Jour. Chem. Soc., (2) 10, 1107 (Abs.); 11, 400 (Abs.); Ann. Phys. u. Chem., 148, 86–99; Chem. News, 27, 133–4.

Spectrum von Kohlensäure.

> Plücker. Ann. Phys. u. Chem., 105, 76.

Ueber die Dauer der spectralanalytische Reaction von Kohlenoxyd.

> Salfeld (E.). Repert. analyt. Chem. (1883), 35–7; Archiv. d. Pharm., (3) 21, 289 (Abs.); Jour. Chem. Soc., 46, 843 (Abs.).

Propriétés optiques d'acide oxalique.

> Sénarmont (H. de). Ann. Chim. et Phys., (3) 41, 336.

Die Zerstreuung der C O₂ durch die Pflanzen im directen Sonnenspectrum.

> Timiriaseff (K.). Mém. Acad. St. Pétersbourg, Sept., 1878; Ber. chem. Ges., 6, 1212 (Abs.); Jour. Chem. Soc., (2) 12, 285 (Abs.).

Recherches sur la décomposition de l'acide carbonique dans le spectre solaire par les parties vertes de végétaux (extrait d'un ouvrage "Sur l'assimilation de la lumière par les végétaux," St. Pétersbourg, 1875.)

> Timiriaseff (C.). Ann. Chim. et Phys., (5) 12, 355–96; Comptes Rendus, 84, 1236–9; Jour. Chem. Soc. (1877), 2, 685 (Abs.).

Ueber die Nachweisung von Kohlenoxydgas.

> Vogel (H. W.). Ber. chem. Ges., 10, 792–5.

Note on the spectrum of carbonic acid.

> Wesendonck (C.). Proc. Royal Soc., 32, 880–2; Chem. News, 44, 42–3; Jour. Chem. Soc., 40, 861 (Abs.).

Ueber die Molecularrefraction der geschwefelten Kohlensäureäther, nebst einigen Bemerkungen über Molecularrefraction im Allgemeinen.

> Wiedemann (E.). Ann. Phys. u. Chem., n. F. 17, 577–80; Jour. Chem. Soc., 44, 762 (Abs.); Jour. de Phys., (2) 2, 139 (Abs.).

Ueber die Brechungsexponenten der geschwefelten Substitutionsproducte des Kohlensäureäthers.

> Wiedemann (E.). Jour. prackt. Chem., (2) 6, 453–5.

Spectrum von Kohlensäure.

> Wüllner (A.). Ann. Phys. u. Chem., 144, 485, 500, 507, 516, 517.

PARATOLUIDINE.

Description and measurement of the spectrum of paratoluidine.
Hartley (W. N.). Jour. Chem. Soc., **47** (1885), 706.

PARAXYLINE.

Description and measurement of the spectrum of Paraxyline (Kahlbaum).
Hartley (W. N.). Jour. Chem. Soc., **47** (1885), 707–10.

PENTACRINUS.

Colouring matter of pentacrinus.
Nature, **21**, 578.

PHENOLS.

On a new class of colouring matters from the phenols.
Meldola (R.). Jour. Chem. Soc., **39**, 37–40

PICOLENE.

Description and measurement of the spectrum of picolene (Dr. Ramsay).
Hartley (W. N.). Jour. Chem. Soc., **47** (1885), 719–21.

PIPERIDINE.

Description and measurement of the spectrum of piperidine (Kahlbaum).
Hartley (W. N.). Jour. Chem. Soc., **47** (1885), 731.

PLANTS.

Zur Theorie des Assimilations-processes in der Pflanzenwelt.
Benkovich (E. von). Ann. Phys. u. Chem., **154**, 468–73.

Zur Frage über die Wirkung des farbigen Lichtes auf die Assimilations-thätigheit der Pflanzen.
Lommel (E.). Ann. Phys. u. Chem., **145**, 442–55; Jour. Chem. Soc., (2) **11**, 292 (Abs.).

Ueber den Einfluss des farbigen Lichtes auf die Assimilation und die damit zusammenhängende Vermehrung der Aschenbestandtheile in Erbsenkeimlingen.
Weber (R.). Landwirthschaftl.-Versuchs-Statistik, **18**, 18–48; Jour. Chem. Soc., (2) **13**, 1211–15 (Abs.).

PURPURIN.

Displacement of the absorption bands of purpurin in solutions of alum.
Morton (H.). Chem. News, **42**, 207; Jour. Chem. Soc., **40**, 488.

Note on the purple of the ancients.

> Schunk (E.).   Jour. Chem. Soc., 37, 612-17.

Die Purpurin-Thonerde-Magnesiareaction

> Vogel (H. W.).   Ber. chem. Ges., 10, 157, 873 ; Bull. Soc. chim. Paris, n. s. 28, 475, 478.

Ueber die Lichtempfindlichkeit des Purpurins.

> Vogel (H. W.).   Ber. chem. Ges., 10, 692.

PYRIDINE.

Description and measurement of the spectrum of pyridine (Kahlbaum).

> Hartley (W. N.).   Jour. Chem. Soc., 47 (1885), 711-16.

QUINOLINE.

Description and measurement of the spectrum of quinoline, specimens I and II.

> Hartley (W. N.).   Jour. Chem. Soc., 47 (1885), 721-7, 728-30.

> (Look below for Tetrahydroquinoline.)

Spectrum of quinoline-red.

> Hoffmann (A. W.).   Ber. chem. Ges., 20, 4-20; Jour. Chem. Soc., 52 (1887), 380 (Abs.).

RASPBERRY.

Ueber die Untersuchungen von Hinbeersaft.

> Vogel (H. W.).   Ber. chem. Ges., 10, 1428-32; Jour. Chem. Soc., 1877, 915 (Abs.).

ROSANILINE.

Ueber Rosolsäure.

> Gräbe (C.) und Caro (H.).   Ann. Phys. u. Chem., 179, 184-203; Jour. Chem. Soc., 1876, 1, 588-91.

Spectrum of rosaniline base.

> Hartley (W. N.).   Jour. Chem. Soc., 51 (1887), 164-6.

Spectrum of rosaniline hydrochloride.

> Hartley (W. N.).   Jour. Chem. Soc., 51 (1887), 169-171.

RUBERINE.

On the colouring matter (ruberine), etc., contained in agaricus ruber.

> Phipson (T. L.).   Chem. News, 46, 199-200; Jour. Chem. Soc., 44, 100 (Abs.); Ber. chem. Ges., 16, 244 (Abs.).

SAFRANIN.

Absorptionsspectrum von safranin.

> Landauer (J.). Ber. chem. Ges., 11, 1772-5; Jour. Chem. Soc., 36, 101 (Abs.); Beiblätter, 3, 195-6.

SODA (CARBONATE).

Propriétés optiques de sous-carbonate de soda.

> Senarmont (H. de). Ann. Chim. et Phys., (8) 41, 336.

SPONGILLA FLUVIATILIS.

Chromatological relations of spongilla fluviatilis.

> Sorby (H. C.). Quar. Jour. Microscop. Sci., 15, 47-52.

CARBON AND SULPHUR.

Note on the absorption spectrum of iodine in solution in carbon disulphide:

> Abney (W. de W.) and Festing (Lieut. Col.). Proc. Royal Soc., 34, 480.

Spectre du sulphure de carbone.

> Becquerel (H.). Comptes Rendus, 85, 1227.

Spectrum von Schwefelkohlenstoff.

> Dibbits (H. C.). Ann. Phys. u. Chem., 122, 531.

Schwefelkohlenspectrum.

> Jahresber. d. Chemie (1875), 122, 125, 126 (Abs.). See Vogel (H. W.), Deutsch. chem. Ges., 1875, 96; Watts (W. M.), Phil. Mag., (4) 48, 369; and Morton (H.), Ann. Phys. u. Chem., 155, 551.

Absorptionsstreifen in Prismen von Schwefelkohlenstoff.

> Lamansky (S.). Ann. Phys. u. Chem., 146, 213, 215.

Ueber das Spectrum der Sell'schen Schwefelkohlenstofflampe.

> Vogel (H. W.). Per. chem. Ges., 8, 96-8; Jour. Chem. Soc., (2) 13, 698 (Abs.).

TEREBINTHENE.

Sur les chlorhydrates liquides de térébinthène.

> Barbier (P.). Comptes Rendus, 96, 1066-9; Jour. Chem. Soc., 44, 809 (Abs.).

Spectre de l'essence de térébinthène.

> Masson (A.). Comptes Rendus, 32, 129.

TERPENES.

Das moleculare Brechungsvermögen der Terpene.
>   Flawitsky (F.).   Ber. chem. Ges., **15**, 15–16.

An examination of terpenes for cymene by means of the ultra-violet spectrum.
>   Hartley (W. N.).   Jour. Chem. Soc., **37**, 676–8.

TETRAHYDROQUINOLINE.

Description and measurement of the spectrum of tetrahydroquinoline.
>   Hartley (W. N.).   Jour. Chem. Soc., **47** (1885), 731–4.

Description and measurement of the spectrum of tetrahydroquinoline hydrochloride (Kahlbaum).
>   Hartley (W. N.).   Jour. Chem. Soc., **47** (1885), 735–8.

TOURMELINE.

On the nature of the light emitted by heated tourmeline.
>   Stewart (Balfour).   Phil. Mag., (4) **21**, 891.

TRIPHENYLMENTHANE.

Spectrum of triphenylmenthane.
>   Hartley (W. N.).   Jour. Chem. Soc., **51** (1887), 162–4.

TROPÆOLIN.

Spectrum of tropæolin *0*.
>   Hartley (W. N.).   Jour. Chem. Soc., **51**, 182–3.

Spectrum of tropæolin *0 0 0*.
>   Hartley (W. N.).   Jour. Chem. Soc., **51**, 184–7.

TURPENTINE.

Spectrum of turpentine vapour.
>   Capron (J. R.).   Photographed Spectra, London, 1877, p. 74.

ULTRAMARINE.

Ueber die Absorptionsspectren verschiedener Ultramarinsorten.
>   Wunder (J.).   Ber. chem. Ges., **9**, 295–9; Jour. Chem. Soc. (1876), **1**, 864.
>   Bemerkungen dazu, Hoffmann (R.).   Ber. chem. Ges., **9**, 494.

URINE.

Researches into the colouring matters of human urine, with an account of the separation of urobilin.

> MacMunn (C. A.). Proc. Royal Soc., **30**, 250-2; **31**, 26-36; Ber. chem. Ges., **14**, 1212-14 (Abs.).

Observations on the colouring matter of the so-called bile of invertebrates, and on some unusual urine pigments, etc.

> MacMunn (C. A.). Proc. Royal Soc., **35**, 370-408; Jour. Chem. Soc., **46**, 194-8 (Abs.).

Ueber das Urorosein, einen neuen Harnfarbstoff.

> Nencki (M.) und Sieber (N.). Jour. prackt. Chemie, **26**, 333-36; Chem. News, **42**, 12 (Abs.); Jour. Chem. Soc., **44**, 101 (Abs.); Ber. chem. Ges., **15**, 3087 (Abs.).

Substances colorantes de l'urine.

> Neusser (E.). Les Mondes, (8) **2**, 468-9; Jour. Chem. Soc., **46**, 93 (Abs.).

WINE.

Recherche et détermination des principales matières colorantes employées pour falsifier les vins.

> Chancel (G.). Comptes Rendus, **84**, 848-51; Jour. Chem. Soc. (1877), **2**, 371 (Abs.); Ber. chem. Ges., **10**, 494.

The detection of foreign colouring matters in wine.

> Dupré (A.). Jour. Chem. Soc., **37**, 572-5; Ber. chem. Ges., **13**, 2004-5 (Abs.).

The detection of the colouring matters of logwood, Brazil-wood, and cochineal in wine.

> Dupré (A.). Analyst, **1**, 26; Jour. Chem. Soc. (1877), **1**, 234 (Abs.).

Zur Weinverfälschung.

> Lepel (F. von). Ber. chem. Ges., **9**, 1906-11; **11**, 1552-6.

WOOD.

Preliminary notes on a blue colouring matter found in certain wood undergoing decomposition in the forest.

> Girdwood (G. P.) and Bemrose (J.). Rept. British Assoc. (1884), 690.

Absorptionsspectrum von Brazilienholtzabkochung.

> Reynolds (J. E.). Jour. prackt. Chemie, **105**, 858.

Absorptionsspectrum von Campecheholtzabkochung.

> Reynolds (J. E.). Jour. prackt. Chemie, **105**, 859.

Notiz über die Strahlen des Lichtes welche das Xantophyll der Pflanzen zerlegen.

> Wiesner (J.). Ann. Phys. u. Chem., **153**, 622–3.

CERIUM.

Contribution to the chemistry of the cerite metals.

> Brauner (B.). Jour. Chem. Soc., **43**, 278–89; Chem. News, **47**, 175 (Abs.).

Sulla diffusione del Cerio, etc.

> Cossa (A.). R. Accad. dei Lincei, (3) **3**, 17–34; Beiblätter, **4**, 43–44 (Abs.).

Le didyme de la cérite est probablement un mélange de plusieurs corps.

> Delafontaine. Comptes Rendus, **87**, 634–5; Jour. Chem. Soc., **36**, 119 (Abs.); Beiblätter, **3**, 197–8 (Abs.).

Sur les terres de la cérite.

> Demarçay (Eug.). Comptes Rendus, **103** (1887), 580.

Contribution to the chemistry of cerium compounds.

> Hartley (W. N.). Jour. Chem. Soc., **41**, 202–9; Chem. News, **45**, 40 (Abs.).

Le didyme de la samarskite diffère-t-il de celui de la cérite?

> Lecoq de Boisbaudran (F.). Comptes Rendus, **88**, 322; Beiblätter, 3. 858 (Abs.).

CHLORINE.

1, CHLORINE ALONE.

Spectre du chlore dans les tubes de Geissler.
> Chautard (J.). Comptes Rendus, **82**, 278.

Spectres appartenant à la famille du chlore.
> Ditte (A.). Comptes Rendus, **73**, 738.

Des spectres d'absorption du chlore.
> Gernez (D.). Bull. Soc. chim. Paris, n. s. **17**, 258; Ber. chem. Ges.,
> **5**, 219; Comptes Rendus, **74**, 465, 660.

Absorptionsspectrum des Chlors.
> Jahresber. d. Chemie (1869), 182 (Abs. See Morren, below).

Réaction spectrale du chlore.
> Lecoq de Boisbaudran (F.). Comptes Rendus, **91**, 902–3; Phil. Mag.,
> (5) **11**, 77–8; Beiblätter, **5**, 118 (Abs.).

Verbindungsspectrum zur Entdeckung von Chlor.
> Mitscherlich. Jour. prackt. Chem., **97**, 218.

Absorptionsspectrum des durch Chlor gegangenen Sonnenlichtes.
> Morren. Ann. Phys. u. Chem., **137**, 165; Comptes Rendus, **68**, 876.

2, CHLORINE COMPOUNDS.

Effect of the spectrum of silver chloride.
> Abney (W. de W.). Rept. British Assoc. (1881), 594.

Sur les chlorhydrates liquides de térébinthène.
> Barbier (P.). Comptes Rendus, **96**, 1066–9; Jour. Chem. Soc., **44**,
> 809 (Abs.).

Spectre du bichlorure de titane.
> Becquerel (H.). Comptes Rendus, **85**, 1227.

Tin chloride spectrum.
> Capron (J. R.). Photographed Spectra, London, 1877, p. 76.

Sur l'indice de réfraction du chlorure d'argent naturel.
> Cloiseux (Des). Bull. Soc. mineral. de France, **5**, 143; Beiblätter, **7**,
> 25 (Abs.).

Spectrum von Kupferchlorid, mit einer Karte.

    Diacon (E.).   Ann. Chim. et Phys., (4) **6**, 1.

Spectres des métalloïdes de la famille du chlore.

    Ditte (A.).   Bull. Soc. chim. Paris, n. s. **16**, 229; Comptes Rendus, **73**, 738.

Ueber Chlorsäure, ein neues Reagens auf Alkaloïde.

    Fraude (G.).   Ber. chem. Ges., **12**, 1558–60.

Spectrum von Chloroxyd und Unterchlorinsäure.

    Gernez (D.).   Ber. chem. Ges., **5**, 218.

Sur les raies d'absorption produites dans le spectre par les solutions des acides chloreux, etc.

    Gernez (D.).   Comptes Rendus, **74**, 465–8; Jour. Chem. Soc., (2) **10**, 280 (Abs.); Ber. chem. Ges., **5**, 218 (Abs.).

Spectre d'absorption du chlorure d'iode.

    Gernez (D.).   Comptes Rendus, **74**, 660; Bull. Soc. chim. Paris, n. s. **17**, 258.

Spectre d'absorption du vapeur de l'acide hypochloreux.

    Gernez (D.).   Comptes Rendus, **74**, 803; Bull. Soc. chim. Paris, n. s. **17**, 257; Ber. chem. Ges., **5**, 219.

Spectre d'absorption du vapeur de protochlorure de tellure.

    Gernez (D.).   Bull. Soc. chim. Paris, n. s. **18**, 172.

On the violet flame of many chlorides.

    Gladstone (J. H.).   Phil. Mag., (4) **24**, 417.

Spectres de chlorure de baryum, de chlorure de cadmium, de chlorure de calcium, de chlorure de cobalt, de chlorure de cuivre, de chlorure de fer, de chlorure de magnésium, de chlorure de platine, de chlorure de strontium.

    Gouy.   Comptes Rendus, **84**, 231; **85**, 439; Chem. News, **35**, 107.

Absorptionsspectrum des Mangansuperchlorids.

    Jahresber. d. Chemie (1869), 184 (Abs. See Luck, below).

Spectra der Chlormetalle.

    Jahresber. d. Chemie (1863), 111 (Abs. See Diacon, above).

Absorptionsspectrum des Chlors und der unterchlorigen Säure.

    Jahresber. d. Chemie (1872), 188, 139 (Abs. See Gernez, above).

Absorptionsspectrum des einfachen Chlorjods.

    Jahresber. d. Chemie (1872), 139 (Abs. See Gernez, above).

Absorptionsspectrum des Chlorselens.

> Jahresber. d. Chemie (1872), 140 (Abs. See Gernez, above).

Absorptionsspectrum des einfachen Chlortellurs.

> Jahresber. d. Chemie (1872), 140 (Abs. See Gernez, above).

Spectrum des Phosphorenzlichts von Chlorophan.

> Kindt. Ann. Phys. u. Chem., 131, 160.

Spectralanalyse des Chlorberylliums.

> Klatzo. Jour. prackt. Chemie, 106, 280.

Protochlorure d'antimoine en solution.

> Lecoq de Boisbaudran (F.). Spectres Lumineux, Paris, 1874, p. 150, planche XXIII.

Chlorure de baryum dans le gaz et en solution, étincelle.

> Lecoq de Boisbaudran (F.). Spectres Lumineux, Paris, 1874, p. 57, 62, planche VII; p. 66, planche IX.

Chlorure de bismuth en solution, étincelle.

> Lecoq de Boisbaudran (F.). Spectres Lumineux, Paris, 1874, p. 145, planche XXII.

Chlorure de cadmium en solution, étincelle.

> Lecoq de Boisbaudran. Spectres Lumineux, p. 139, planche XX.

Chlorure de calcium dans le gaz chargé de H Cl; et en solution, étincelle.

> Lecoq de Boisbaudran (F.). Spectres Lumineux, Paris, 1874, p. 79, planche XI; p. 81, planche XII.

Sesquichlorure de chrome en solution, étincelle.

> Lecoq de Boisbaudran (F.). Spectres Lumineux, Paris, 1874, p. 106, planche XVI.

Chlorure de cobalt en solution, étincelle.

> Lecoq de Boisbaudran (F.). Spectres Lumineux, Paris, 1874, p. 129, planche XIX.

Chlorure de cuivre en solution, étincelle; et dans le gaz.

> Lecoq de Boisbaudran (F.). Spectres Lumineux, Paris, 1874, p. 152, planche XXIV; p. 156, planche XXIV.

Chlorure de didyme en solution concentrée, absorption; et en solution étendue, absorption.

> Lecoq de Boisbaudran. Spectres Lumineux, Paris, 1874, p. 87, planche XIII; p. 90, planche XIII.

Chlorure de l'erbium en solution, absorption.

> Lecoq de Boisbaudran. Spectres Lumineux, Paris, 1874, p. 100, planche XV.

Spectre de chlorure d'or.

> Lecoq de Boisbaudran (F.). Comptes Rendus, **77**, 1152–4; Jour. Chem. Soc., (2) **12**, 217 (Abs.); Ber. chem. Ges., **6**, 1418 (Abs.); Bull. Soc. chim. Paris, n. s. **21**, 125.

Chlorure d'or en solution, étincelle; et dans le gaz.

> Lecoq de Boisbaudran (F.). Spectres Lumineux, Paris, 1874, p. 172, planche XXVI; p. 176, planche XXVI.

Perchlorure de fer en solution, étincelle.

> Lecoq de Boisbaudran (F.). Spectres Lumineux, Paris, 1874, p. 122, planche XVIII.

Chlorure de magnésium en solution, étincelle.

> Lecoq de Boisbaudran. Spectres Lumineux, Paris, 1874, p. 85, planche XII.

Chlorure de manganèse en solution, dans le gaz, étincelle courte, étincelle moyenne.

> Lecoq de Boisbaudran (F.). Spectres Lumineux, Paris, 1874, p. 110, 114, 120, planches XVII, XVIII.

Bichlorure de mercure en solution, étincelle.

> Lecoq de Boisbaudran (F.). Spectres Lumineux, Paris, 1874, p. 169, planche XXV.

Chlorure de nickel en solution, étincelle.

> Lecoq de Boisbaudran. Spectres Lumineux, Paris, 1874, p. 188, planche XIX.

Chlorure de palladium en solution, étincelle.

> Lecoq de Boisbaudran. Spectres Lumineux, Paris, 1874, p. 184, planche XXVII.

Chlorure de platine en solution, étincelle.

> Lecoq de Boisbaudran (F.). Spectres Lumineux, Paris, 1874, p. 181, planche XXVII.

Chlorure de potassium dans le gaz.

> Lecoq de Boisbaudran (F.). Spectres Lumineux, Paris, 1874, p. 47, planche IV.

Chlorure de rubidium dans le gaz.

> Lecoq de Boisbaudran (F.). Spectres Lumineux, Paris, 1874, p. 46, planche IV.

Chlorure de strontium dans le gaz chargé de H Cl; et en solution, étincelle.

Lecoq de Boisbaudran. Spectres Lumineux, Paris, 1874, p. 72, 75, planche X; p. 69, planche IX.

Bichlorure de l'étain en solution, étincelle.

Lecoq de Boisbaudran. Spectres Lumineux, Paris, 1874, p. 148, planche XXII.

Chlorure de zinc en solution, étincelle.

Lecoq de Boisbaudran. Spectres Lumineux, Paris, 1874, p. 188, planche XX.

Absorptionsspectrum des Mangansuperchlorids.

Luck (E.). Zeitschr. analyt. Chemie, 8, 405.

Verbindungspectrum zur Entdeckung von Chlor.

Mitscherlich (A.). Jour. prackt. Chemie, 97, 218.

Entdeckung sehr geringer Mengen von Chlor in Verbindungen.

Mitscherlich (A.). Ann. Phys. u. Chem., 125, 629.

Spectroscopic anomalies, especially in chlorides.

Palmieri (L.). Chem. News, 47, 247.

Absorption spectra of bromine and of iodine monochloride.

Roscoe (H. E.) and Thorpe (T. E.). Proc. Royal Soc., 25, 4.

Spectroscopic observations on dissolved cobaltous chloride.

Russell (W. J.). Chem. News, 51, 250.

Spectren organischer Chlorverbindungen.

Salet (G.). Ber. chem. Ges., 5, 222; Bull. Soc. chim. Paris, 1 mars 1872.

Recent discoveries with the spectroscope, especially in the absorption spectrum of chromochloric anhydride.

Stoney (Johnstone). Chem. News, 23, 104.

Ueber die verschiedenen Modificationen des Chlorsilbers.

Vogel (H. W.). Ber. chem. Ges., 16, 1170–9.

Ueber die Brechung und Dispersion des Lichtes in Chlorsilber.

Wernicke (W.). Ann. Phys. u. Chem., 142, 560–73; Jour. Chem. Soc., (2) 9, 653 (Abs.); Ann. Chim. et Phys., (4) 26, 287 (Abs.)

CHLOROPHYLL.

Propriétés optiques de la chlorophylle.

Ann. Chim. et Phys., (4) 26, 277-9.

Recherches sur les raies de la chlorophylle.

Chautard (J.). Comptes Rendus, 75, 1836.

Examen spectroscopique de la chlorophylle dans les résidus de la digestion.

Chautard (J.). Comptes Rendus, 76, 103-5; Jour. Chem. Soc., (2) 11, 521.
Observations par M. Millardet. Comptes Rendus, 76, 105-7.

Modifications du spectre de la chlorophylle sous l'influence des alcalis.

Chautard (J.). Comptes Rendus, 76, 570; Bull. Soc. chim. Paris, 20, 89; Jour. Chem. Soc., (2) 11, 582 (Abs.).

Influence des rayons de diverses couleurs sur le spectre de la chlorophylle.

Chautard (J.). Comptes Rendus, 76, 1031-3; Jour. Chem. Soc., (2) 11, 718 (Abs.).

Examen des différences presentées par le spectre de la chlorophylle, selon la nature du dissolvant.

Chautard (J.). Comptes Rendus, 76, 1066-9; Jour. Chem. Soc., (2) 11, 996-7.

Classification des bandes d'absorption de la chlorophylle; raies accidentales.

Chautard (J.). Comptes Rendus, 76, 1278.

(Look below under Pocklington.)

Spectre de la chlorophylle.

Chautard (J.). Comptes Rendus, 77, 596.

Nouvelles bandes surnuméraires produites dans les solutions de chlorophylle sous l'influence des agents sulfurés.

Chautard (J.). Comptes Rendus, 78, 414-16; Jour. Chem. Soc., (2) 12, 643 (Abs.).

Recherches sur le spectre de la chlorophylle.

Chautard (J.). Ann. Chim. et Phys., (5) 3, 5-56.

Note sur la chlorophylle.

Filhol (E.). Comptes Rendus, 79, 612-14; Jour. Chem. Soc., (2) 13, 371-2 (Abs.).

Recherches sur.la chlorophylle et quelques uns de ses dérivés.

> Gerland (E.) et Rauwenhoff (W. H.). Arch. Neerlandaises, **6**, 97–116; Ann. Phys. u. Chem., **143**, 231–9; Jour. Chem. Soc., (2) **9**, 1201–2 (Abs.).

Ueber die Einwirkung des Lichtes auf das Chlorophyll.

> Gerland (J.). Ann. Phys. u. Chem., **143**, 585–610; Jour. Chem. Soc., (2) **10**, 160 (Abs.).

Ueber die Rolle des Chlorophylls bei der Assimilationsthätigkeit der Planzen und das Spectrum der Blätter.

> Gerland (J.). Ann. Phys. u. Chem., **148**, 99–115; Jour. Chem. Soc., (2) **11**, 401 (Abs.).

Purpurophyll, ein neues (?) Derivat des Chlorophylls.

> Hartsen (T. A.). Ann. Phys. u. Chem., **146**, 158–60.

Absorptionsspectrum des Chlorophylls.

> Jahresber. d. Chemie (1872), 136 (Abs. See Chautard, above).

Spectroscopische Untersuchungen des Chlorophylls.

> Jahresber. d. Chemie (1873), 154–7 (Abs. See Chautard, above).

Zur Kenntniss der Chlorophyll-farbstoffe.

> Krauss (G.). Archives de Genève, (2) **46**, 359 (Abs.).

Untersuchungen über das Chlorophyll, den Blumenfarbstoff und deren Beziehungen zum Blutfarbstoffe.

> Liebermann (L.). Sitzungsber. d. Wiener Akad., **72** II, 599–618; Chem. Centralblatt, (3) **7**, 615–16; Jour. Chem. Soc., 1877, **2**, 208 (Abs.).

Ueber das Verhalten des Chlorophylls zum Licht.

> Lommel (E.). Ann. Phys. u. Chem., **143**, 568–85; Jour. Chem. Soc., (2) **10**, 150–60 (Abs.).

Observations sur l'examen spectroscopique de la chlorophylle par M. Chautard.

> Millardet (A.). Comptes Rendus, **76**, 105–7; Jour. Chem. Soc., (2) **11**, 996 (Abs.).

Spectroscopic study of chlorophyll.

> Nature, **26**, 636.

M. Chautard's classification of the absorption-bands of chlorophyll.

> Pocklington (H.). Pharmaceutical Trans., (3) **4**, 61–3.

Ueber die Absorptionsspectra der Chlorophyllfarbstoffe.

> Pringsheim. Monatsber. d. Berliner Akad. (1874), 628–59.

13 T

Ueber natürliche Chlorophyllmodificationen und die Farbstoffe der Florideen.

> Pringsheim. Monatsber. d. Berliner Akad. (1875), 745–59.

Spectroscopic study of chlorophyll.

> Russell (W. J.) and Lapraik (W.). Jour. Chem. Soc., 41, 334–41; Nature, 26, 636–9; Ber. chem. Ges., 15, 2746 (Abs.); Chem. News, 45, 250.

Ueber die Bedeutung des Chlorophylls.

> Sachsse (R.). Sitzungsber. d. Naturforsch. Ges. zu Leipzig, 2, 120–55; Chemisches Centralblatt, (3) 7, 550–2; Jour. Chem. Soc. (1877), 2, 208 (Abs.).

Ueber eine neue Reaction des Chlorophylls.

> Sachsse (R.). Chemisches Centralblatt, (3) 9, 121–5; Jour. Chem. Soc., 34, 516 (Abs.).

Die Reindarstellung des Chlorophyllfarbstoffes.

> Tschirch (A.). Ber. chem. Ges., 16, 2731–6; Jour. Chem. Soc., 45, 57–62.

Untersuchungen über das Chlorophyll und einige seiner Derivate.

> Tschirch (A.). Ann. Phys. u. Chem., n. F. 21, 870–83.

Beziehungen des Lichtes zum Chlorophyll.

> Wiesner (J.). Sitzungsber. d. Wiener Akad., 59 I, 827; Ann. Phys. u. Chem., 152, 497; Jour. Chem. Soc., (2) 12, 999 (Abs.).

CHROMIUM.

On the colour properties and relations of chromium.
> Bayley (T.). Jour. Chem. Soc., **37**, 828–36.

The chromium arc spectrum, photographed.
> Capron (J. R.). Photographed Spectra, London, 1877, p. 26

On the optical properties of a new chromic oxalate.
> Hartley (W. N.). Proc. Royal Soc., **21**, 499–507 ; Ber. chem. Ges.. **6**, 1425 (Abs.).

Distribution of heat in green oxide of chromium.
> Jacques (W. W.). Proc. American Acad., **14**, 142.

Sesquichlorure de chrome en solution, étincelle.
> Lecoq de Boisbaudran (F.). Spectres Lumineux, Paris, 1874, p. 106, planche XVI.

Absorptionsspectra der Alkalichromate und der Chromsäure.
> Sabatier (P.). Beiblätter, **11**, 228.

COBALT.

On the colour, properties, and relations of cobalt, etc.
> Bayley (T.).   Jour. Chem. Soc., **37**, 828–36.

Cobalt arc spectrum, photographed.
> Capron (J. R.).   Photographed Spectra, London, 1877, p. 27.

Spectre de chlorure de cobalt.
> Gouy.   Comptes Rendus, **84**, 281 ; Chem. News, **35**, 107.

Spectra of some cobalt compounds in blowpipe chemistry.
> Horner (C.).   Chem. News, **27**, 241 ; Jour. Chem. Soc., (2) **11**, 1161–2
> (Abs.).

Spectrum von Kobalt.
> Jahresber. d. Chemie (1872), 145.   (See Lockyer, below.)

Spectrum von Kobaltverbindungen.
> Jahresber. d. Chemie (1873), 150.   (See Horner, above.)

Spectre des sels de cobalt.
> Lallemand (A.).   Comptes Rendus, **78**, 1272.

Chlorure de cobalt en solution, étincelle.
> Lecoq de Boisbaudran (F.).   Spectres Lumineux, Paris, 1874, p. 129,
> planche XIX.

On the spectrum of cobalt.
> Lockyer (J. N.).   Proc. Royal Soc., **17**, 289.

Absorption spectra of cobalt salts.
> Russell (W. J.).   Proc. Royal Soc., **31**, 51 ; **32**, 258 ; Chem. News,
> **43**, 27.

Spectroscopic observations on dissolved cobaltous chloride.
> Russell (W. J.).   Chem. News, **51**, 259.

Erkennung des Kobalts neben Eisen und Nickel.
> Vogel (H. W.).   Ber. chem. Ges., **12**, 2318–16 ; Beiblätter, **4**, 278
> (Abs.); **5**, 118 (Abs.).

Methods for the determination of cobalt by spectral analysis.
> Wolff.   Chem. News, **39**, 124.

COLOUR.

Metachromism, or colour-change.
> Ackroyd (W.). Chem. News, **34**, 75–7.

Ueber die Aenderung des Farbentones von Spectralfarben bei abnehmender Lichtstärke.
> Albert (E.). Ann. Phys. u. Chem., n. F. **16**, 129–60; Jour. Chem. Soc., **42**, 1158 (Abs.).

Influence de la lumière sur les animaux.
> Béclard. Comptes Rendus, **46**, 441.

Influence des rayons colorés du spectre sur le développement des animaux.
> Béclard. Comptes Rendus, **73**, 1487.

Nouvelles recherches sur les impressions colorées produites lors de l'action chimique de la lumière.
> Becquerel (Éd.). Comptes Rendus, **39**, 65.

Ueber die Entstehung von farbigem Licht durch elective Reflection.
> Behrens (H.). Ann. Phys. u. Chem., **150**, 808–11.

Action of various coloured bodies on the spectrum.
> Brewster (Sir D.). Phil. Mag., (4) **24**, 441.

Étude expérimentale de la réflexion des rayons actiniques; influence du poli speculaire.
> Chardonnet (E. de). Comptes Rendus, **96**, 441; Jour. de Phys., **12**, 219.

La perception des couleurs.
> Charpentier (Aug.). Comptes Rendus, **96**, 859.

Recherches expérimentales sur les anneaux colorés de Newton.
> Desains (P.). Comptes Rendus, **78**, 219–21; Phil. Mag., (4) **47**, 286–7.

Farbe und Assimilation.
> Engelmann (T. W.). Onderzoekingen physiol. Lab. Utrecht, (8) **7**, 209–83; Beiblätter, **7**, 878–80 (Abs.); Centralblatt f. Agriculturchemie (1883), 174–8 (Abs.); Jour. Chem. Soc., **44**, 819 (Abs.).

Bacterium photometricum.
> Engelmann (T. W.). Onderzoekingen physiol. Lab. Utrecht, (8) **7**, 252–90; Pflüger's Arch. f. physiol., **30**, 95–124; Proc. Verb. K. Akad. v. Wetenschappen, Amsterdam, Mar. 25, 1882, 3–6 (Abs.); Beiblätter, **7**, 381 (Abs.).

Das Verhalten verschiedener Wärmefarben bei der Reflexion polarisirten Strahlen von Metallen.

> Knoblauch (H.).   Ann. Phys. u. Chem., n. F. 10, 654.

Ueber den neutralen Punckt im Spectrum der Farbenblinden.

> König (A.).   Verhandl. d. physischen Ges. in Berlin (1883), 20-23.

Influence of colour upon reduction by light.

> Lea (M. Carey).   Amer. Jour. Sci., (3) 7, 200-207.

Influence of colour upon the refraction of Light.[1]

> Lea (M. Carey).   Amer. Jour. Sci., (3) 9, 855-7.

Dr. Vogel's colour theory.

> Lea (M. Carey).   Amer. Jour. Sci., (3) 12, 48-50.

On the development of the colour sense.

> Lubbock (Dr. Montague).   Rept. British Assoc. (1881), 715.

On the relations of the colours of the spectrum.

> Maxwell (J. Clerk).   Proc. Royal Soc., 10, 484.

On the duration of colour impressions upon the retina.

> Nichols (E. L.).   Amer. Jour. Sci., (3) 28, 248-52.

Eine Beziehung zwischen der Farbe gewisser Flammen und den durch das Licht gefärbten heliographischen Bildern.

> Niepce de Saint Victor.   Ann. Phys. u. Chem., Ergänzungsband, 3 (1858), 442; Ann. Chim. et Phys., (3) 32, 373.

On the sensitiveness of the eye to slight differences of colour.

> Peirce (B. O., Jr.).   Amer. Jour. Sci., (3) 26, 299-302; Z. Instrumentenkunde, 4, 67-8 (Abs.); Beiblätter, 8, 120.

Sur l'achromatisme chimique.

> Prazmowski.   Comptes Rendus, 79, 107-110; Jour. Chem. Soc., (2) 12, 1125 (Abs.).

Experiments in colour.

> Rayleigh (Lord).   Nature, 25, 64-6.

Sur l'application de la succession anomale des couleurs dans le spectre de plusieurs substances.

> Sellmeier.   Jour. de Phys., 1, 104.
> Bemerkungen hiezu, A. Levistal.   Ann. Phys. u. Chem., 143, 272.

Colour in practical astronomy, spectroscopically examined.

> Smyth (C. Piazzi).   Trans. Roy. Soc. Edinburgh, 28, 779-843; Beiblätter, 4, 548 (Abs.).

·Comparative vegetable chromatology.

> Sorby (H. C.). Proc. Royal Soc., **21**, 442–83; Jour. Chem. Soc., (2) **12**, 279–85 (Abs.).

Sur la transparence des milieux de l'œil pour les rayons ultra-violets.

> Soret (J. L.). Comptes Rendus, **88**, 1012–15; Beiblätter, **3**, 620 (Abs.).

On combinations of colour by means of polarized light.

> Spottiswoode (W.). Proc. Royal Soc., **22**, 354–8.

Farbenwahrnehmung.

> Weinhold (A.). Ann. Phys. u. Chem., n. F. **2**, 631.

De l'influence de différentes couleurs du spectre sur la dévellopement des animaux.

> Yung (E.). Comptes Rendus, **87**, 998–1000.

CONE–SPECTRUM.

The blowpipe cone-spectrum and the distribution of the intensity of light in the prismatic and diffraction spectra.

> Draper (J. W.). Nature, **20**, 301.

CONSTANTS.

Beziehungen zwischen physikalischen Constanten chemischer Verbindungen.

> Brühl (J. W.). Ber. chem. Ges., **15**, 467.

Spectroscopische Untersuchung der Constanten von Lösungen.

> Bürger (H.). Ber. chem. Ges., **11**, 1876.

On a new optical constant.

> Gibbs (Wolcott). Proc. Amer. Acad., **10**, 401-16; Ann. Phys. u. Chem., **156**, 120-44.

Optische Constanten.

> Janowsky (J. V.). Ber. chem. Ges., **13**, 2272-77.

Ueber die Refractionsconstante.

> Lorenz (L.). Ann. Phys. u. Chem., n. F. **11**, 70-103.

Experimentelle Untersuchungen über die Refractionsconstante.

> Prytz (K.). K. Dän. Ges. d. Wiss. 1880, **6**, 8-22; Ann. Phys. u. Chem., n. F. **11**, 104-20.

Ueber einige von den Herrn J. W. Brühl und V. Zenger aufgestellte Beziehungen zwischen physikalischen Constanten chemischer Verbindungen.

> Wiedemann. Ber. chem. Ges., **15**, 464-70; Beiblätter, **6**, 370 (Abs.), 877 (Abs.).

COPPER.

On the colour, properties, and relations of the metals copper, nickel,
cobalt, iron, manganese, and chromium.

Bayley (T.). Jour. Chem. Soc., **37**, 828–36.

On the colour relations of copper and its salts.

Bayley (T.). Phil. Mag., (5) **5**, 222–4.

On the analysis of alloys containing copper.

Bayley (T.). Phil. Mag., (5) **6**, 14–19.

On the colour properties and colour relations of the metals of the iron-
copper group.

Bayley (T.). Jour. Chem. Soc., **39**, 862–70.

Copper spark spectrum; copper arc spectrum; copper and silver arc
spectrum; copper, gold, and silver (alloy) arc spectrum; copper
and iron spark spectrum.

Capron (J. R.). Photographed Spectra, London, 1877, p. 27, 31, 43.

Spectrum of nitrate of copper.

Chem News, **35**, 107.

Renversement des raies spectrales de cuivre.

Cornu (A.). Comptes Rendus, **73**, 332.

Spectre du cuivre.

Debray. Comptes Rendus, **54**, 169.

Spectre du bromure de cuivre, et du chlorure de cuivre.

Diacon (E.). Ann. Chim. et Phys., (4) **6**, 1

Spectre de l'azotate de cuivre.

Gouy. Comptes Rendus, **84**, 231; Chem. News, **35**, 107.

Caractères des flammes chargées de l'oxyde de cuivre et de l'acetate de
cuivre.

Gouy. Comptes Rendus, **85**, 439.

Black oxide of copper.

Jacques (W. W.). Proc. Royal Soc., **14**, 159.

Spectrum des Kupfers.

> Jahresber. d. Chemie, **15**, 80.   (See Debray, above.)

Spectre de l'oxyde de cuivre.

> Lallemand (A.).   Comptes Rendus, **78**, 1272.

Sur la diffusion lumineuse du sulfure et du phosphure de cuivre obtenus sans précipitation.

> Lallemand (A.).   Comptes Rendus, **79**, 698.

Chlorure de cuivre en solution, étincelle; chlorure de cuivre dans le gaz.

> · Lecoq de Boisbaudran, Paris, 1874, p. 152, 156, planche XXIV.

Erkennung von Chlor, Brom und Iod durch das Spektrum der Kupfer-verbindung.

> Mitscherlich (A.).   Ann. Phys. u. Chem., **125**, 629.

Spectrum von Kupfer.

> Simmler (R. Th.).   Ann. Phys. u. Chem., **115**, 249.

Methods for the determination of copper by spectral analysis.

> Wolff.   Chem. News, **39**, 124.

CRYSTALS.

Sur le pouvoir rotatoire du quartz dans le spectre ultra-violet.
> Croullebois. Comptes Rendus, 81, 666.

Action rotatoire du quartz sur le plan de polarization des rayons calorifiques obscurs d'un spectre.
> Desains (P.). Comptes Rendus, 84, 1056.

Anwendung des Spectroskops zur optischen Untersuchung der Krystalle.
> Ditscheiner (L.). Sitzungsber. d. Wiener Akad., 58 II, 4, 15–29.

Indices de réfraction ordinaire et extraordinaire du quartz, pour les
rayons de différentes longueurs d'onde jusqu'à l'extrême ultra-
violet.
> Sarasin (E.). Arch. de Genève, (2) 61, 109–19; Comptes Rendus, 85,
> 1230–2 (Abs.); Beiblätter, 2, 77 (Abs.).

Indices de réfraction ordinaire et extraordinaire du spath d'Islande pour
les rayons de diverses longueurs d'onde jusqu'à l'extrême ultra-
violet.
> Sarasin (E.). Comptes Rendus, 95, 680.

Indices de réfraction du spath-fluor pour les rayons de différentes longueurs
d'onde, jusqu'à l'extrême ultra-violet.
> Sarasin (E.). Comptes Rendus, 97, 850.

Propriétés optiques de quelques cristaux; acide oxalique, hyposulfite de
soude, sous-carbonate de soude, borax.
> Senarmont (H. de). Ann. Chim. et Phys., (3) 41, 336.

Sur la polarization rotatoire du quartz.
> Soret (J. L.). Arch. de Genève, (3) 8, 5–59, 97–132, 201–28; Jour. de
> Phys., (2) 2, 281–6 (Abs.).

Sur la polarization rotatoire du quartz.
> Soret (J. L.) et Sarasin (E.). Comptes Rendus, 83, 818; 95, 635.

D LINE.

Dark double line D in the spectrum from the electric arc.
> Foucault.   L'Institut (1848), 45.

Darstellung der dunklen Fraunhofer'schen Linie D.
> Kirchhoff (G.).   Ann. Phys. u. Chem., **109**, 148.

Die Ursache der dunklen Linie D nicht in dem Atmosphäre.
> Kirchhoff (G.).   Ann. Phys. u. Chem., **109**, 297.

Détermination de la valeur absolue de la longueur d'onde de la raie D.
> Macé de Lépinay (J.).   Ann. Chim. et Phys., (6) **10** (1887), 170-199.

Détermination de la longueur d'onde de la raie D_2.
> Macé de Lépinay (J.).   Jour. de Phys., (2) **5**, 411-16.

Indice du quartz pour la raie D.
> Sarasin (Ed.).   Comptes Rendus, **85**, 1280.

D line spectra.
> Stokes (G. G.).   Nature, **13**, 247.

Monographie du groupe D du spectre solaire.
> Thollon (L.).   Jour. de Phys., **13**, 5.

DARK LINES.

Étude des bandes froides des spectres obscurs.

Dessains (P.) et Aymonnet. Comptes Rendus, **81**, 423.

Die brechbarsten oder unsichtbaren Lichtstrahlen im Beugungsspectrum, und ihre Wellenlänge.

Eisenlohr (W.). Ann. Phys. u. Chem., **98**, 353.

Dark double line D in the spectrum from the electric arc.

Foucault. L'Institut (1849), 45.

Anwendung der dunklen Linien des Spectrums als Reagens auf Uran und Mangansäure.

Jahresber. d. Chemie, **5**, 125. (See Stokes in L'Institut, 1852, p. 392.)

Umwandlung heller Linien in Dunkle.

Jahresber. d. Chemie, **14**, 44. (See Kirchhoff, below.)

Dunkle Spectrallinien der Elemente.

Jahresber. d. Chemie, **17**, 108. (See Hinrichs (G.) in Amer. Jour. Sci., [2] **38**, 81.)

Umkehrung der hellen Spectrallinien der Metalle, insbesondere des Natriums, in Dunkle.

Jahresber. d. Chemie, **18**, 90. (See Madan (H. G.) in Phil. Mag., [4] **29**, 388.)

Die Ursache der dunklen Linie D nicht in dem Atmosphäre.

Kirchhoff (G.). Ann. Phys. u. Chem., **109**, 297.

Umkehrung der hellen und dunklen Linien.

Kirchhoff (G.) und Bunsen (R.). Ann. Phys. u. Chem., **110**, 187.

Spectrum des Phosphorescenzlichtes von Chlorophan, etc., mit dunklen Linien.

Kindt. Ann. Phys. u. Chem., **131**, 160; Phil. Mag., Dec., 1867.

Absorptionsspectren dunkler Wärmestrahlen in Gasen und Dämpfen.

Lecher und Pernter. Sitzungsber. d. Wiener Akad., **82** II, 265.

Dunkle Linien in den Spectren einiger Fixsterne.

Merz (L.). Ann. Phys. u. Chem., **117**, 654.

Dunkle Linien in dem photographirten Spectrum weit über dem sichtbaren Theil hinaus.

> Müller (J.).   Ann. Phys. u. Chem., **97**, 185.

Wellenlänge und Brechungsexponent der äussersten dunklen Wärmestrahlen des Sonnenspectrums.

> Müller (J.).   Ann. Phys. u. Chem., **116**, 543; Berichtigung dazu, **116**, 644.

A method of examining refractive and dispersive powers by prismatic reflection.   (Contains the first discovery of the dark solar lines.)

> Wollaston (W. H.).   Phil. Trans. (1802), 365.

Ursache der ungleichen Intensität der dunklen Linien im Spectrum der Sonne und der Fixsterne.

> Zöllner (F.).   Ann. Phys. u. Chem., **141**, 873.

DAVYUM.

Spectre du davyum.

> Kern (S.).   Comptes Rendus, **85**, 667; Nature, **17**, 245; Chem. News, **36**, 114, 155, 164; Beiblätter, **1**, 619.

DECIPIUM.

Sur le décipium, métal nouveau de la samarskite.

Delafontaine. Comptes Rendus, **87**, 632–4; Jour. Chem. Soc., **36**, 117–8; Amer. Jour. Sci., (3) **17**, 61-2 (Abs.); Beiblätter, **3**, 197–8 (Abs.).

Remarques sur le décipium et ses principaux composés.

Delafontaine. Comptes Rendus, **90**, 221–3; Arch. de Genève, (8) **3**, 250–60; Beiblätter, **4**, 549 (Abs.).

Spectre du nitrate de décipium.

Lecoq de Boisbaudran (F.). Comptes Rendus, **89**, 212.

DENSITY.

Ueber den Einfluss der Dichte und der Temperatur auf die Spectren von Dämpfen und Gasen.

Ciamician (G.). Wiener Anzeigen (1878), 158–60; Chemisches Centralblatt (1878), 689–90; Jour. Chem. Soc., **36**, 101 (Abs.).

Ueber den Einfluss der Dichte und der Temperatur auf die Spectren von Dämpfen und Gasen, 1879.

Ciamician (G.). Sitzungsber. d. Wiener Akad., **78** II, 867–90; Chemisches Centralblatt (1879), 507–9, 537–42, 555–7; Nature, **20**, 90 (Abs.); Beiblätter, **3**, 609–11.

Ueber den Einfluss der Dichtigkeit eines Körpers auf die Menge des von ihm absorbirten Lichtes.

Glan (P.). Ann. Phys. u. Chem., n. F. **3**, 54–82.

De l'intensité lumineuse des couleurs spectrales.

Parinaud (H.). Comptes Rendus, **99**, 937.

De l'influence qu'exerce l'intensité de la lumière colorée, etc.
    Prillieux.   Comptes Rendus, **69**, 294, 408, 412.

Ueber die Abhängigkeit der Brechungsexponenten anomal dispergirender
Medien von der Concentration der Lösung und der Temperatur.
    Sioben (G.).   Ann. Phys. u. Chem., **23**, 312.

Note sur un procédé destiné à mesurer l'intensité relative des éléments
constitutifs des différentes scources lumineuses.
    Trannin (H.).   Comptes Rendus, **77**, 1495.

Aenderung der Lage und Breite der Linien in Salpetergas und anderen
Substanzen mit der Dicke und Schicht.
    Weiss (A.).   Ann. Phys. u. Chem., **112**, 153.

Ueber den Einfluss der Dichtigkeit und Temperatur auf die Spectra
glühender Gase.
    Zöllner (F.).   Ber. Sächs. Ges. d. Wiss., **22**, 233–58; Ann. Phys. u.
Chem., **142**, 88–111; Phil. Mag., (4) **41**, 190–205.

DIDYMIUM.

Sur les variations des spectres d'absorption du dídyme.

> Becquerel (H.). Comptes Rendus, **103** (1887), 777–80; Chem. News, **55**, 148 (Abs.).

Sur le didyme.

> Brauner (B.). Comptes Rendus, **94**, 1718–19; Chem. News, **46**, 16–17; Jour. Chem. Soc., **44**, 18 (Abs.); Ber. chem. Ges., **15**, 2281 (Abs.).

Das Absorptionsspectrum des Didyms.

> Bührig (H.). Jour. prackt. Chemie, (2) **12**, 209–15; Amer. Jour. Sci., (3) **11**, 142 (Abs.).

Erscheinungen beim Absorptionsspectrum des Didyms; Aenderung bei Anwendung polarisirten Lichtes.

> Bunsen (R.). Ann. Phys. u. Chem., **128**, 100.

On the inversion of the bands in the didymium absorption spectra.

> Bunsen (R.). Phil. Mag., (4) **28**, 246; **32**, 177. (See Roscoe's Spectrum Analysis, Lecture 4, Appendix F, Third Edition.)

Photograph of the didymium arc spectrum.

> Capron (J. R.). Photographed Spectra, London, 1877, p. 28.

Note préliminaire sur le didyme.

> Clève (P. T.). Comptes Rendus, **94**, 1528–30; Chem. News, **45**, 273; Jour. Chem. Soc., **44**, 18 (Abs.); Ber. chem. Ges., **15**, 1750 (Abs.); Beiblätter, **6**, 771–2 (Abs.).

Quelques remarques sur le didyme.

> Clève (P. T.). Comptes Rendus, **95**, 83; Jour. Chem. Soc., **42**, 1165 (Abs.); Beiblätter, **6**, 772 (Abs.).

Note on the absorption spectrum of didymium.

> Crookes (W.). Chem. News, **54** (1886), 27.

Vergleich der Absorptionsspectra von Didym, etc.

> Delafontaine. Ann. Phys. u. Chem., **124**, 685.

Sur les spectres du didyme et du samarium.

> Demarçay (Eug.). Comptes Rendus, **102** (1886), 1551–2.

Absorptionslinien der Didymlösungen.

> Erdmann. Jour. prackt. Chemie, **85**, 394; **94**, 803.

14 T

On an optical test for didymium.

> Gladstone (J. H.).   Jour. Chem. Soc. (1858), **10**, 219.

Absorptionsspectrum des Didymnitrats.

> Jahresber. d. Chemie (1870), 321.

Chlorure de didyme en solution concentrée, absorption; do. en solution étendue, absorption.

> Lecoq de Boisbaudran (F.).   Spectres Lumineux, Paris, 1874, p. 87, 90, XIII.

The didymium absorption spectrum.

> Rood (O. N.).   Amer. Jour. Sci., (2) **34**, 129; Ann. Phys. u. Chem., **118**, 350.

Sur le spectre du nitrate de didyme.

> Smith (Lawrence) et Lecoq de Boisbaudran (F.).   Comptes Rendus, **88**, 1167.

Recherches sur l'absorption des rayons ultra-violets par diverses substances; spectre du didyme.

> Soret (J. L.).   Arch. de Genève, (2) **63**, 89–112; Comptes Rendus, **86**, 1062–4; Beiblätter, **2**, 410–11; **3**, 196–7.

Recherches sur les spectres d'absorption du didyme et de quelques autres substances extraites de la samarskite.

> Soret (J. L.).   Comptes Rendus, **88**, 422–4.

Om de lysande spectra hos Didym och Samarium (Sur les spectres brilliants du didyme et du samarium).

> Thalen (R.).   Ofversigt K. Svensk. Vetensk. Akad. Forhandl., **40**, No. 7, 3–16; Jour. de Phys., (2) **2**, 446–49; Ber. chem. Ges., **16**, 2760 (Abs.); Beiblätter, **7**, 893 (Abs.).

Om spectra tillhörande didym, yttrium, erbium och lanthan.

> Thalen (R.).   K. Svensk. Vetenskaps Akad. Förhandlingar, **12**, No. 4, 24; Bull. Soc. chim. Paris, (2) **22**, 350 (Abs.); Jour. de Phys., **4**, 33, avec une planche.

Note on the spectrum of didymium.

> Thompson (Claude M.).   Chem. News, **55** (1887), 227.

DIFFRACTION.

Spectrum der brechbarsten Strahlen.

Crookes. Cosmos, **8**, 90; Ann. Phys. u. Chem., **97**, 621.

Krümmung der Spectrallinien.

Ditscheiner (L.). Sitzungsber. d. Wiener Akad., **51** II, 341, 368–388.

On diffraction spectrum photography.

Draper (H.). Amer. Jour. Sci., **106**, 401–9; Phil. Mag., (4) **46**, 417–25; Nature, **9**, 224–6; Ann. Phys. u. Chem., **151**, 887–50.

Beugungsspectrum auf fluorescirenden Substanzen.

Eisenlohr (W.). Ann. Phys. u. Chem., **99**, 163.

Albertotypie eines photographirten Diffractionsspectrums.

Jahresber. d. Chemie (1873), 166. (See Draper, above.)

Diffraction bands in the spectrum.

Moreland. Amer. Jour. Sci., (3) **29**, 5.

Wärmevertheilung im Diffractionsspectrum.

Müller (J.). Ann. Phys. u. Chem., **105**, 355.

Comparison of prismatic and diffraction spectra.

Pickering (E. C.). Proc. Amer. Acad., **11**, 273.

On diffraction spectra.

Quincke (G.). Phil. Mag., (4) **45**, 365–71.

Beugungserscheinungen im Spectrum.

Rosiky. Sitzungsber. d. Wiener Akad., **71** I, 391.

Reduction for diffraction in spectrum observation.

Rosenberg (E.). Jour. Franklin Inst., **106**, 95.

Sur les phénomènes de diffraction produits par les réseaux circulaires.

Soret (J. L.). Archives de Genève, (2) **52**, 320–87; Ann. Phys. u. Chem., **156**, 99–113; Ann. Chim. et Phys., (5) **7**, 409–24.

Einige Bermerkungen über die Diffractionsspectra.

Spée (E.). Bull. de l'Acad. de Belgique, (3) **12**, 82–4; Beiblätter, **11** (1887), 99 (Abs.).

Imitation des spectres de diffraction par dispersion.

Zenger (Ch. V.). Comptes Rendus, **96**, 521.

DISCONTINUOUS SPECTRA.

On discontinuous spectra in high vacua.

> Crookes (W.). Proc. Royal Soc., **32**, 206–18; Nature, **24**, 89–91; Chem. News, **43**, 287–9; Ber. chem. Ges., **14**, 1696–7.

DISPERSION SPECTRA.

Experimentelle Prüfung der aelteren und neueren Dispersionsformeln.

> Brühl (J. W.). Ber. chem. Ges., **19** (1886), 2821–87; Beiblätter, **11**, 244–8; Jour. Chem. Soc., **52**, 195–8 (Abs.).

Note on the curvature of lines in the dispersion spectrum, and the method of correcting it.

> Christie (W. H. M.). Monthly Notices Astronom. Soc., **34**, 268–5. Note on this by Simms, same vol., 868–4.

Specific refraction and dispersion of light by liquids.

> Gladstone (J. H.). Rept. British Assoc. (1881), 591; Nature, **24**, 468 (Abs.); Beiblätter, **6**, 21 (Abs.).

Specific refraction and dispersion of isomeric bodies.

> Gladstone (J. H.). Proc. Royal Soc., **4**, 94–100; Phil. Mag., (5) **11**, 54–60; Ber. chem. Ges., **14**, 835 (Abs.); Jour. Chem. Soc., **40**, 213 (Abs.); Beiblätter, **5**, 276 (Abs.).

Zur Theorie der anomalen Dispersion.

> Helmholtz (H.). Monatsber. d. Berliner Akad. (1874), 667–80; Ann. Phys. u. Chem., **154**, 582–96.

Untersuchungen über das Dispersionsgesetz.

> Hesse (O.). Ann. Phys. u. Chem., n. F. **11**, 871–908.

Sur la dispersion anomale.

> Hurion. ·Jour. de Phys., **7**, 181; Ann. de l'École normale, (2) **6**, 367–412; Beiblätter, **2**, 79 (Abs.).

Zusammenhang zwischen Absorption und Dispersion.

> Ketteler (E.). Ann. Phys. u. Chem., **160**, 466–86.

Das specifische Gesetz der sogenannten anomalen Dispersion.

> Ketteler (E.). Ann. Phys. u. Chem., Jubelband, 166–82.

Notiz, betreffend die Dispersionscurve der Mittel mit mehr als einem Absorptionsstreifen.

> Ketteler (E.). Ann. Phys. u. Chem., n. F. **1**, 340–51.

Einige Anwendungen des Dispersionsgesetzes auf durchsichtige, halb-durchsichtige und undurchsichtige Mittel.

> Ketteler (E.). Ann. Phys. u. Chem., n. F. **12**, 368.

Attempt at a theory of the (anomalous) dispersion of light in singly and doubly refracting media.

> Ketteler (E.). Verhandl. d. naturhist. Vereinsd. preuss. Rheinlande und Westphalens, **33** (1876); Phil. Mag., (5) **2**, 332–45, 414–22, 508–22.

Zur Handhabung der Dispersionsformel.

> Ketteler (E.). Ann. Phys. u. Chem., (2) **30**, 299–31

Recherches sur la dispersion prismatique de la lumière.

> Klercker (C. E. de). Bihang till k. Svensk. Vet. Akad. Handl., **7**, 1–55; Comptes Rendus, **97**, 707 (Abs.).

Ueber anomale Dispersion der Körper mit Oberflächenfarben.

> Kundt (A.). Ann. Phys. u. Chem., **142**, 163–171; **143**, 149–52, 259–79; **144**, 128–87; **145**, 67–80; Nachtrag, **145**, 164–66; Ann. Chim. et Phys., (4) **25**, 404–10 (Abs.), 413–19 (Abs.), 419–21 (Abs.).

Ueber einige Beziehungen zwischen der Dispersion und Absorption des Lichtes.

> Kundt (A.). Ann. Phys. u. Chem., Jubelband, 615–24.

Ueber anomale Dispersion in glühendem Natriumdampf.

> Kundt (A.). Ann. Phys. u. Chem., n. F. **10**, 321–5; Phil. Mag., '5 **10**, 53–57.

Ueber die Dispersion des Aragonits nach arbiträrer Richtung.

> Zang (V. von). Sitzungsber. d. Wiener Akad.. **83** II, 671–6; Wiener Anzeigen (1881), 84 (Abs.).

On the dispersion of a solution of mercuric iodide.

> Liveing (G. D.). Proc. Philosoph. Soc. Cambridge, **3**, 258-60; Beiblätter, **4**, 610 (Abs.).

Theorie der normalen und anomalen Dispersion.

> Lommel (E.). Ann. Phys. u. Chem., n. F. **3**, 329-56.

Ueber einige zweiconstantige Dispersionsformel.

> Lommel (E.). Ann. Phys. u. Chem., n. F. **8**, 628-634.

Ueber das Dispersionsgesetz.

> Lommel (E.). Ann. Phys. u. Chem., n. F. **13**, 353-60.

Das Gesetz der Rotationsdispersion.

> Lommel (E.). Ann. Phys. u. Chem., n. F. **20**, 578.

Theorie der Dispersion.

> Lorenz (L.). Ann. Phys. u. Chem., n. F. **10**, 1-21.

Einige Versuche über totale Reflexion und anomale Dispersion.

> Mach (E.) und Arbes (J.). Ann. Phys. u. Chem., (2) **27**, 436-44.

Sur la dispersion des gaz.

> Mascart. Comptes Rendus, **78**, 679-82; Amer. Jour. Sci., (3) **7**, 591-2 (Abs.).

Versuch einer Erklärung der anomalen Farbenzerstreuung.

> Meyer (O. E.). Ann. Phys. u. Chem., **145**, 80-86; Ann. Chim. et Phys., (4) **43**, 821-38.

Quelques phénomènes de décomposition produits par la lumière.

> Morren. Comptes Rendus, **69**, 899.

Une méthode pour mesurer la dispersion dans les différentes parties du spectre fourni par un prisme ou un spectroscope quelconque.

> Mousson. Arch. de Genève, (2) **45**, 13; Ann. Phys. u. Chem., **148**, 660.
> (See Mach in Ann. Phys. u. Chem., **149**, 270.)

Sur les lois de la dispersion.

> Mouton. Comptes Rendus, **88**, 1189-92; Beiblätter, **3**, 616 (Abs.): Ann. Chim. et Phys., (5) **18**, 145-89.

Dispersion de la lumière.

> Ricour (Th.). Comptes Rendus, **69**, 1281; **70**, 115.

Ueber eine neue Flüssigkeit von hohem specifischen Gewicht, ..onem Brechungsexponenten und grosser Dispersion.

> Rohrbach (C.). Ann. Phys. u. Chem., n. F. **1**, 169-174; Amer. Jour. Sci., (3) **26**, 406 (Abs.); Jour. Chem. Soc., **46**, 145 (Abs.).

Recherches concernant la dispersion électromagnétique sur une spectre de grande étendue.

> Schaik (W. C. L. von). Arch. Neerlandaises, **17**, 378–90; Beiblätter, **7**, 919 (Abs.).

Ueber das Dispersionsäquivalent von Diamant.

> Schrauf (A.). Ann. Phys. u. Chem., n. F. **22**, 424–9; Jour. Chem. Soc., **48**, 14 (Abs.).

Ueber die durch die Aetherschwingungen erregten Mitschwingungen der Körpertheilchen und deren Rückwirkung auf die erstern, besonders zur Erklärung der Dispersion und ihrer Anomalien.

> Sellmeier (W.). Ann. Phys. u. Chem., **145**, 399–421, 520–49; **147**, 386–403, 525–54.

Untersuchungen über die anomale Dispersion des Lichtes.

> Sieben (G.). Ann. Phys. u. Chem., n. F. **8**, 137–57.

Micrometrical measures of gaseous spectra under high dispersion.

> Smyth (C. Piazzi). Trans. Royal. Soc. Edinburgh, **32** III, 415–60, 1884, with plates.

Sur la dispersion anormale de quelques substances.

> Soret (J. L.). Arch. de Genève, (2) **40**, 280–3; Ann. Phys. u. Chem., **143**, 825–7; Phil. Mag., (4) **44**, 395–6; Ann. Chim. et Phys., (4) **25**, 412 (Abs.).

Sur la réfraction et la dispersion des aluns crystallisés.

> Soret (C.). Arch. de Genève, (3) **10**, 300–2; Beiblätter, **8**, 874 (Abs.).

On an easy and at the same time accurate method of determining the ratio of the dispersions of glasses intended for objectives.

> Stokes (G. G.). Proc. Royal Soc., **27**, 485–94; Beiblätter, **3**, 185–7 (Abs.).

Minimum de dispersion des prismes; achromatisme de deux lentilles de mêmes substances.

> Thollon (L.). Comptes Rendus, **89**, 93–6; Beiblätter, **4**, 82–4.

Ueber die Beziehung zwischen chemischer Wirkung des Sonnenspectrums und anomaler Dispersion.

> Vogel (H.). Ber. chem. Ges., **7**, 976–9; Jour. Chem. Soc., (2) **12**, 1121–2.

Theorie der Dispersion.

> Voigt (W.). Göttinger gelehrten Nachr. (1884), 262.

Zur Dispersion farblos durchsichtiger Medien.

> Wüllner (A.).   Ann. Phys. u. Chem., n. F. 17, 580–7; Jour. de Phys.,
> (2) 2, 281 (Abs.).

Ausdehnung der Dispersionstheorie auf die ultra-rothen Strahlen.

> Wüllner (A.).   Ann. Phys. u. Chem., n. F. 23, 806; Jour. de Phys,
> (2) 4, 824 (Abs.).

Sur la dispersion du chromate de soude à 4 H₂ O.

> Wyrouboff (G.).   Bull. Soc. mineral. de France, 5, 160–1.

DISSOCIATION.

Dissociation of the elements.

> Crookes (W.).   Chem. News, 39, 65–6.

Ueber die neuen Wasserstofflinien und die Dissociation des Calciums.

> Vogel (H. W.).   Ber. chem. Ges., 13, 274–6; Jour. Chem. Soc., 33,
> 597 (Abs.); Beiblätter, 4, 274.

Ueber Lockyer's Dissociationstheorie.

> Vogel (H. W.).   Sitzungsber. d. Berliner Akad. (1882), 905–7; Nature,
> 27, 233; Ann. Phys. u. Chem., n. F. 19, 284–287; Phil. Mag., (5)
> 15, 28–30; Jour. Chem. Soc., 44, 762 (Abs.); Chem. News, 49, 201
> (Abs.).

DISTRIBUTION IN THE SPECTRUM.

The distribution of heat in the visible spectrum.
Conroy (Sir J.). Proc. Phys. Soc., 3, 106-12; Phil. Mag., (5) 8, 208-9; Beiblätter, 4, 44 (Abs.).

On the distribution of lines in spectra.
Hinrichs. Amer. Jour. Sci., July, 1864.

Vertheilung der chemischen Wirkung im Spectrum.
Jahresber. d. Chemie (1873), 160.

Distribution de l'energie dans le spectre normal.
Langley (S. P.). Ann. de Chim. et de Phys., (5) 25, 211.

Wärmevertheilung im Normalspectrum.
Lundquist (G.). Ann. Phys. u. Chem., 155, 146.

Sur la distribution des bandes dans les spectres primaires.
Salet (G.). Comptes Rendus, 79, 1229-30; Ber. chem. Ges., 7, 1788 (Abs.); Bull. Soc. chim. Paris, 22, 543.

DOUBLE SPECTRA.

Secondary Spectrum.
Rood (O. N.). Amer. Jour. Sci., 106, 172.

Sur les spectres doubles.
Salet (G.). Jour. de Phys., 4, 225.

On double spectra.
Watts (W. M.). Quar. Jour. Sci., Jan., 1871.

DYSPROSIUM.

Spectre du dysprosium.

> Lecoq de Boisbaudran (F.). Comptes Rendus, **102**, 1005–6; Jour.
> Chem. Soc., **50**, 667 (Abs.).

ELECTRIC SPECTRA.

Relation between electric energy and radiation in the spectrum of incandescence lamps.

> Abney and Festing. Proc. Royal Soc., **37**, 157.

Continuirliches Spectrum des electrischen Funkens.

> Abt (A.). Ann. Phys. u. Chem., n. F. **7**, 159; K. Ungar. Acad. d.
> Wiss. in Buda-Pest, Dec. 11, 1878; Jour. Chem. Soc., **36**, 765;
> Amer. Jour. Sci., (3) **18**, 68–9.

Spectrum des electrischen Lichtes.

> Angström (A. J.). Ann. Phys. u. Chem., **94**, 145; Phil. Mag., (4) **9**,
> 327.

Pouvoir phosphorescent de la lumière électrique.

> Becquerel (E.). Comptes Rendus, **8**, 217; **101**, 205–10; Jour. Chem.
> Soc., **48**, 1098 (Abs.).

Nouvelles expériences sur les effets électriques produits sous l'influence des rayons solaires.

> Becquerel (E.). Comptes Rendus, **9**, 561; remarques par M. Biot, 569.

Nouvelles expériences sur le même sujet.

> Becquerel (E.). Comptes Rendus, **9**, 711; nouvelles remarques par M.
> Biot, 713, 719.

Sur le rayonnement chimique qui accompagne la lumière solaire et la lumière électrique.

> Becquerel (E.). Comptes Rendus, 11, 702; rapport de M. Biot à propos de ce mémoire, 12, 101.

Effets électro-chimiques produits sous l'influence de la lumière.

> Becquerel (E.). Comptes Rendus, 32, 85.

A new form of absorption-cell.

> Bostwick (A. E.). Amer. Jour. Sci., Dec., 1885; Phil. Mag., (5) 21, 80 (Abs.).

Einfluss des Drucks auf das Spectrum des electrischen Funkens in Gasen.

> Cailletet. Ber. chem. Ges., 5, 482.

Kleinste im Inductionsfunken durch die Spectralanalyse noch erkennbare Gewichtsmenge verschiedener Metalle.

> Cappel (E.). Ann. Phys. u. Chem., 139, 681–6.

Wolfram arc spectrum, photographed.

> Capron (J. R.). Photographed Spectra, London, 1877, 50.

Sur la photographie du spectre de l'étincelle électrique.

> Cazin (A.). Bull. Soc. philom. de Paris, 1877, (7) 1, 6–7; Beiblätter, 1, 287–8 (Abs.).

Sur le spectre de l'étincelle électrique dans les gaz soumis à une pression croissante.

> Cazin (A.). Comptes Rendus, 84, 1151–4; Phil. Mag., (5) 4, 153–6; Beiblätter, 1, 620 (Abs.); Jour. Chem. Soc., 34, 857 (Abs.); Jour. de Phys., 6, 271; Amer. Jour. Sci., (3) 15, 148 (Abs.).

Phénomènes observés dans les spectres produits par la lumière des courants d'induction traversant les gaz raréfiés.

> Chautard (J.). Comptes Rendus, 59, 383.

Action exercée par un électro-aimant sur les spectres des gaz raréfiés, traversés par des décharges électriques.

> Chautard (J.). Comptes Rendus, 79, 1123–4.

Action des aimants sur les gaz raréfiés renfermés dans les tubes capillaires et illuminés par un courant induit.

> Chautard (J.). Comptes Rendus, 80, 1161–4.

Phénomènes magnéto-chimiques produits au sein des gaz raréfiés dans les tubes de Geissler.

> Chautard (J.). Comptes Rendus, 81, 75–7; 82, 272–274; Jour. Chem. Soc., 1876, 1, 29 (Abs.).

Observations of the spectrum of lightning.

> Clark (J. W.). Chem. News, **30**, 28; **32**, 65; **35**, 2; Beiblätter, **1**, 192.

Den Einfluss welchen die Natur der electrischen Stromquelle auf das Aussehen von Gasspectren ausübt.

> Czechowicz. Versammlung russischer Naturforscher und Aertzte in Warschau, Sept., 1876; Ber. chem. Ges., **9**, 1598 (Abs.).

Analyse spectrale de l'étincelle électrique produite dans les liquides et les gaz.

> Daniel. Comptes Rendus, **57**, 98.

Notice sur la constitution de l'univers. Première partie, analyse spectrale.

> Delaunay. Ann. du Bureau des Longitudes, Paris, 1869.

Sur les spectres des étincelles des bobines à gros fil.

> Demarçay (E.). Comptes Rendus, **103** (1887), 678.

Spectre du pôle négatif de l'azote.

> Deslandes (H.). Comptes Rendus, **103** (1886), 375–9; Jour. Chem. Soc., **50**, 957.

Recherches sur l'influence des éléments électro négatifs sur le spectre des métaux.

> Diacon (E.). Ann. Chim. et Phys., (4) **6**, 5.

Ueber den Unterschied der prismatischen Spectra des am positiven und negativen Pol im luftverdünnten Raume hervortretenden electrischen Lichtes.

> Dove (H. W.). Ann. Phys. u. Chem., **104**, 184.

Over de zamenstelling van zonlicht, gaslicht en het von Edison's lamp, vergelijkend onderzocht met behulp der bacterien-methode.

> Engelmann (T. W.). Proc. verb. k. Akad. v. Wetensch. te Amsterdam, Nov. 25, 1882, No. 5, 4–5; Beiblätter, **7**, 880 (Abs.).

Sur les changements de réfrangibilité observés dans les spectres électriques de l'hydrogène et du magnésium.

> Fiévez (C.). Bull. Acad. do Bélgique, (3), **7**, 245–7; Beiblätter, **8**, 506 (Abs.).

Spectrum of lightning.

> Gibbons (J.). Chem. News, **24**, 96; **40**, 65.

Spectrum of lightning.

> Grandeau (L.). Chem. News, **9**, 66.

Note of an experiment on the spectrum of the electric discharge.
> Grove (Sir W. R.). Proc. Royal Soc., **28**, 181–4; Beiblätter, **3**, 860 (Abs.).

Das Stokes'sche Gesetz.
> Hagenbach (E.). Ann. Phys. u. Chem., n. F. **8**, 369.

The investigation by means of photography of the ultra-violet spark spectra emitted by metallic elements and their combinations under varying conditions.
> Hartley (W. N.). Chem. News, **48**, 195–6; Nature, **29**, 89–90; Jour. Chem. Soc., **46**, 187 (Abs.); Beiblätter, **8**, 802 (Abs.).

Spectrum of lightning.
> Herschel (Lieut. John). Proc. Royal Soc., **16**, 418; **17**, 61.

Spectra of lightning.
> Hoh (Th.). Chem. News, **30**, 258; Ann. Phys. u. Chem., **152**, 178.

Spectrum of lightning.
> Holden (E. S.). Amer. Jour. Sci., (3) **4**, 474–5.

Spectrum of the electric light.
> Hopkins-Walters (J.). Nature, **25**, 103.

Electric spectra in various gases and with electrodes of various substances.
> Huggins (W.). Phil. Trans., 1864; Ann. Phys. u. Chem., **124**, 275–292, 621.

Photographische Wirkung electrischer Metallspectren.
> Jahresber. d. Chemie, (1862) 33, (1863) 104, 106, 107, 113, (1864) 109, 110, 115, (1865) 90, 91, 92, (1868) 126–7, (1872) 148, (1873) 150–2, (1875) 123.

Spectrum des Blitzes.
> Jahresber. d. Chemie, (1864) 109, (1868) 126, 127, (1872) 148.

Spectralanalyse mittelst des Inductionsstroms.
> Jahresber. d. Chemie, (1865) 91, 92, (1873) 150, 151–2, (1864) 110.

Spectrum of lightning.
> Joule (J. P.). Nature, **6**, 161.

Spectra of two hundred and fourteen flashes of lightning observed at the astrophysical observatory in Herény, Hungary.
> Konkoly (N. von). Observatory (1883), 267–8; Beiblätter, **7**, 862 (Abs.).

Wärmevertheilung im Spectrum des Kalklichtes bei Flintglas-und Stein-
salz-prismen.

> Lamansky (S.).   Ann. Phys. u. Chom., **146**, 227.

Sur la loi de Stokes.

> Lamansky (S.).   Jour. de Phys., **8**, 367; Ann. Phys. u. Chem., n. F.
> **8**, 624.

Observations sur quelques points d'analyse spectrale et sur la constitution
des étincelles d'induction.

> Lecoq de Boisbaudran (F.).   Comptes Rendus, **73**, 943.

Spectre de l'ammoniaque par renversement du courant induit.

> Lecoq de Boisbaudran (F.).   Comptes Rendus, **101** (1885), 42–5; Jour.
> Chem. Soc., **48**, 1025 (Abs.).

Sur un spectre électrique particulier aux terres rares du groupe terbique.

> Lecoq de Boisbaudran.   Comptes Rendus, **102** (1886), 153–5.

Fluorescence des composés du manganèse, soumis à l'effluve électrique
dans le vide.

> Lecoq de Boisbaudran.   Comptes Rendus, **103** (1886), 468–71, 629–31,
> 1064–7, 1107; Jour. Chem. Soc., **52** (Abs.); Amer. Jour. Sci., (3)
> **33**, 149–51 (Abs.); Beiblätter, **11**, 87, 89 (Abs.).

An arrangement of the electric arc for the study, with the spectroscope,
of the radiation of vapours, together with preliminary results.

> Liveing (G. D.) and Dewar (J.).   Proc. Royal Soc., **34**, 119.

Note on some phenomena attending the reversal of lines in the arc produced by a Siemens machine.

> Lockyer (J. N.).   Proc. Royal Soc., **28**, 428.

Ueber die Glüherscheinungen an Metallectroden innerhalb einer Wasser-
stoffatmosphäre von verschiedenen Drucke.

> Lohse (O.).   Ann. Phys. u. Chom., n. F. **12**, 109–114.

Das Stokes'sche Gesetz.

> Lommel (E.).   Ann. Phys. u. Chom., n. F. **8**, 244.

Die weitausgedehnten ultravioletten Strahlen im Spectrum des electrischen Funkens mit dem Auge wahrnehmbar.

> Mascart.   Ann. Phys. u. Chom., **137**, 163.

Spectre de la lumière des piles dans l'air.

> Masson (A.).   Comptes Rendus, **32**, 128; Ann. Chim. et Phys., (3)
> **31**, 295.

On the photographic effects of metallic and other spectra obtained by means of the electric spark.

Miller (W. Allen). Proc. Royal Soc., **12**, 159; Phil. Trans. (1862), 861.

Spectre de la lumière électrique dans le vide.

Du Moncel. Comptes Rendus, **49**, 40.

Spectre fluorescent de l'étincelle électrique.

Müller (J.). Ann. Chim. et Phys., (4) **13**, 465.

Report on spark spectra, from the British Association Report on the Present State of our Knowledge of Spectrum Analysis.

Nature, **26**, 459. (By A. Schuster.)

Ueber das Sauerstoffspectrum und über die electrischen Lichterscheinungen verdünnter Gaze in Röhren mit Flüssigkeitselectroden.

Paalzow. Monatsber. d. Berliner Akad. (1878), 705–9; Phil. Mag., (5) **7**, 297–300; Ann. Phys. u. Chem., n. F. **7**, 130–5; Jour. Chem. Soc., **36**, 861.

Photographing spark spectra.

Parry (J.). Chem. News, **36**, 140.

Experimentelle Untersuchung über das electrische Lichtspectrum in Beziehung auf die Farben der Doppelsterne.

Petzval (Jos.). Sitzungsber. d. Wiener Akad., **41**, 561, 581–9.

Spectra der electrischen Lichtströmungen.

Plücker. Ann. Phys. u. Chem., **104**, 122; **105**, 67; **107**, 497, 505, 506, 518–642; **116**, 27.

Spectrum of lightning.

Proctor (H. R.). Nature, **6**, 161, 220.

Spectra negativer Electroden und lange gebrauchter Geissler'schen Röhren.

Reitlinger (Edm.) und Kuhn (M.). Sitzungsber. d. Wiener Akad., **51** II, 405, 408–16; Ann. Phys. u. Chem., **141**, 185–6.

Electric spectra.

Robinson (Dr.). Phil. Trans. (1868).

Recherches sur les raies du spectre solaire et des différentes spectres électriques.

Robiquet. Comptes Rendus, **49**, 606.

Spectrum des electrischen Glimmlichts in atmosphärischer Luft.

Schimkow (A.). Ann. Phys. u. Chem., **129**, 513.

On the spectra of lightning.

Schuster (A.).   Phil. Mag., (5) 7, 816–21; Beiblätter, 3, 872 (Abs.).

Sur les spectres de l'étincelle électrique dans les gaz composés et en particulier dans le fluorure de silicium.

Seguin (J. M.).   Comptes Rendus, 54, 988.

Spectrum des Inductionsfunken.

Simmler (R. Th.).   Ann. Phys. u. Chem., 115, 268.

Beiträge zur Electricitätsleitung der Gase.

Stenger (F.).   Ann. Phys. u. Chem., (2) 25, 31–48; Jour. Chem. Soc., 48, 1028 (Abs.).
(See Phil. Trans., 171, 65.)

On the long spectrum of the electric light.

Stokes (G. G.).   Proc. Royal Soc., 12, 166; Phil. Trans. (1862), 599; Ann. Phys. u. Chem., 123, 30, 37, 472.

Effluviography.

Tomassi (D.).   Bull. Soc. chim. Paris, 45, 878; Jour. Chem. Soc., 50, 959 (Abs.).

Ueber die Spectra der Blitze.

Vogel (H.).   Ann. Phys. u. Chem., 143, 653–4.

Chemische Intensität des magnesium und electrischen Lichtes.

Vogel (H. W.).   Photographische Mittheilungen, 16, 187–8; Beiblätter, 4, 49 (Abs.).

Spectrum of the electric (Jablochkoff) light.

Walker (E.).   Nature, 18, 884; Beiblätter, 3, 505 (Abs.).

Spectra des electrischen Funkenstroms in verdünnten Gasen.

Waltenhofen (A. von).   Dingler's Jour., 177, 88.

Spectrum of the electric light.

Walters (J. Hopkins).   Nature, 25, 108.

The prismatic decomposition of the electric, voltaïc, and electro-magnetic sparks.

Wheatstone (C.).   Chem. News, 3, 198.

Das Leuchten der Gase durch electrische Entladungen.

Wiedemann (E.).   Ann. Phys. u. Chem., n. F. 6, 298.

Das thermische und optische Verhalten von Gasen unter dem Einfluss electrischer Entladungen.

Wiedemann (E.).   Ann. Phys. u. Chem., n. F. 10, 202.

Das electrische Leuchten der Gase.

Wiedemann (E.). Ann. Phys. u. Chem., n. F. **18**, 509–10.

Note au sujet d'un mémoire de M. Lagarde.

Wiedemann (E.). Ann. Chim. et Phys., (6) **7**, 143; Amer. Jour. Sci., (3) **31**, 218 (Abs.).

Das electrische Spectrum.

Willigen (S. M. von der). Ann. Phys. u. Chem., **106**, 615, 619, 621, 622, 624, 628; **107**, 473.

Sur le spectre de l'étincelle électrique dans les gaz soumis à une pression croissante.

Wüllner (A.). Comptes Rendus, **85**, 280–1; Ann. Chim. et Phys., (5) **12**, 143–4; Beiblätter, **1**, 620.

Das Linienspectrum gehört dem Funken, das Bandenspectrum gehört der Lichthülle an.

Wüllner (A.). Ann. Phys. u. Chem., **147**, 324–48.

15 T

EMISSION SPECTRA.

Sur la variation des spectres d'absorption et des spectres d'émission par
phosphorescence d'un même corps.

    Becquerel (H.).   Comptes Rendus, **102**, 106–10.

Notes on photographs of the ultra-violet emission spectra of certain elements.

    Hartley (W. N.).   Chem. News, **43**, 289; Ber. chem. Ges., **15**, 1432a,
    2924b.

Das Verhältniss zwischen Emission und Absorption ist bei allen Körpern
dasselbe.

    Kirchhoff (G.).   Ann. Phys. u. Chem., **109**, 299.

Ueber den Zusammenhang zwischen Emission und Absorption von Licht
und Wärme.

    Kirchhoff (G.).   Monatsber. d. Berliner Akad., Oct. 27, 1859; Phil.
    Mag., (4) **19**, 168.

ENERGY IN THE SPECTRUM.

Étude expérimentale de la réflexion des rayons actiniques.

De Chardonnet. Jour. de Phys., 11, 549.

Distribution of chemical force in the spectrum.

Draper (J. W.). Amer. Jour. Sci., 105, 25, 91–8; Phil. Mag., (4) 44, 422–48; Jour. Chem. Soc., (2) 11, 282–5.

Actinometry.

Duclaux (E.). Comptes Rendus, 103, 1010–12; Jour. Chem. Soc., 52, 189 (Abs.).

Einführung des Princips der Erhaltung der Energie in die Theorie der Diffraction.

Fröhlich (J.). Ann. Phys. u. Chem., n. F. 3, 876.

The Bolometer and radiant energy.

Langley (S. P.). Proc. Amer. Acad., 16, 342–58; Zeitschr. Instrumentenkunde, 4, 27-82 (Abs.).

Distribution de l'énergie dans le spectre normal.

Langley (S. P.). Comptes Rendus, 93, 140; Ann. Chim. et Phys., (5) 25, 211.

Distribution of energy in the spectrum.

Rayleigh (Lord). Nature, 27, 559.

La distribution de l'énergie dans le spectre solaire et la chlorophylle.

Timiriaseff. Comptes Rendus, 96, 875.

ERBIUM.

Erbinerdelösungen coïncidirend mit den hellen Streifen leuchtender Erbinerde.

>Bahr und Bunsen. Jour. prackt. Chemie, **97**, 277; Ann. f. Chem. u. Pharm., **137**, 1.

Aenderung des Absorptionsspectrums von Erbium bei Anwendung polarisirten Lichtes.

>Bunsen (R.). Ann. Phys. u. Chem., **128**, 100.

Erbium arc spectrum.

>Capron (J. R.). Photographed Spectra, London, 1877, p. 29.

Sur deux nouveaux éléments dans l'erbine.

>Clève (P. T.). Comptes Rendus, **89**, 478–80; Amer. Jour. Sci., (3) **18**, 400–1; Beiblätter, **4**, 43 (Abs.).

Spectre de l'erbine.

>Clève (P. T.). Comptes Rendus, **89**, 708; **91**, 381.

Sur les combinaisons de l'yttrium et de l'erbium.

>Clève (P. T.) et Hoegland (O.). Bull. Soc. chim. Paris, **18**, 193–201; 289–97; Jour. Chem. Soc., (2) **11**, 136.

Note on the spectra of erbia.

>Crookes (W.). Chem. News, **53** (1886), 75, 154, 179; Proc. Royal Soc., **40**, 77–9, Jour. Chem. Soc., **50**, 749 (Abs.); Comptes Rendus, **102**, 506.

Absorptionsspectrum von Erbiumlösungen.

>Delafontaine. Jour. prackt. Chemie, **94**, 303.

Vergleich der Absorptionsspectra von Didym, Erbium und Terbium.

>Delafontaine. Ann. Phys. u. Chem., **124**, 635; Chem. News, **11**, 253; Ann. Chim. et Phys., **135**, 194.

Note on the spectra of erbia and of some other earths.

>Huggins (W.). Chem. News, **22**, 175.

Spectren der Erbinerde.

>Jahresber. d. Chemie (1878), 150.

Phosphate de l'erbine, émission; erbine, émission; chlorure de l'erbium en solution, absorption.

> Lecoq de Boisbaudran (F.). Spectres Lumineux, Paris, 1874, p. 92, 97, planche XIV; p. 100, planche XV.

Spectre d'émission de l'erbine.

> Lecoq de Boisbaudran (F.). Comptes Rendus, **76**, 1080.

Spectre du nitrate de l'erbium.

> Lecoq de Boisbaudran (F.). Comptes Rendus, **88**, 1107.

Examen spectral de l'erbine.

> Lecoq de Boisbaudran (F.). Comptes Rendus, **88**, 1842–44; Jour. Chem. Soc., **36**, 861 (Abs.); Amer. Jour. Sci., (3) **18**, 216–7; Beiblätter, **3**, 871 (Abs.).

Spectre de l'erbine.

> Lecoq de Boisbaudran (F.). Comptes Rendus, **89**, 516; Beiblätter, **4**, 43 (Abs.); Chem. News, **40**, 147.

Remarques à M. P. T. Clève "Sur deux nouveaux éléments dans l'erbine."

> Smith (L.). Comptes Rendus, **89**, 480-1; Beiblätter, **4**, 43 (Abs.).

Om spectra tillhörande yttrium, erbium, didym och lanthan.

> Thalén (R.). K. Svensk. Vetenskaps. Akad. Forhandlinger, **12**, No. 4, 24; Bull. Soc. chim. Paris, (2) **22**, 850 (Abs.).

Spectrum of erbium.

> Thalén (R.). Chem. News, **42**, 184; Comptes Rendus, **91**, 326; Jour. de Phys., (2) **4**, 33.

Spektralundersökningar rörande skandium, ytterbium, erbium och thulium.

> Thalén (R.). Ofversigt af Kongl. Vetensk. Acad. Förhandlingar, **38**, No. 6, 13–21; Jour. de Phys., (2) **2**, 85–40; Chem. News, **47**, 217 (Abs.); Jour. Chem. Soc., **44**, 954 (Abs.).

EXCHANGES.

On the Theory of Exchanges.

> Stewart (Balfour). Trans. Royal Soc. Edinburgh (1858), Vol. **22**
> part I, 1; Rept. British Assoc. (1861), 97.

EXPLOSIONS.

Spectroscopic studies on gaseous explosions.

> Liveing (G. D.) and Dewar (J.). Proc. Royal Soc., **36**, 471–8; Chem.
> News, **49**, 227–9; Nature, **29**, 614–15; Beiblätter, **8**, 644–5 (Abs.).

Spectral lines of the metals developed by exploding gases

> Liveing (G. D.) and Dewar (J.). Phil. Mag., (5) **18**, 161–78; Jour.
> Chem. Soc., **48** (1885), 317 (Abs.).

Spectroscopic studies of explosions.

> Liveing (G. D.) and Dewar (J.). Rept. British Assoc. (1884), 672;
> Jour. de Phys., (2) **4**, 51 (Abs.).

Spectrum des Lichtes explodirender Schiessbaumwolle.

> Vogel (H. W.). Ann. Phys. u. Chem., n. F. **3**, 615.

FLAME AND GAS SPECTRA.

The dichroism of the vapour of iodine.

> Andrews (T.). Chem. News, **24**, 75; Jour. Chem. Soc., (2) **9**, 973
> (Abs.).

Spectres des gaz simples.

> Angström (A. J.). Comptes Rendus, **73**, 860; Bull. Soc. chim. Paris
> n. s. **16**, 228.

Recherches expérimentales sur la polarization rotatoire magnétique dans
les gaz.

> Becquerel (H.). Comptes Rendus, **90**, 1407.

Spectres d'émission infra-rouges des vapeurs métalliques.

> Becquerel (H.). Comptes Rendus, **97**, 71-4; Chem. News, **48**, 46
> (Abs.); Nature, **28**, 287 (Abs.); Beiblätter, **7**, 701-2 (Abs.); Amer.
> Jour. Sci., (3) **26**, 821 (Abs.); Ber. chem. Ges., **16**, 2487 (Abs.);
> Jour. Chem. Soc., **46**, 1 (Abs.); Zeitschr. analyt. Chem., **23**, 49
> (Abs.).

Spectres d'émission infra-rouges des vapeurs métalliques.

> Becquerel (H.). Comptes Rendus, **99**, 374; Amer. Jour. Sci., (3) **28**,
> 459; Phil. Mag., Oct., 1884.

Spectres de quelques corps composés dans les systèmes gazeux en équilibre.

> Berthelot et Richard. Comptes Rendus, **68**, 1546.

Experimentaluntersuchung zur Bestimmung der Brechungsexponenten
verflüssigter Gase.

> Bleekrode (L.). Ann. Phys. u. Chem., n. F. **8**, 400

Experiments on Flame.

> Burch (G. J.). Nature, **31**, 272_5; Jour. Chem. Soc., **48**, 466 (Abs.).

Einfluss des Drucks auf das Spectrum des electrischen Funkens in Gazen.

> Cailletet. Ber. chem. Ges., **5**, 482.

Spectrum of coal gas.

> Capron (J. R.). Photographed Spectra, London, 1877, p. 24, 61, 62,
> 71, 72.

Relative intensity of the spectral lines of gases.

> Capron (J. R.). Phil. Mag., (5) **9**, 329-80; Jour. Chem. Soc., **38**, 685
> (Abs.); Beiblätter, **4**, 613-14 (Abs.).

Spectre de l'étincelle électrique dans les gaz soumis à une pression croissante.

    Cazin (A.).   Comptes Rendus, **84**, 1151–4; Phil. Mag., (5) **4**, 153–6.

Action des ainmants sur les gaz raréfiés renfermés dans les tubes capillaires et illuminés par un courant induit.

    Chautard (J.).   Comptes Rendus, **59**, 883; **79**, 1123; **80**, 1161; **81**,
    75; Phil. Mag., Nov., 1864.

Ueber den Einfluss des Drucks und der Temperatur auf die Spectren
von Dämpfen und Gasen.

    Ciamician (G.).   Sitzungsber. d. Wiener Akad., **77** II, 829–41; Jour.
    Chem. Soc., **36**, 685 (Abs.); Nature, **23**, 160; Beiblätter, **3**, 193–4.

Viscosity of gases at high exhaustions.

    Crookes (W.).   Phil. Trans., **173**, 387–434; Chem. News, **43**, 85–9
    (Abs.); Nature, **23**, 421–3, 443–6 (Abs.); Beiblätter, **5**, 836–46 (Abs.).

Position of the chemical rays in the spectra of sunlight and gaslight.

    Crookes (W.).   Cosmos, **8**, 90; Ann. Phys. u. Chem., **97**, 619; Bull.
    London Photogr. Soc., 21 Jan., 1856.

Étude des radiations émises par les corps incandescents.

    Crova (A.).   Ann. Chim. et Phys., (5) **19**, 472–550; Beiblätter, **5**, 117
    (Abs.).

Spectre du pôle négatif de l'azote.

    Deslandres (H.).   Comptes Rendus, **103**, 375–9; Beiblätter, **11**, 36.

Spectra zusammengesetzter Gase.

    Dibbits (H. C.).   Ann. Phys. u. Chem., **122**, 538.

Essai d'analyse spectrale appliquée à l'examen de gaz simples et de leurs
mélanges.

    Dubrumfaut.   Comptes Rendus, **69**, 1245; Ber. chem. Ges., **2**, 745.

Flame-spectra.

    Fielding (G. F. M.).   Chem. News, **54**, 212.

Preliminary note of researches on gaseous spectra in relation to the physical constitution of the Sun, fixed stars and nebulæ.

    Franckland (E.) and Lockyer (J. N.).   Proc. Royal Soc., **17**, 288;
    18; 79.

Sur les spectres d'absorption des vapeurs de sélénium, de protochloruro et
de bromure de sélénium, de tellure, de protochlorure et de protobromure de tellure, protobromure d'iode et d'alizarine.

    Gernez (D.).   Comptes Rendus, **74**, 1190–2; Jour. Chem. Soc., (2) **10**,
    665 (Abs.); Phil. Mag., (4) **43**, 473–5; Amer. Jour. Sci., **4**, 59–60.

Blue flame from common salt.

 Gladstone (J. H.). Proc. Royal Soc., **19**, 582.

Note on the atmospheric lines of the solar spectrum, and on certain spectra of gases.

 Gladstone (J. H.). Proc. Royal Soc., **11**, 805.

Beobachtungen an Gasspektris.

 Goldstein (E.). Monatsber. d. Berliner Akad. (1874), 593–610; Ann. Phys. u. Chem., **154**, 128–149; Jour. Chem. Soc., (2) **13**, 527 (Abs.); Phil. Mag., (4) **49**, 388–45; Bemerkungen dazu, von A. Wüllner, Monatsber. d. Berliner Akad. (1874), 755–61; Phil. Mag., (4) **49**, 448–53.

Recherches photométriques sur les flammes colorées.

 Gouy. Comptes Rendus, **83**, 269–72; Phil. Mag., (5) **2**, 817–19.

Recherches sur les spectres des métaux à la base des flammes.

 Gouy. Comptes Rendus, **84**, 231.

Recherches photométriques sur les flammes colorées; sodium, lithium, strontium, calcium, etc.

 Gouy. Comptes Rendus, **85**, 70.

Sur le caractères des flammes chargées de calcium, de poussières salines, de chlorure de cuivre, de l'azotate et du chlorure de calcium, du chlorure de strontium, du chlorure de baryum, de l'oxyde de cuivre, de l'acetate de cuivre.

 Gouy. Comptes Rendus, **85**, 439.

Sur la transparence des flammes colorées, spectres continus du potassium, du sodium, des sels de l'alumine et de magnésie, du strontium, du calcium et du baryum.

 Gouy. Comptes Rendus, **86**, 878.

Transparence des flammes colorées pour leurs propres radiations; la double raie du sodium, la double raie du potassium; lithium, strontium, rubidium, calcium.

 Gouy. Comptes Rendus, **86**, 1078.

Du pouvoir émissif des flammes colorées.

 Gouy. Comptes Rendus, **88**, 418.

Ueber ein einfaches Verfahren die Umkehrung der farbigen Linien der Flammenspectra, insbesondere der Natriumlinie, subjectiv darzustellen.

 Günther (E.). Ann. Phys. u. Chem., n. F. **2**, 477.

De la recherche des composés gazeux et de l'étude de quelques-unes de
leur propriétés à l'aide du spectroscope.

> Hautefeuille (P.) et Chappuis (J.). Comptes Rendus, 92, 80–2; Jour.
> Chem. Soc., 40, 221–222 (Abs.); Beiblätter, 5, 317 (Abs.).

Bemerkungen zu dem Aufsatze von W. Siemens: Über das Leuchten der
Flamme.

> Hittorf (W.). Ann. Phys. u. Chem., n. F. 19, 78–7; Jour. Chem.
> Soc., 44, 697 (Abs.).

Prismatische Zerlegung des Lichtes glühender oder brennender Körper.

> Jahresber. d. Chemie, 1, 161; 3, 155.

Verschiedene Spectren desselben Gases.

> Jahresber. d. Chemie (1868), 125.

Spectra der Flammen grünfärbender Substanzen.

> Jahresber. d. Chemie, 14, 43.

Gas Spectra.

> Jahresber. d. Chemie, (1864) 109, (1868) 125, (1869) 176–80, (1870) 176,
> (1872) 143, (1873) 143, (1875) 122.

Sur le spectre de la vapeur de l'eau.

> Janssen (J.). Ann. Chim. et Phys., (4) 24, 215–7; Jour. Chem. Soc.,
> (2) 10, 280 (Abs.).

Flamme bleue du gaz d'éclairage.

> Lecoq de Boisbaudran (F.). Spectres Lumineux, Paris, 1874, p. 41,
> planche III.

Spectra kohlenstoffhaltiger Gase.

> Lielegg. Jour. prackt. Chemie, 103, 507; Phil. Mag., (4) 37, 203.

Untersuchungen über die Spectra gasförmiger Körper.

> Lippich (F.). Sitzungsber. d. Wiener Akad., 82 II, 15–88; Ann.
> Phys. u. chem., n. F. 12, 380.

Erklärung der Verbreiterung der Spectrallinien in den Gazen.

> Lippich (F.). Ann. Phys. u. Chem., 139, 465.

Origin of the spectrum of the hydrocarbon flame.

> Liveing (G. D.) and Dewar (J.). Nature, 27, 257.

On the reversal of the lines of metallic vapours.

> Liveing (G. D.) and Dewar (J.). No. I in Proc. Royal Soc., 27, 132–6;
> No. II in do., 27, 350–4; No. III in do., 27, 494–6; No. IV in do.,
> 28, 352–8; No. V in do., 28, 367–72; No. VI in do., 28, 471–5;
> No. VII in do., 29, 402–6; Beiblätter, 2, 261–3 (Abs.), 490 (Abs.);
> 3, 502 (Abs.), 710 (Abs.); 4, 364 (Abs.).

Disappearance of some spectral lines and the variation of metallic spectra due to mixed vapours.

> Liveing and Dewar. Proc. Royal Soc., **33**, 428.

An arrangement of the electric arc for the study, with the spectroscope, of the radiation of vapours, together with preliminary results.

> Liveing and Dewar. Proc. Royal Soc., **34**, 119.

Spectral lines of metals developed by exploding gases.

> Liveing (G. D.) and Dewar (J.). Phil. Mag., (5) **18**, 161–73; Jour. Chem. Soc., **48**, 817 (Abs.); Jour. de Phys., (2) **4**, 51.

Spectroscopic studies on gaseous explosions.

> Liveing (G. D.) and Dewar (J.). Proc. Royal Soc., **36**, 471–8; Jour. Chem. Soc., **48**, 465.

Spectroscopic Notes. Note I, on the absorption of great thicknesses of metallic and metalloidal vapours; Note II, on the evidence of variation in molecular structure; Note III, on the molecular structure of vapours in connection with their densities; Note IV, on a new class of absorption phenomena.

> Lockyer (J. N.). Proc. Royal Soc., **22**, 371–8.

On a new method of studying metallic vapours.

> Lockyer (J. N.). Proc. Royal Soc., **29**, 266–72; Beiblätter, **4**, 86 (Abs.).

On the spectra of metals volatilized by the oxyhydrogen flame.

> Lockyer (J. N.) and Roberts (W. C.). Proc. Royal Soc., **23**, 844–9; Phil. Mag., (5) **1**, 284–9; Jour. Chem. Soc., 1876, **2**, 156 (Abs.).

Sur les spectres des vapeurs, aux températures élévées; hydrogène, nitrogène, potassium, carbone, sodium, zinc, cadmium, antimoine, phosphore, soufre, arsénic, bismuth, iode, mercure, lithium.

> Lockyer (J. N.). Comptes Rendus, **78**, 1790; Nature, **30**, 178.

On the indices of refraction of certain compound ethers.

> Long (J. H.). Amer. Jour. Sci., (3) **21**, 279–86.

Comparaison des spectres des flammes éclairantes et des flammes pâles.

> Magnus (G.). Ann. Chim. et Phys., (4) **6**, 159.

Réfraction des gaz.

> Mascart. Comptes Rendus, **78**, 417; Ann. Phys. u. Chem., **153**, 153.

Sur la comparaison des gaz et des vapeurs.

> Mascart. Comptes Rendus, **86**, 321–3; Jour. Chem. Soc., **34**, 859 (Abs.).

Sur la réfraction des corps organiques considérées à l'état gazeux.

>Mascart. Comptes Rendus, **86**, 321-3, 1182-5; Jour. Chem. Soc., **34**, 693 (Abs.); Ann. de l'École normale (2) **6**, 9-78; Beiblätter, **1.** 257-70.

Examination of coloured flames by the prism.

>Melvill (T.). Edinburgh Physical and Literary Essays, **2**, 12, 1752.

Experiments and observations on some cases of lines in the prismatic spectrum produced by the passage of light through coloured vapours and gases, and from certain coloured flames.

>Miller (W. A.). Phil. Mag., (3) **27**, 81.

Flame spectra.

>Milne (G. A.). Chem. News, **54**, 225.

Spectra von Flammen im Allgemeinen.

>Mitscherlich (A.). Ann. Phys. u. Chem., **121**, 487.

Ueber die Beziehung der chemischen Beschaffenheit zu der lichtbrechenden Kraft der Gaze.

>Mohr (F.). Ber. chem. Ges., **4**, 149-55; Jour. Chem. Soc., (2) **9**, 183 (Abs.).

Sur les moyens propres à la réproduction photographique des spectres ultra-violets des gaz.

>Monckhoven (van). Bull. de l'Acad. de Belgique, (2) **43**, 187-92; Beiblätter, **1**, 286 (Abs.).

De la flamme de quelques gaz carburés.

>Morren (M. A.). Ann. Chim. et Phys., (4) **4**, 305; Chem. News, 9, 135.

Das Sauerstoffspectrum und die electrischen Erscheinungen verdünnter Gase in Röhren mit Flüssigkeitselectroden.

>Paalzow (A.). Ann. Phys. u. Chem., n. F. **7**, 180.

The spectroscopic examination of the vapours evolved on heating iron, etc., at atmospheric pressure.

>Parry (J.). Chem. News, **49**, 241-2; **50**, 303-4; Ber. chem. Ges., **17**, Referate, 387 (Abs.); Jour. Chem. Soc., **46**, 801 (Abs.); Beiblätter, **8**, 646 (Abs.).

Comparaison des indices de réfraction dans quelques éthers composés isomères.

>Pierre (Is.) et Puchat (E.). Comptes Rendus, **76**, 1566-8.

Spectrum von Fluorborgas.

>Plücker (J.). Ann. Phys. u. Chem., **104**, 125.

Spectra der verschiedenen Gase wenn durch dieselben bei starker Verdünnung die electrische Entladung hindurchgeht.

> Plücker (J.). Ann. Phys. u. Chem., 105, 67.

Constitution der electrischen Spectra der verschiedenen Gase und Dämpfe.

> Plücker (J.). * Ann. Phys. u. Chem., 107, 497.

Zusammengesetzte Gase haben wie die einfachen ihr eigenthümliches Spectrum.

> Plücker (J.). Ann. Phys. u. Chem., 113, 276.

Recurrente Ströme und ihre Anwendung zur Darstellung von Gasspectren.

> Plücker (J.). Ann. Phys. u. Chem., 116, 27.

On the spectra of ignited gases and vapours, with especial regard to the different spectra of the same elementary gaseous substance.

> Plücker (J.) and Hittorf (S. W.). Proc. Royal Soc., 13, 153; Phil. Trans., 1865, p. 1.

De la flamme du soufre et des diverses lumières utilisables en photographie.

> Riche (A.) et Bardy (C.). Comptes Rendus, 80, 238-41; Ber. chem. Ges., 8, 182-3.

Sur le spectre d'absorption de la vapeur du soufre.

> Salet (G.). Comptes Rendus, 74, 865-6; Jour. Chem. Soc., (2) 10, 382 (Abs.); Ber. chem. Ges., 5, 828 (Abs.).

Coloration of the hydrogen flame.

> Santini (S.). Gazzetta, XIV, 274-6; Jour. Chem. Soc., 48, 465 (Abs.).

Veränderlichkeit der Spectra glühender Gase.

> Schenck (O.). Zeitschr. analyt. Chem., 12, 886-90; Jour. Chem. Soc., (2) 12, 1122-3 (Abs.).

Notiz über das Flammenspectrum der Schiessbaumwolle.

> Schöttner (F.). Carl's Repert., 14, 55-6; Beiblätter, 3, 279.

Harmonic ratios in the spectra of gases.

> Schuster (A.). Nature, 20, 533; 31, 337-47; Beiblätter, 4, 37; 5, 435-8 (Abs.).

Spectrum der Bunsen'schen Gasflamme, oder Spectrum des inneren Flammenkegels.

> Simmler (R. Th.). Ann. Phys. u. Chem., 115, 247.

Spectra der verschiedenen grünen Flammen.

> Simmler (R. Th.). Ann. Phys. u. Chem., 115, 249.

Blue flame from common salt.

> Smith (A. P.). Nature, 19, 488; 20, 5; Chem. News, 39, 141; Jour. Chem. Soc., 36, 497 (Abs.).

Gaseous spectra in vacuum tubes.

> Smyth (C. Piazzi). Proc. Royal Soc. Edinburgh, 10, 711–12 (Abs.); Trans. Royal Soc. Edinburgh, 32, Part III, 415–60, with plates.

Observations sur la note de M. M. Stoney et Reynolds sur les spectres des gaz.

> Soret (G. L.). Arch. de Genève, 42, 82–4; Phil. Mag., 42, 464–5; Ann. Chim. et Phys., (4) 26, 269.

Spectres d'absorption ultra-violets des éthers azotiques et azoteux.

> Soret (J. L.) et Rilliet (Alb. A.). Comptes Rendus, 89, 747.

On the effect of pressure on the character of the spectra of gases.

> Stearn (C. H.) and Lee (G. H.). Proc. Royal Soc., 21, 282–3; Jour. Chem. Soc., (2) 11, 996 (Abs.); Ber. chem. Ges., 6, 973 (Abs.); Phil. Mag., (4) 46, 406–7.

Zur Spectralanalyse gefärbter Flüssigkeiten, Gläser und Dämpfe.

> Stein (W.). Jour. prackt. Chemie, 10, 368–84; Jour. Chem. Soc., (2) 13, 412–14 (Abs.).

On the cause of the interrupted spectra of gases.

> Stoney (G. J.). Phil. Mag., (4) 41, 291–6; 42, 41–52; Ann. Chim. et Phys., (4) 26, 265–6 (Abs.), 266–8 (Abs.). (Look under Soret, above.)

On the blue lines of the spectrum of the non-luminous gas-flame.

> Swan (W.). Edinburgh Philosoph. Trans., 3, 376; 21, 353.

Prismatic spectra of the flames of carbon and hydrogen.

> Swan (W.). Edinburgh Philosoph. Trans., 21 (1857), 411–29; Ann. Phys. u. Chem., 100, 306.

Some experiments on coloured flames.

> Talbot (H. Fox). Brewster's Jour. Sci., 5, 1826.

Ueber die photographische Aufnahme von Spectren der in Geisslerrohren eingeschlossenen Gase.

> Vogel (H. W.). Monatsber. d. Berliner Akad. (1879), 115–19; Beiblätter, 4, 125–30 (Abs.).

Spectroscopische Notizen. Die Wasserstoffflamme in der Spectralanalyse.

> Vogel (H. W.). Ber. chem. Ges., 12, 2313–16; Beiblätter, 4, 278 (Abs.); 5, 118 (Abs.).

Gasspectra in Geissler'schen Röhren; bei zunehmender Verdünnung der Gase verschwinden die minder brechbaren Streifen zuerst.

    Waltenhofen (A. von). Ann. Phys. u. Chem., 126, 527-87.

On the spectrum of the Bessemer flame.

    Watts (W. M.). Phil. Mag., (4) 45, 81-90; Jour. Chem. Soc., (2) 11, 460 (Abs.).

Untersuchungen über die Natur der Spectra: 1, Theorie; 2, Spectra gemischter Gase.

    Wiedemann (E.). Ann. Phys. u. Chem., n. F. 5, 500-24; Phil. Mag., (5) 7, 77-95; Amer. Jour. Sci., (3) 17, 250-1.

Das Leuchten der Gase durch electrische Entladungen; Nachtrag zu der Arbeit über die Natur der Spectra.

    Wiedemann (E.). Ann. Phys. u. Chem., n. F. 6, 298.

Das thermische und optische Verhalten von Gasen unter dem Einfluss electrischer Entladungen.

    Wiedemann (E.). Ann. Phys. u. Chem., n. F. 10, 202.

Ueber die Dissociationswärme des Wasserstoffmoleculs und das electrische Leuchten der Gasen.

    Wiedemann (E.). Ann. Phys. u. Chem., n. F. 18, 509-10.

Spectroscopic examination of gases from meteoric iron.

    Wright (A. W.). Amer. Jour. Sci., (3) 9, 294-302; Jour. Chem. Soc., 1876, 1, 27 (Abs.).

Spectra der Gase unter hohem Druck.

    Wüllner (A.). Ann. Phys. u. Chem., 137, 337-56; Phil. Mag., (4) 37, 405; 39, 865.

Ueber die Spectra einiger Gase in Geissler'schen Röhren.

    Wüllner (A.). Ann. Phys. u. Chem., 144, 481-525; 147, 321-53; 149, 103-12; Ann. Chim. et Phys., (4) 26, 258-63 (Abs.); Bull. Soc. chim. Paris, n. s. 12, 445.

Ueber die Spectra der Gase.

    Wüllner (A.). Verhandl. d. naturwiss. Ges. zu Aachen, Dec., 1874; Ann. Phys. u. Chem., 154, 149-56; Jour. Chem. Soc., (2) 13, 527 (Abs.).

Reinheit der Spectren von Gasen.

    Wüllner (A.). Ber. chem. Ges., 3, 100.

Spectres des Gaz simples.

    Wüllner (A.). Comptes Rendus, 70, 125, 890.

Sur le spectre de l'étincelle électrique dans les gaz soumis à une pression croissante.

> Wüllner (A.). Comptes Rendus, 85, 280–1; Ann. Chim. et Phys., (5) 12, 148–4; Beiblätter, 1, 620 (Abs.).

Des transformations que subissent les spectres des gaz incandescents avec la pression et la température.

> Wüllner (A.). Arch. de Genève, (2) 40, 305–10.

Bemerkungen zu Herrn Goldstein's Beobachtungen an Gasspectris.

> Wüllner (A.). Monatsber. d. Berliner Akad., 1874, 755–61; Phil. Mag., (4) 49, 448–53.

Ueber den Einfluss der Dichtigkeit und Temperatur auf die Spectra glühender Gase.

> Zöllner (F.). Ber. chem. d. k. Sächs. Ges. d. Wiss., 22, 233–73; Ann. Phys. u. Chem., 142, 88–111; Phil. Mag., (4) 41, 190–205.

FLUORESCENCE.

Observations relatives à une note de M. Lamansky ayant pour titre " Sur
la loi de Stokes."
>Becquerel (E.). Comptes Rendus, **88**, 1287-9; Beiblätter, **3**, 619;
>Jour. Chem. Soc., **36**, 862 (Abs.).
>(Look below, under Lamansky.)

Sur la phosphorescence du sulfure de calcium.
>Becquerel (E.). Comptes Rendus, **103**, 551-3; Chem. News, **55**, 123.

Action du manganèse sur le pouvoir de phosphorescence du carbonate de
chaux.
>Becquerel (E.). Comptes Rendus, **103**, 1098-1101.

Zur Geschichte der Fluorescenz.
>Berthold (G.). Ann. Phys. u. Chem., **158**, 623.

Ueber die Fluorescenz der lebenden Netzhaut.
>Bezold (M. von) und Engelhardt (G.). Sitzungsber. d. Münchener
>Akad., **7**, 226-38; Phil. Mag., (5) **4**, 397-400.

On the crimson line of phosphorescent alumina.
>Crookes (W.). Proc. Royal Soc., **42**, 25-30; Chem. News, **55**, 25;
>Nature, **35**, 310; Amer. Jour. Sci., (3) **33**, 304 (Abs.).

Beugungsspectrum auf fluorescirenden Substanzen.
>Eisenlohr (W.). Ann. Phys. u. Chem., **99**, 163.

Les vibrations de la matière et les ondes de l'éther dans la phosphorescence et la fluorescence.
>Favé. Comptes Rendus, **86**, 289-94.

Action des fluorures sur l'alumine.
>Frémy et Varneuil. Comptes Rendus, **103** (1887), 788-40.

De la fluorescence.
>Gripon (E.). Jour. de Phys., **2**, 199, 246.

Versuche über Fluorescenz.
>Hagenbach (E.). Ann. Phys. u. Chem., **146**, 65-89, 232-57, 375-405,
>508-38; Jour. Chem. Soc., (2) **10**, 1058-61 (Abs.); Phil. Mag., (4)
>**45**, 57-64 (Abs.); Chem. News, **26**, 173 (Abs.).

Fernere Versuche über Fluorescenz.
>Hagenbach (E.). Ann. Phys. u. Chem., Jubelband, 303-18.

16 T

Das Aufleuchten, die Phosphorescenz und Fluorescenz des Flussspaths.

Hagenbach (E.). Naturforscherversammlung in München, 1877; Ber. chem. Ges., **10**, 2282 (Abs.).

Fluorescenz nach Stokes's Gesetz.

Hagenbach (E.). Ann. Phys. u. Chem., n. F. **18**, 45–56; Jour. Chem. Soc., **44**, 587–8 (Abs.).

Das Stokes'sche Gesetz.

Hagenbach (E.). Ann. Phys. u. Chem., n. F. **8**, 369–400.

Note on the behavior of certain fluorescent bodies in castor oil.

Horner (C.). Phil. Mag., (4) **48**, 165–6.

Herstellung des Spectrums fluorescirender Substanzen.

Jahresber. d. Chemie (1867), 105.

Bemerkungen zu den Arbeiten der Herrn Lommel, Glazebrook und Matthieu.

Ketteler (E.). Ann. Phys. u. Chem., n. F. **15**, 618.

Ueber Fluorescenz.

Lamansky (S.). Ann. Phys. u. Chem., n. F. **11**, 908–12; Jour. Chem. Soc., **40**, 214 (Abs.).

Ueber das Stokes'sche Gesetz.

Lamansky (S.). Ann. Phys. u. Chem., n. F. **8**, 624–8; Comptes Rendus, **88**, 1192–4, 1851; Jour. Chem. Soc., **36**, 862 (Abs.); Beiblätter, **3**, 619.

(Look above, under Becquerel, and below, under Lubarsch.)

Sur la fluorescence des terres rares.

Lecoq de Boisbaudran. Comptes Rendus, **101** (1885), 552, 588; Jour. Chem. Soc., **48**, 1174 (Abs.).

Les fluorescences Z α et Z β appartiennent-elles à des terres différentes?

Lecoq de Boisbaudran. Comptes Rendus, **102**, 899–902; Jour. Chem. Soc., **50**, 666 (Abs.).

Identité d'origine de la fluorescence Z β par renversement et des bandes obtenus dans le vide par M. Crookes.

Lecoq de Boisbaudran. Comptes Rendus, **103**, 113–17; Jour. Chem. Soc., **50**, 958.

Fluorescence des composés du manganèse soumis à l'effluve électrique dans le vide.

Lecoq de Boisbaudran. Comptes Rendus, **103**, 468–71, 629–31, 1064–7, 1107; Jour. Chem. Soc., **52**, 189, 191; Amer. Jour. Sci., (3) **33**, 149–51.

Fluorescence rouge de l'alumine.
>Lecoq de Boisbaudran (F.). Comptes Rendus, **104**, 830–4; Jour.
>Chem. Soc., **52**, 409 (Abs.).

Ueber die Fluorescenz in der Anthracenreihe.
>Liebermann (C.). Ber. chem. Ges., **13**, 913–16.

Ueber Fluorescenz.
>Lommel (E.). Sitzungsber. d. phys. med. Ges. Erlangen, 1871, 39–60;
>Ann. Phys. u. Chem., **143**, 26–51; Ann. Chim. et Phys., (4) **26**,
>288 (Abs.).

Ueber Fluorescenz.
>Lommel (E.). Ann. Phys. u. Chem., **159**, 514–36; Jour. Chem. Soc.,
>1877, **1**, 676; Amer. Jour. Sci., (3) **13**, 380 (Abs.).

Intensität des Fluorescenzlichtes.
>Lommel (E.). Ann. Phys. u. Chem., **160**, 75–96.

Fluorescenz.
>Lommel (E.). Naturforscherversammlung in München, 1877; Ber.
>chem. Ges., **10**, 2282 (Abs.); Ann. Phys. u. Chem., n. F. **3**, 113–25;
>Jour. Chem. Soc., **34**, 858 (Abs.).

Theorie der Absorption und Fluorescenz.
>Lommel (E.). Ann. Phys. u. Chem., n. F. **3**, 251–88.

Zwei neue fluorescirende Substanzen, Anthracenblau und bisulfobichlo-
ranthracenige Säure.
>Lommel (E.). Ann. Phys. u. Chem., n. F. **6**, 115–118.

Ueber das Stokes'sche Gesetz.
>Lommel (E.). Ann. Phys. u. Chem., n. F. **8**, 244.

Die dichroïtische Fluorescenz des Magnesiumplatincyanürs.
>Lommel (E.). Ann. Phys. u. Chem., n. F. **8**, 634; **9**, 108.

Ueber Fluorescenz.
>Lommel (E.). Ann. Phys. u. Chem., n. F. **10**, 449-72, 681-54.

Die Fluorescenz des Ioddampfes.
>Lommel (E.). Ann. Phys. u. Chem., n. F. **19**, 856.

Die Fluorescenz des Kalkspathes.
>Lommel (E.). Ann. Phys. u. Chem., n. F. **21**, 422; Jour. Chem.
>Soc., **46**, 649 (Abs.).

Beobachtungen über Fluorescenz, Didymglas und Aescorcin.
>Lommel (E.). Ann. Phys. u. Chem., (2) **24**, 288–92.

Zur Theorie der Fluorescenz.

Lommel (E.). Ann. Phys. u. Chem.,(2) 25, 648–55; Jour. de Phys., (2) 5, 516 (Abs.).

Ueber Fluorescenz.

Lubarsch (O.). Ann. Phys. u. Chem., 153, 420–40; n. F. 6, 248–67; Jour. Chem. Soc., (2) 13, 528 (Abs.).

Das Stokes'sche Gesetz.

Lubarsch (O.). Ann. Phys. u. Chem., n. F. 9, 665–71.

Neue Experimentaluntersuchungen über Fluorescenz.

Lubarsch (O.). Ann. Phys. u. Chem., n. F. 11, 46–69; Jour. Chem. Soc., 40, 70 (Abs.).

Bemerkungen zu den Arbeiten des Hernn Lamansky über Fluorescenz.

Lubarsch (O.). Ann. Phys. u. Chem., n. F. 14, 575–80.

Observations on the colour of fluorescent solutions.

Morton (H.). Chem. News, 24, 77; Jour. Chem. Soc., (2) 9, 992–3 (Abs.); (2) 10, 27; Amer. Jour. Sci., (3) 2, 198, 855.

Fluorescent relations of certain solid hydrocarbons found in coal-tar and petroleum distillates.

Morton (H.). Phil. Mag., (4) 44, 845–9; Ann. Phys. u. Chem., 148, 292–7; Chem. News, 26, 199–201, 272–4; Jour. Chem. Soc., (2) 11, 285 (Abs.).

Fluorescenzverhältnisse gewisser Kohlenwasserstoffverbindungen in den Steinkohlen-und Petroleum-Destillaten.

Morton (H.). Ann. Phys. u. Chem., 155, 551–79.

Fluorescence and the violet end of a projected spectrum.

Morton (H.). Chem. News, 27, 88.

Investigation of the fluorescent and absorption spectra of the uranium salts.

Morton (H.) and Bolton (H. C.). Chem. News, 28, 47–50, 113–16, 164–7, 288–4, 244–6, 257–9, 268–70; Jour. Chem. Soc., (2) 12, 12 (Abs.).

Fluorescent relations of the basic salts of uranic oxide.

Morton (H.). Chem. News, 29, 17–18; Jour. Chem. Soc., (2) 12, 642 (Abs.).

Fluorescent relations of chrysene and pyrene.

Morton (H.). Chem. News, 31, 35–6, 45–7.

On the connection between fluorescence and absorption.

Sorby (H. C.). Monthly Microscop. Jour., 13, 161–4.

Sur la fluorescence des sels des métaux terreux.

> Soret (J. L.). Comptes Rendus, **88**, 1077–8; Jour. Chem. Soc., **36**, 862 (Abs.); Beiblätter, **3**, 620 (Abs.).

Zur Kenntniss der Fluorescenzerscheinungen.

> Stenger (Fr.). Ann. Phys. u. Chem., (2) **28**, 201–30; Berichtigung dazu, do., 868.

On the change of refrangibility of light.

> Stokes (G. G.). Phil. Trans. (1852), 463–562.
> (His discovery of what has since been known as fluorescence.)

Sur la fluorescence de la matière colorante des champignons.

> Weiss (A.). Acad. de Vienne, Wiener Anzeiger (1885), 111; Jour. de Phys., (2) **5**, 240; Chem. Centralblatt (1886), 670–1; Jour. Chem. Soc., **52**, 814.

Fluorescence des Naphthalinrothes.

> Wesendonck (K.). Ann. Phys., (2) **26**, 521–7; Jour. Chem. Soc., **50**, 585; Jour. de Phys., (2) **5**, 517.

Berichtigung zu einer Notiz des Herrn Lommel betreffend die Theorie der Fluorescenz.

> Wüllner (A.). Ann. Phys. u. Chem., Ergänzungsband, 1878, **8**, 474–8.

FLUORINE.

Silicic fluoride spectrum.

> Capron (J. R.). Photographed Spectra, London, 1877, p. 75, 76.

Spectre du fluorure de silicium dans les tubes de Geissler.

> Chautard (J.). Comptes Rendus, **82**, 273.

Das Aufleuchten, die Phosphorescenz und die Fluorescenz des Fluss-spaths.

> Hagenbach (E.). Naturforscherversammlung in München, 1877; Ber. chem. Ges., **10**, 2232 (Abs.).

Spectrum des Fluors.

> Jahresber. d. Chemie, **15** (1862), 88.

Spectrum des Phosphorescenzlichtes von Flusspath.

> Kindt. Ann. Phys. u. Chem., **131**, 160.

Note on the spectra of calcium fluoride.

> Liveing (G. D.). Proc. Cambridge Philosoph. Soc., **3**, 96–8; Beiblätter, **4**, 611 (Abs.).

Spectrum von Fluorborgas.

> Plücker. Ann. Phys. u. Chem., **104**, 125.

Indices de réfraction du spath fluor.

> Sarasin (E.). Arch. de Genève, (8) **10**, 808–4.

Spectre du fluorure de silicium.

> Séguin (J. M.). Comptes Rendus, **54**, 998.

Ueber die Spectra des Fluorsiliciums und des Siliciumwasserstoffs.

> Wesendonck (K.). Ann. Phys. u. Chem., n. F. **21**, 427–87; Jour. Chem. Soc., **46**, 649 (Abs.).

GADOLINITE.

New elements in gadolinite and samarskite.

> Crookes (W.). Proc. Royal Soc., **40**, 502–9; Jour. Chem. Soc., **52**, 384.

Remarques sur la gadolinite.

> Delafontaine. Comptes Rendus, **90**, 221.

Gadolinium, le Ya de Marignac.

> Lecoq de Boisbaudran (F.). Comptes Rendus, **102**, 902; Jour. Chem. Soc., **50**, 667 (Abs.).

Sur les terres de la gadolinite.

> Marignac (C.). Ann. Chim. et Phys., (5) **14**, 247–258; Jour. Chem. Soc., **36**, 113 (Abs.).

Sur l'ytterbine, nouvelle terre contenue dans la gadolinite.

> Marignac (C.). Comptes Rendus, **87**, 578–81; Amer. Jour. Sci., (3) **17**, 62–8 (Abs.); Jour. Chem. Soc., **36**, 118–19 (Abs.).

Notice sur les nouveaux métaux obtenus du gadolinite.

> Mendelejeff. Jour. Soc. phys. chim. russe, **13**, 517–20; Bull. Soc. chim. Paris, **38**, 189–43.

Recherches sur l'absorption des rayons ultra-violets par diverses substances. II, Sur les spectres d'absorption des terres de la gadolinite.

> Soret (J. L.). Arch. de Genève, (2) **63**, 89–112; Comptes Rendus, **86**, 1062–4; Beiblätter, **3**, 196 (Abs.); **2**, 410–11; Jour. Chem. Soc., **2**, 410 (Abs.).

Ueber die Erden des Gadolinits von Ytterby.

> Welsbach (C. Auer von). Sitzungsber. d. Wiener Akad., **88** II, 382–44, 1237–51; Zeitschr. analyt. Chem., **23**, 520 (Abs.); Chem. News **51**, 25 (Abs.).

GALLIUM.

Caractères chimiques et spectroscopiques d'un nouveau métal, le gallium, découvert dans une blende de la mine de Pierrefitte, vallée d'Argelès (Pyrénnées).

Lecoq de Boisbaudran (F.).　Comptes Rendus, **81**, 493–5; **82**, 168, 1086, 1098; Bull. Soc. chim. Paris, n. s. **24**, 870; Jour. Chem. Soc., 1876, **1**, 190 (Abs.); Amer. Jour. Sci., (3) **11**, 320 (Abs.); Ann. Chim. et Phys., (5) **10**, 117; Ann. Phys. u. Chem., **159**, 650; Chem. News, **32**, 159, 294.

Remarques à propos de la découverte du gallium.

Mendelejef (D.).　Comptes Rendus, **81**, 969.

GERMANIUM.

Ueber das Spectrum des Germaniums.

Kobb (G.).　Ann. Phys. u. Chem., (2) **29** (1886), 670–2; Jour. Chem. Soc., **52**, 818 (Abs.); Amer. Jour. Sci., (3) **33**, 151 (Abs.).

Spectre du germanium.

Lecoq de Boisbaudran (F.).　Comptes Rendus, **102**, 1291–5; Jour. Chem. Soc., **50**, 768 (Abs.).

GLASS.

Prüfung des gelben Glases für Dunkelzimmer der Photographen.

Foster (Le Neve). Dingler's Journal, **207**, 427; Jour. Chem. Soc., (2) **11**, 948 (Abs.).

Phasenveränderung des Lichtes bei Reflexion an Glas.

Glan (P.). Ann. Phys. u. Chem., **155**, 14.

On the influence of temperature on the optical constants of glass.

Hastings (C. S.). Amer. Jour. Sci., (3) **15**, 269–75; Beiblätter, **2**, 838 (Abs.).

Refractive indices of glass.

Hopkinson (J.). Proc. Royal Soc., **26**, 290–7; Beiblätter, **1**, 680 (Abs.).

Vertheilung der Wärme im Flintglasspectrum.

Lamansky (S.). Ann. Phys. u. Chem., **146**, 207, 209.

The yellow glass of commerce lets through portions of nearly the whole spectrum.

Lea (M. Carey). Amer. Jour. Sci., (3) **33**, 363.

On the refractive and dispersive powers of various samples of glass.

Lohse (J. G.). Monthly Notices Astronom. Soc., **40**, 563–4; Beiblätter, **4**, 891 (Abs.).

Spectra produced in glass by scratching.

Love (E. J. J.). Nature, **32**, 270.

Spectrale Untersuchung eines longitudinaltönenden Glasstabes.

Mach (E.). Ann. Phys. u. Chem., **146**, 316–17.

Ueber die Dispersionsverhältnisse optischer Gläser.

Merz (S.). Zeitschr. f. Instrumentenkunde, **2**, 176–80; Beiblätter, **6**, 673 (Abs.).

Zur Spectralanalyse gefärbter Flüssigkeiten, Gläser und Dämpfe.

Stein (W.). Jour. prackt. Chemie, **10**, 868–84; Jour. Chem. Soc., (3) **13**, 412 (Abs.).

Methoden zur Bestimmung der Brechungsexponenten von Flüssigkeiten und Glasplatten.

Wiedemann (E.). Ann. Phys. u. Chem., **158**, 875–86.

GOLD.

Gold arc spectrum.

> Capron (J. R.).   Photographed Spectra, London, 1877, p. 30.

L'or n'a donné aucune apparence de renversement.

> Cornu (A.).   Comptes Rendus, **73**, 382.

Spectrum des Goldchlorids.

> Jahresber. d. Chemie (1873), 152.

Chlorure d'or en solution, étincelle; chlorure d'or dans le gaz.

> Lecoq de Boisbaudran (F.).   Spectres Lumineux, Paris, 1874, p. 172, 176, planche XXVI.

Spectre de chlorure d'or.

> Lecoq de Boisbaudran (F.).   Bull. Soc. chim. Paris, n. s. **21**, 125.

Sur quelques spectres métalliques, chlorure d'or.

> Lecoq de Boisbaudran (F.).   Comptes Rendus, **77**, 1152–4; Jour. Chem. Soc., (2) **12**, 217 (Abs.); Ber. chem. Ges., **6**, 1418 (Abs.).

HEAT SPECTRA.

Measurement of the so-called thermospectrum.

Abney (W. de W.).   Chem. News, **40**, 21.

Sur un moyen d'isoler les radiations calorifiques des radiations lumineuses et chimiques.

Assche (F. von).   Comptes Rendus, **97**, 888.

Spectres calorifiques.

Aymonnet.   Comptes Rendus, **82**, 1158.

Pouvoirs absorbants des corps pour la chaleur.

Aymonnet.   Comptes Rendus, **83**, 971.

Nouvelle méthode pour étudier les spectres calorifiques.

Aymonnet.   Comptes Rendus, **83**, 1102.

Ein einfacher Versuch zur Versinnlichung des Zusammenhanges zwischen der Temperatur eines glühenden Drahtes und der Zusammensetzung des von ihm ausgehenden Lichtes.

Bezold (W. von).   Ann. Phys. u. Chem., n. F. **21**, 175–8.

Verschiebung der Spectrallinien unter Wirkung der Temperatur des Prismas.

Blaserna (P.).   Ann. Phys. u. Chem., **143**, 655.

Einfluss der Temperatur auf die Empfindlichkeit der Spectralreaction.

Cappel (E.).   Ann. Phys. u. Chem., **139**, 628.

Einfluss des Druckes und der Temperatur auf die Spectren von Dämpfen und Gasen.

Ciamician.   Sitzungsber. d. Wiener Akad., **77** II, 839; **78** II, 867.

Distribution of heat in the visible spectrum.

Conroy (Sir J.).   Proc. Royal Soc., **3**, 106–12; Phil. Mag., (5) **8**, 203–9; Beiblätter, **4**, 44 (Abs.).

Étude des radiations émises par les corps incandescents.   Mesure optique des hautes températures.

Crova (A.).   Ann. Chim. et Phys., (5) **19**, 472–550; Beiblätter, **5**, 117–18 (Abs.).

Mesure spectrométrique des hautes températures.

> Crova (A.). Comptes Rendus, **87**, 979; **90**, 252; Jour. de Phys., **8**, 196–8.

Recherches sur les spectres calorifiques obscurs.

> Desains (P.). Comptes Rendus, **67**, 296–7, 1097; **70**, 985; **84**, 285; **88**, 1047; **89**, 189; **94**, 1144; **95**, 433; Jour. Chem. Soc., **36**, 864 (Abs.); Beiblätter, **3**, 869 (Abs.).

Détermination des longueurs d'onde des rayons calorifiques à basse température dans le spectre.

> Desains (P.) et Curie (P.). Comptes Rendus, **90**, 1506.

Measurement of high temperatures.

> Dewar (J.). Chem. News, **28**, 174.

Distribution of heat in the spectrum.

> Draper (J. W.). Amer. Jour. Sci., (8) **4**, 161–75; Phil. Mag., (4) **44**, 104–17; Jour. Chem. Soc., (2) **10**, 968 (Abs.).

Absorption of light at different temperatures.

> Feussner. Phil. Mag., (4) **29**, 471; Monatsber. d. Berliner Akad., März, 1865.

De l'influence de la température sur les caractères des raies spectrales.

> Fiévez (C.). Bull. de l'Acad. de Belgique, (3) **7**, 348–55; Beiblätter, **8**, 645 (Abs.); Les Mondes, (3) **8**, 481–3; Chem. News, **50**, 128 (Abs.).

Influence of temperature on the optical constants of glass.

> Hastings (C. S.). Amer. Jour. Sci., (8) **15**, 269–75; Beiblätter, **2**, 388 (Abs.).

Distribution of heat in the spectra of various sources of radiation.

> Jacques (W. W.). Dissertations of the Johns Hopkins University, 1879; Proc. Amer. Acad., **14**, 142–61; Beiblätter, **3**, 865 (Abs.).

Einfluss der Temperatur der Flamme auf das Spectrum.

> Jahresber. d. Chemie, **15** (1862), 29; **21** (1868), 80; **23** (1870), 148, 175; **26** (1873), 54.

Durchgang der strahlenden Wärme durch polirtes und berüsstes Steinsalz; Diffusion der Wärmestrahlen; Lage des Wärmemaximums im Sonnenspectrum.

> Knoblauch (H.). Ann. Phys. u. Chem., **120**, 177.

Einfluss der Temperatur auf spectroscopische Beobachtungen.

> Krüss (G.). Ber. chem. Ges., **17**, 2782b; Jour. Chem. Soc., **48**, 209 (Abs.).

Geschichtliches über das Wärmespectrum der Sonne; Vertheilung der Warme im Flintglasspectrum.

Lamansky (S.). Ann. Phys. u. Chem., **146**, 200–80.

Abhängigkeit des Brechungsquotienten der Luft von der Temperatur.

Lang (V. von). Ann. Phys. u. Chem., **153**, 450.

Observations on invisible heat-spectra and the recognition of hitherto unmeasured wave-lengths, made at the Alleghany Observatory, Alleghany, Pa.

Langley (S. P.). Amer. Jour. Sci., (3) **31** (1886), 1–12; **32**, 83–106; Phil. Mag., (5) **21**, 394–409: **22**, 149–178; Jour. de Phys., (2) **5**, 377–80; Ann. Chim. et Phys., (6) **9**, 433–506; Beiblätter, **11**, 245.

Ueber die spectrale Vertheilung der strahlenden Wärme.

Lecher (E.). Wiener Anzeigen (1881), 193–4.

Spectra of vapours at elevated temperatures.

Lockyer (J. N.). Chem. News, **30**, 98.

Nothwendigkeit bei spectroscopische Messungen die Temperatur zu berücksichtigen.

Lommel (E.). Ann. Phys. u. Chem., **143**, 656.

Om Värmefördelningen i Normalspektrum (Ueber die Wärmevertheilung im Normalspectrum).

Lundquist (G.). Oefversigt af K. Vetensk. Acad. Hand., 1874, **31**, X, 19–27; Ann. Phys. u. Chem., **155**, 146–55.

Maximum de température.

Magnus (G.). Ann. Chim. et Phys., (4) **6**, 155.

Sur l'identité des diverses radiations lumineuses, calorifiques et chimiques.

Melloni. Comptes Rendus, **15**, 454.

Température des différentes parties du spectre solaire.

Melloni. Comptes Rendus, **18**, 39.

Recherches sur la réflexion métallique des rayons calorifiques obscurs et polarisés.

Mouton. Comptes Rendus, **84**, 650.

Spectre calorifique normal du Soleil et de la lampe à platine incandescent Bourbouze.

Mouton. Comptes Rendus, **89**, 295.

Wärmevertheilung im Spectrum eines Glas-und Steinsalzprismas.

Müller (J.). Ann. Phys. u. Chem., **105**, 347.

Wärmevertheilung im Diffractionsspectrum.

    Müller (J.). Ann. Phys. u. Chem., **105**, 855.

Untersuchungen über die thermischen Wirkungen des Sonnenspectrums.

    Müller (J.). Ann. Phys. u. Chem., **115**, 387.

Wellenlänge und Brechungsexponent der äussersten dunklen Wärmestrahlen des Sonnenspectrums.

    Müller (J.). Ann. Phys. u. Chem., **115**, 543; Berichtigung dazu,
    116, 644.

Effect of increased temperature upon the nature of the light emitted by
the vapour of certain metals or metallic compounds.

    Roscoe and Clifton. Chem. News, **5**, 283.

On spectral lines of low temperature.

    Salisbury (The Marquis of). Phil. Mag., (4) **45**, 241–5; Jour. Chem.
    Soc., (2) **11**, 711 (Abs.); Amer. Jour. Sci., (3) **6**, 141 (Abs.).

Stickstoff gibt je nach der Temperatur drei Spectra.

    Schimkow (A.). Ann. Phys. u. Chem., **129**, 513.

Ueber die Abhängigkeit der Brechungsexponenten anomal dispergirender Medien von Concentration der Lösung und der Temperatur.

    Sieben (G.). Ann. Phys. u. Chem., n. F. **23**, 312.

Einfluss der Temperatur auf das optische Drehvermögen des Quartzes
und des chlorsauren Natrons.

    Sohnke (L.). Ann. Phys. u. Chem., n. F. **3**, 516.

Rapport sur un travail de M. Fiévez concernant l'influence de la température sur les caractères des raies spectrales.

    Stas. Bull. de l'Acad. de Belgique, (3) **7**, 290–4.

Ueber den Einfluss der Wärme auf die Brechung des Lichtes in festen
Körpern.

    Stefan (J.). Sitzungsber. d. Wiener Akad., **63** II, 223–45.

Ueber den Einfluss der Dichtigkeit und Temperatur auf die Spectra glühender Gase.

    Zöllner (F.). Ber. d. k. Sächs. Ges. d. Wiss., **22**, 288–53; Ann. Phys.
    u. Chem., **142**, 88–111; Phil. Mag., (4) **41**, 190–205.

HELIUM.

Sur la raie dite de l'hélium.

> Spée (E.). Bull. de l'Acad. de Belgique, (3) **49**, 879–96; Beiblätter, **4**, 614 (Abs.).

SPECTRA AT HIGH ALTITUDES.

Notes on some recent astronomical experiments at high altitudes on the Andes.

> Copeland (R.). Nature, **28**, 606; Beiblätter, **8**, 220 (Abs.).

Ascension scientifique à grande hauteur, exécutée le 22 mars 1874.

> Crocé-Spinelli (J.) et Sivel. Comptes Rendus, **78**, 946–50; Amer Jour. Sci., (3) **8**, 36 (Abs.).
> (Look below under Janssen and Pecchi.)

Note sur des observations spectroscopiques, faites dans l'ascension du 24 Spet. 1874, pour étudier les variations des couleurs du spectre.

> Fonvielle (W. de). Comptes Rendus, **89**, 816–17.

Die Fraunhofer'schen Linien auf grossen Höhen dieselben wie in der Ebne.

> Heusser (J. C.). Ann. Phys. u. Chem., **90**, 819.

Remarques sur le spectre d'eau à l'occasion du voyage aérostatique de M. M. Crocé-Spinelli et Sivel.

> Janssen (J.). Comptes Rendus, **78**, 995–8.

Sunlight and skylight at high altitudes.

> Langley (S. P.). Nature, **26**, 586–9; Amer. Jour. Sci., (3) **24**, 898–8; Beiblätter, **7**, 28 (Abs.); Jour. de Phys., (2) **3**, 47 (Abs.).

Observations relatives à une communication de M. Crocé-Spinelli sur les bandes de la vapeur d'eau dans le spectre solaire.

> Secchi (A.). Comptes Rendus, **78**, 1080–81.

HOLMIUM.

Spectre de holmium.

   Clève (P. T.).   Comptes Rendus, **89**, 478.

Remarques sur le holmium ou philippine.

   Delafontaine.   Comptes Rendus, **90**, 221.

Holmium, ou l'**x** de M. Soret.

   Lecoq de Boisbaudran (F.).   Comptes Rendus, **102**, 1003–4; Jour. Chem. Soc., **50**, 667 (Abs.).

HOMOLOGOUS SPECTRA.

On homologous spectra.

   Hartley (W. N.).   Jour. Chem. Soc., **43**, 390–400; Nature, **27**, 522 (Abs.); Chem. News, **47**, 188 (Abs.); Amer. Jour. Sci., (8) **26**, 401 (Abs.); Ber. chem. Ges., **16**, 2659 (Abs.); Beiblätter, **8**, 217 (Abs.).

HYDROGEN.

Spectrum von Wasserstoff.

Angström (A. J.). Ann. Phys. u. Chem., **94**, 157.

Wasserstoff hat nur ein Spectrum; die vielfachen Spectren rühren bei Bemengungen her.

Angström (A. J.). Ann. Phys. u. Chem., **144**, 302, 304.

Spectres des gaz simples; l'hydrogène, etc.

Angström (A. J.). Comptes Rendus, **73**, 869.

Notiz über die Spectrallinien des Wasserstoffs.

Balmer (J. J.). Ann. Phys. u. Chem., (2) **25**, 80–7; Jour. Chem. Soc., **48**, 1025 (Abs.); Jour. de Phys., (2) **5**, 515 (Abs.).

Absorptionsspectrum des durch Wasserstoffsuperoxyd gebräunten blausäurehaltigen Blutes.

Buchner. Jour. prackt. Chemie, **105**, 845.

Hydrogen tube spectrum.

Capron (J. R.). Photographed Spectra, London, 1877, p. 61, 62, 63.

Sur le spectre ultra-violet de l'hydrogène.

Cornu (J.). Jour. de Phys., (2) **5**, 341–54.

Continuous spectra of hydrogen observed by combustion of hydrogen in oxygen and chlorine.

Dibbits. Ann. Phys. u. Chem., **122**, 497.

Recherches sur l'intensité relative des raies spectrales de l'hydrogène et de l'azote en rapport avec la constitution des nébuleuses.

Fiévez (C.). Bull. de l'Acad. de Belgique, (2) **49**, 107–118; Phil. Mag., (5) **9**, 309–12; Beiblätter, **4**, 461 (Abs.); Ann. Chim. et Phys., (5) **20**, 179–85; Jour. Chem. Soc., **40**, 69 (Abs.).

Sur l'élargissement des raies de l'hydrogène.

Fiévez (C.). Comptes Rendus, **92**, 521–2; Beiblätter, **5**, 281 (Abs.); Jour. Chem. Soc., **40**, 955 (Abs.).

Combustion of hydrogen and carbonic oxide under great pressure.

Franckland. Proc. Royal Soc., **16**, 419.

17 T

The refraction equivalents of carbon, hydrogen, nitrogen, and oxygen in organic compounds.

> Gladstone (J. H.). Proc. Royal Soc., **31**, 827–30; Ber. chem. Ges., **14**, 1553 (Abs.).

Untersuchungen über das zweite Spectrum des Wasserstoffes.

> Hasselberg (B.). Mem. Acad. imp. St. Pétersbourg, **30**, No. 7, 24; **31**, No. 14, 80; Beiblätter, **8**, 881–4 (Abs.); Mem. Spettr. ital., **13**, 97 (Abs.); Phil. Mag., (5) **17**, 329–52; Jour. Chem. Soc., **48**, 317 (Abs.); Jour. de Phys., (2) **4**, 241 (Abs.).

Bemerkungen zu Hrn. Wüllner's Aufsatz; "Ueber die Spectra des Wasserstoffs und des Acetylens."

> Hasselberg (B.). Ann. Phys. u. Chem., n. F. **15**, 45–9.

Zusatz zu meinen Untersuchungen über das zweite Spectrum des Wasserstoffs.

> Hasselberg (B.). Mélanges phys. et chim. tirés du Bull. de l'Acad. de St. Pétersbourg, **12**, 208–14; Beiblätter, **9**, 519 (Abs.).

Die Spectralerscheinungen des Phosphorwasserstoffs und des Ammoniaks.

> Hofmann (K. B.). Ann. Phys. u. Chem., **147**, 92–5.

On the spectrum of the flame of hydrogen.

> Huggins (W.). Proc. Royal Soc., **80**, 576; Amer. Jour. Sci., (3) **20**, 121–8; Beiblätter, **4**, 658 (Abs.).

L'intensité relative des raies spectrales de l'hydrogène et de l'azote en rapport avec la constitution des nébuleuses.

> Huggins (W.). Bull. de l'Acad. de Belgique, (2) **49**, 266–7; Beiblätter, **4**, 658 (Abs.).

Spectrum des Wasserstoffs.

> Jahresber. d. Chemie, **16** (1863), 111.

Absorptionsspectrum des Phosphorwasserstoffs.

> Jahresber. d. Chemie, **25** (1872), 142.

Absorptionsspectra von Kohlenwasserstoffen.

> Jahresber. d. Chemie, **28** (1875), 126.

Absorptionsspectrum des Wasserstoffs.

> Jahresber. d. Chemie, **25** (1872), 141, 143–6.

Recherches photométriques sur le spectre de l'hydrogène.

> Lagarde (H.). Ann. Chim. et Phys., (6) **4**, 248–369, avec 1 planche; Jour. de Phys., (2) **5**, 186 (Abs.); note par Wiedemann (E.), Ann. Chim. et Phys., (6) **7**, 143–4.

Spectre de l'hydrogène phosphoré.

Lecoq de Boisbaudran (F.). Spectres Lumineux, Paris, 1874, p. 187, planche XXVII.

Action de la lumière sur l'acide iodhydrique.

Lemoine (G.). Comptes Rendus, 85, 144_7; Beiblätter, 1, 510 (Abs.).

Spectra of compounds of carbon with hydrogen.

. . Liveing (G. D.) ånd Dewar (J.). Nature, 22, 620.

Note on the reversal of hydrogen lines, and on the outburst of hydrogen lines when water is dropped into the arc.

Liveing (G. D.) and Dewar (J.). Proc. Royal Soc., 35, 74-6; Chem. News, 47, 122; Nature, 28, 21 (Abs.); Beiblätter, 7, 371 (Abs.); Jour. de Phys., (2) 4, 51.

Note on the spectrum of hydrogen.

Lockyer (J. N.). Proc. Royal Soc., 30, 81-2; Beiblätter, 4, 868 (Abs.).

Sur les spectres des vapeurs aux températures élévées; hydrogène.

Lockyer (J. N.). Comptes Rendus, 78, 1790; Chem. News, 30, 98. (Original in French.)

De l'élargissement des raies spectrales de l'hydrogène.

Monckhoven (D. von). Comptes Rendus, 95, 878.

Spectrum von Wasserstoff in der Geissler'schen Röhre.

Plücker. Ann. Phys. u. Chem., 104, 122; 105, 76.

Spectrum von Wasserstoff.

Plücker. Ann. Phys. u. Chem., 105, 81.

Spectra am negativen Pol in Stickstoff-und Wasserstoff-röhren; Modification beider Röhren nach langer Gebrauch.

Reitlinger (E.). Ann. Phys. u. Chem., 141, 135_6.

Coloration of the hydrogen flame.

Santini (S.). Gazzetta chim. ital., 14, 142-6; Jour. Chem. Soc., 48, 209 (Abs.); Beiblätter, 9, 82 (Abs.).

On the spectrum of hydrogen at low pressure.

Seabroke (G. M.). Monthly Notices Astronom. Soc., 32, 68-4; Phil. Mag., (4) 43, 155_7; Chem. News, 25, 111; Ann. Chim. et Phys., (4) 26, 264 (Abs.).

Remarques sur la relation entre les protubérances et les taches solaires; intérêt qu'auraient les expériences sur la lumière spectrale de l'hydrogène brûlant sous une très forte pression.

Secchi (A.). Comptes Rendus, 68, 237-8.

Hydrogène et la raie D$_s$ dans le spectre de la chromosphère solaire.
 Secchi (A.). Comptes Rendus, **73**, 1800.

Prismatic spectra of the flames of compounds of carbon and hydrogen.
 Swan. Phil. Trans. Edinburgh, **21**, 411; Ann. Phys. u. Chem., **100**, 306.

Spectres de l'hydrogène, etc., sur la surface du Soleil.
 Vicaire (E.). Comptes Rendus, **76**, 1540.

Spectrum von Wasserstoff.
 Vogel (H. C.). Ann. Phys. u. Chem., **146**, 576.

Ueber die Spectra des Wasserstoffs.
 Vogel (H. C.). Monatsber. d. Berliner Akad. (1879), 586–604; Beiblätter, **4**, 125–30; Amer. Jour. Sci., (8) **19**, 406 (Abs.).

Die Wasserstoffflamme in der Spectralanalyse.
 Vogel (H. W.). Ber. chem. Ges., **12**, 2818; Beiblätter, **4**, 278 (Abs.); **5**, 118 (Abs.).

Ueber die neuen Wasserstofflinien.
 Vogel (H. W.). Ber. chem. Ges., **13**, 274–6; Jour. Chem. Soc., **38**, 597–8 (Abs.); Beiblätter, **4**, 274 (Abs.).

Die Photographie des Wasserstoffspectrums.
 Vogel (H. W.). Photographische Mittheilungen, **16**, 276–8.

Ueber die Spectra des Fluorsiliciums und des Siliciumwasserstoffs.
 Wesendonck (K.). Ann. Phys. u. Chem., n. F. **21**, 427–37; Jour. Chem. Soc., **46**, 649 (Abs.).

Ueber die Dissociationswärme des Wasserstoffmoleculs.
 Wiedemann (E.). Ann. Phys. u. Chem., n. F. **18**, 509–10.

Electrische Spectra in Wasserstoff.
 Willigen (S. M. van der). Ann. Phys. u. Chem., **106**, 622.

Drei Spectra bei Wasserstoff.
 Wüllner (A.). Ann. Phys. u. Chem., **135**, 499.

Spectra der Gase unter hohem Druck; Wasserstoff gibt dabei ein continuirliches Spectrum; vier Spectra beim Wasserstoff.
 Wüllner (A.). Ann. Phys. u. Chem., **137**, 337–47.

Spectra des Wasserstoffs.
 Wüllner (A.). Ann. Phys. u. Chem., n. F. **14**, 855.
 (Look above, under Hasselberg.)

INDIGO (THE).

The indigo color in the spectrum.
> Rood (O. N.).   Amer. Jour. Sci., (3) **19**, 185

INDIUM.

Indium arc spectrum.
> Capron (J. R.).   Photographed Spectra, London, 1877, p. 80, 45.

Spectra of indium.
> Clayden (A. W.) and Heycock (C. T.).   Phil. Mag., (5) **2**, 387-9;
> Amer. Jour. Sci., (3) **13**, 57 (Abs.); Beiblätter, **1**, 90-2.

Sels d'indium en solution, étincelle.
> Lecoq de Boisbaudran.   Spectres Lumineux, Paris, 1874, p. 142,
> planche XXI.

Vorläufige Notiz über ein neues Metall (Indium).
> Reich (F.) und Richter (Th.).   Jour. prackt. Chemie, **89**, 441.

Ueber das Indium.
> Reich (F.) und Richter (Th.).   Jour. prackt. Chemie, **90**, 172; Phil.
> Mag., (4) **26**, 488.

Spectrum des Indiums.
> Schrötter.   Jour. prackt. Chemie, **95**, 446.

Spectrum des Indiums.
> Winkler.   Jour. prackt. Chemie, **94**, 1.

Zur spectralanalytische Ermittelung des Indiums.
> Wleugel (S.).   Correspondenzblatt d. Vereins analytischer Chemiker,
> **3**, 89; Beiblätter, **5**, 281 (Abs.); Zeitschr. analyt. Chemie, **20**, 115
> (Abs.).

INTERFERENCE.

Beobachtungen dunkler Interferenzstreifen im Spectrum des weissen Lichtes.

    Abt (A.). Math. naturwiss. Ber. aus Ungarn, **1**, 852–4.

Interferenzstreifen im Spectrum.

    Arons (L.). Ann. Phys. u. Chem., (2) **24**, 669–71.

Sur les phénomènes d'interférence produits par les réseaux parallèles.

    Crova (A.). Comptes Rendus, **72**, 855–8; **74**, 982–86.

Ueber Interferenzstreifen welche durch zwei getrübte Flächen erzeugt werden.

    Exner (K.). Sitzungsber. d. Wiener Akad., **72** II, 675.

Sur les conditions d'achromatisme dans les phénomènes d'interférence.

    Hurion (A.). Comptes Rendus, **94**, 1845; **95**, 75.

Projection der Interferenz der Flüssigkeitswellen.

    Lommel (L.). Ann. Phys. u. Chem., (2) **26**, 156.

Sur l'application du spectroscope à l'observation des phénomènes d'interférence.

    Mascart. Jour. de Phys., **1**, 17; **3**, 310.

Bedeutung von Newton's Construction der Farbenordnungen dünner Blättchen für die Spectraluntersuchung der Interferenzfarben.

    Rollett (Alex.). Sitzungsber. d. Wiener Akad., **75** III, 17.

Graphische Darstellung der Spectren der Interferenzfarben für einen Gypskeil.

    Rollett (Alex.). Sitzungsber. d. Wiener Akad., **77** III, 177.

Ueber die an bestaubten und unreinen Spiegeln sichtbare Interferenzerscheinung.

    Sekulic. Ann. Phys. u. Chem., **154**, 808.

Prismatisches und Beugungsspectrum, Interferenzerscheinungen in demselben.

    Stefan (J.). Sitzungsber. d. Wiener Akad., **50** II, 127, 188–42; Ann. Phys. u. Chem., **123**, 509.

Interferenzstreifen im prismatischen und im Beugungsspectrum.

    Weinberg (M.). Carl's Repertorium, **18**, 600–608.

INVERSION.

Reversal of the sodium lines.

>   Ackroyd (W.). Chem. News, **36**, 164–5.

Renversement des raies spectrales des vapeurs métalliques.

>   Cornu (A.). Comptes Rendus, **73**, 882.

Sur les raies spontanément renversables.

>   Cornu (A.). Comptes Rendus, **100**, 1181–1188; Jour. Chem. Soc.,
>   **48**, 858 (Abs.), 1885.

Sur le renversement des raies du spectre.

>   Duhem. Jour. de Phys., (2), **4**, 221–4.

Ueber ein einfaches Verfahren die Umkehrung der farbigen Linien der
Flammenspectra, insbesondere der Natriumlinie, subjectiv darzustellen.

>   Günther (O.). Ann. Phys. u. Chem., n. F. **2**, 477.

Umkehrung der hellen Spectrallinien der Metalle, insbesondere des
Natriums in dunkle.

>   Jahresber. d. Chemie (1865), 90.

Umkehrung der Spectra.

>   Kirchhoff (G.). Ann. Phys. u. Chem., **109**, 275, 295; **110**, 187; Jour.
>   praekt. Chemie, **80**, 480–8.

Wandlung der Spectren.

>   Lepel (F. von). Ber. chem. Ges., **11**, 1146.

Reversal of the lines of metallic vapours.

>   Liveing (G. D.) and Dewar (J.). Nature, **24**, 206; **26**, 466.

Note on some phenomena attending the reversal of lines.

>   Lockyer (J. N.). Proc. Royal Soc., **28**, 428–82; Beiblätter, **3**, 608
>   (Abs.).

Wandlung der Spectren.

>   Moser (J.). Ber. chem. Ges., **11**, 1416.

Umkehrung der Spectra.

>   Tyndall. Jour. prackt. Chemie, **85**, 261.

Wandlung der Spectren.

Vogel (H. W.).   Ber. chem. Ges., **11**, 622, 918, 1468, 1562.

Leichte Umkehrung der Natriumlinie.

Weinhold (A.).   Ann. Phys. u. Chem., **142**, 321.

Re-reversal of sodium lines.

Young (C. A.).   Nature, **21**, 274–5; Beiblätter, **4**, 370.

IODINE.

Note on the absorption spectrum of iodine in solution in carbon disulphide.

Abney and Festing. Proc. Royal Soc., 24, 480.

The dichroïsm of the vapour of iodine.

Andrews (T.). Chem. News, 24, 75; Jour. Chem. Soc., (2) 9, 998 (Abs.).

Action des rayons différemment réfrangible sur l'iodure et le bromure d'argent.

Becquerel (E.). Comptes Rendus, 79, 185–90; Jour. Chem. Soc., (2) 13, 80 (Abs.).

Iodine vapour; spark in iodine vapour.

Capron (J. R.). Photographed Spectra, London, 1877, p. 76.

Spectre de l'iode dans les tubes de Geissler.

Chautard (J.). Comptes Rendus, 82, 278.

Absorption spectra of iodine.

Conroy (Sir John). Proc. Royal Soc., 25, 46.

Wellenlänge der auf Iodsilber chemisch wirkenden Strahlen.

Eisenlohr (W.). Ann. Phys. u. Chem., 99, 162.

Spectre d'absorption du chlorure d'iode.

Gernez (D.). Comptes Rendus, 74, 660.

Spectre d'absorption des vapeurs de protobromure d'iode, etc.

Gernez (D.). Comptes Rendus, 74, 1190–92; Jour. Chem. Soc., (2) 10, 665 (Abs.); Phil. Mag., (4) 43, 473–5; Amer. Jour. Sci., (3) 4, 59–60.

Spectre d'absorption du chlorure d'iode.

Gernez (D.). Bull. Soc. chim. Paris, n. s. 17, 258; Ber. chem. Ges., 5, 219.

Iodure.

Gouy. Comptes Rendus, 85, 70.

Spectrum des Iods.

Jahresber. d. Chemie, 16, 109.

Absorptionsspectrum des Ioddampfes

Jahresber. d. Chemie, 23, 174.

Absorptionsspectrum des einfachen Chlorjods.

> Jahresber. d. Chemie, **25**, 139.

Absorptionsspectrum des Bromjods.

> Jahresber. d. Chemie, **25**, 140.

Absorptionsspectrum des Iods.

> Jahresber. d. Chemie, **25**, 141.

On the action of the less refrangible rays of light on silver iodide.

> Lea (M. Carey). Amer. Jour. Sci., (8) **9**, 269–78; Jour. Chem. Soc. 1876, **1**, 28 (Abs.).

Iodure de baryum dans le gaz chargé d'iode.

> Lecoq de Boisbaudran (F.). Spectres Lumineux, Paris, 1874, p. 63, 65, planche VIII.

Action de la lumière sur l'acide iodhydrique.

> Lemoine (G.). Comptes Rendus, **85**, 144–7; Beiblätter, 510 (Abs.).

On the dispersion of a solution of mercuric iodide.

> Liveing (G. D.). Proc. Philosoph. Soc. Cambridge, **3**, 258–60; Beiblätter, **4**, 610 (Abs.).

Sur les spectres des vapeurs aux températures elévees; iode.

> Lockyer (J. N.). Comptes Rendus, **78**, 1790; Nature, **30**, 78; Chem. News, **30**, 98.

Die Fluorescenz des Ioddampfes.

> Lommel (E.). Ann. Phys. u. Chem., n. F. **19**, 356.

Verbindungsspectren zur Entdeckung von Iod.

> Mitscherlich (A.). Jour. prackt. Chemie, **97**, 218.

Entdeckung sehr geringer Mengen von Chlor, Brown und Iod in Verbindungen.

> Mitscherlich (A.). Ann. Phys. u. Chem., **125**, 629.

Lo spettro di assorbimento del vapore di jodio.

> Morghen (A.). Mem. Spettr. ital., **13**, 127–31; Beiblätter, **8**, 822 (Abs.); Atti R. Accad. Lincei, Transunti, (8) **3**, 327–30.

Absorption-spectra of bromine and of iodine-monochloride.

> Roscoe (H. E.) and Thorpe (T. E.). Proc. Royal Soc., **25**, 4.

Sur la lumière émise par la vapeur d'iode.

> Salet (G.). Comptes Rendus, **74**, 1249.

Le spectre primaire de l'iode.

> Salet (G.). Comptes Rendus, 75, 76 ; Bull. Soc. chim. Paris, n. s. 18, 216.

Absorptionsspectrum des Ioddampfes.

> Thalén (R.). Ann. Phys. u. Chem., 139, 508.

Ueber die Brechung und Dispersion des Lichtes in Iod-Silber.

> Wernicke (W.). Ann. Phys. u. Chem., 142, 560–73; Jour. Chem. Soc., (2) 9, 658 (Abs.); Ann. Chim. et Phys., (4) 26, 287 (Abs.).

Uebereinstimmung des Absorptionsspectrums und des ersten Iodspectrums mit dem Spectrum dessen Dampfes.

> Wüllner (A.). Ann. Phys. u. Chem., 120, 159, 161.

IRIDIUM.

Iridium arc spectrum.

> Capron (J. R.). Photographed Spectra, London, 1877, p. 30.

IRON.

On the estimation of small quantities of phosphorus in iron and steel by spectrum analysis.

> Alleyne (Sir J. G. N.).   Jour. Iron and Steel Inst. (1875), 62-72.

Iron spark spectrum, and iron arc spectrum; iron meteoric spectrum.

> Capron (J. R.).   Photographed Spectra, London, 1877, p. 31-3.

Le fer n'à donné aucune apparence de renversement.

> Cornu (A.).   Comptes Rendus, 73, 332.

Spectre du chlorure de fer.

> Gouy.   Comptes Rendus, 84, 231; Chem. News, 35, 107.

Ueber-phosphorhaltigen Stahl.

> Greiner (A.).   Dingler's Jour., 217, 33-41; Jour. Chem. Soc., 1876, 1, 454 (Abs.).

Distribution of heat in the various scources of radiation; black oxide of iron, etc.

> Jacques (W. W.).   Proc. Amer. Acad., 14, 161.

Spectrum der Bessemerflamme.

> Jahresber. d. Chemie, (1867) 105, (1873) 150.

Perchlorure de fer en solution, étincelle.

> Lecoq de Boisbaudran (F.).   Spectres Lumineux, Paris, 1874, p. 122, planche XVIII.

Spectrum der Bessemerflamme.

> Lielegg (A.).   Sitzungsber. d. Wiener Akad., 55 II, 150, 153-81; 56 II. 8, 24-80; Jour. prackt. Chemie, 100, 333; Phil. Mag., (4) 34, 302.

On the iron lines widened in solar spots.

> Lockyer (J. N.).   Proc. Royal Soc., 31, 348.

On the examination of the Bessemer flame with colored glasses and with the spectroscope.

> Parker (J. Spear).   Chem. News, 23, 25.

The spectroscopic examination of the vapours evolved on heating iron at atmospheric pressure.

> Parry (J.).   Chem. Soc., 49, 241-2; 50, 303; Ber. chem. Ges., 17, Referate, 337 (Abs.); Jour. Chem. Soc., 46, 301 (Abs.); Beiblätter, 8, 646 (Abs.).

The spectroscope applied to the Bessemer Process.

> Roscoe (H. E). Chem. News, **22**, 44; **23**, 174; Phil. Mag., (4) **25**, 818.

Employment of spectrum analysis in the Bessemer Process.

> Roscoe (H. E.). Jour. Iron and Steel Inst., 1871, **2**, 38–62; Ber. chem. Ges., **4**, 419–21 (Abs.).

Spectre du fer dans l'arc voltaïque.

> Secchi (A.). Comptes Rendus, **77**, 173.

Examination of the Bessemer Flame with colored glasses and with the spectroscope.

> Silliman (J. M.). Chem. News, **22**, 213; **23**, 5.

Ueber das Eisenspectrum, erhalten mit dem Flammenbogen.

> Thalén (Rob.). Nova Acta. Roy. Soc. Upsala, (3) 1884; Beiblätter, 9 (1885), 520 (Abs.).

Spectre du fer sur la surface du Soleil.

> Vicaire (E.). Comptes Rendus, **76**, 1540.

Ueber die Absorptionsspectren einiger Salze der Eisengruppe.

> Vogel (H. W.). Ber. chem. Ges., **8**, 1538–40.

Ueber eine empfindliche spectralanalytische Reaction auf Thonerde.

> Vogel (H. W.). Ber. chem. Ges., **9**, 1641.

Erkennung von Thonerde neben Eisensalzen.

> Vogel (H. W.). Ber. chem. Ges., **10**, 373; Jour. Chem. Soc., 1877, **2**, 269 (Abs.).

Ueber die Erkennung des Kobalts, neben Eisen und Nickel.

> Vogel (H. W.). Ber. chem. Ges., **12**, 2318–16; Beiblätter, **4**, 278 (Abs.); **5**, 118 (Abs.).

Spectrum of the Bessemer flame.

> Watts (W. M.). Phil. Mag., (4) **34**, 437; **45**, 81; Chem. News, **23**, 49; Jour. prackt. Chemie, **104**, 420.

Coïncidence of the spectrum lines of iron, calcium, and titanium.

> Williams (W. M.). Nature, **8**, 46.

Methods for the determination of metallic iron by spectral analysis.

> Wolff. Chem. News, **39**, 124.

Spectroscopic examination of gases from meteoric iron.

> Wright (A. W.). Amer. Jour. Sci., (3) **9**, 294–302; Jour. Chem. Soc., 1876, **1**, 27 (Abs.).

JARGONIUM.

Jargonium, a new element accompanying zirconium.

> Sorby (H. C.).   Chem. News, **19**, 121 ; Proc. Royal Soc., **17**, 511.

LANTHANUM.

Sur le poids atomique du lanthane.

> Clève (P. T.).   Bull. Soc. chim. Paris, **39**, 151–5 ; Chem. News, **47**,
> 154–5 ; Amer. Jour. Sci., (3) **25**, 381 (Abs.).

Spectre du lanthane, avec une planche.

> Thalén (Rob.).   Jour. de Phys., **4**, 33.

LEAD.

Ueber den Einfluss der Temperatur auf die Brechungsexponenten der natürlichen Sulfate des Baryum, Strontium und Blei.

> Arzruni (A.). Zeitschr. f. Krystallogr. u. Mineral., 1, 165–92; Jahrb. f. Mineral. (1877), 526 (Abs.); Jour. Chem. Soc., 34, 189 (Abs.).

Lead arc spectrum, lead and antimony spark spectrum, lead and magnesium spark spectrum.

> Capron (J. R.). Photographed Spectra, London, 1877, p. 84, 35.

Renversement des raies spectrales du plomb.

> Cornu (A.). Comptes Rendus, 73, 332.

Spectre de l'azotate de plomb.

> Gouy. Comptes Rendus, 84, 231; Chem. News, 35, 707.

Spectren zwischen Bleielectroden.

> Jahresber. d. Chemie (1873), 152.

Spectre du sulfure de plomb.

> Lallemand (A.). Comptes Rendus, 78, 1272.

Spectre du plomb.

> Lecoq de Boisbaudran (F.). Comptes Rendus, 77, 1152; Chem. News, 24, 10.

Plomb métallique, étincelle.

> Lecoq de Boisbaudran (F.). Spectres Lumineux, Paris, 1874, p. 147, planche XXIII.

LIGHT.

Vitesse de la lumière fait que les bords du spectre sont diffus.

>Arago. Comptes Rendus, **36**, 43.

Sur la rayonnement chimique qui accompagne la lumière, et sur les effets électriques en résultent.

>Becquerel (Ed.). Comptes Rendus, **13**, 198.

Note accompagnant la presentation du II. volume de son ouvrage intitulé "Lumière, ses Causes et ses Effets."

>Becquerel (Ed.). Comptes Rendus, **67**, 8.

Étude sur la part de la lumière dans les actions chimiques.

>Chastaing (P.). Ann. Chim. et Phys., (5) **11**, 145–223 ; Jour. Chem. Soc., 1877, **2**, 818 (Abs.); Beiblätter, **1**, 515–20 (Abs.). (Look below, under Vogel.)

Lage der chemischen Strahlen im Spectrum des Sonnen-und Gas-Lichts.

>Crookes (W.). Ann. Phys. u. Chem., **97**, 619; Cosmos, **8**, 90; Bull. Lond. Photographical Soc., 21 Jan., 1856.

Sur l'emploi de la lumière monochromatique, produite par les sels de soude.

>Henry (L. d'). Comptes Rendus, **76**, 222–4 (Abs.); Ann. Chem. u. Pharm., **169**, 272; Dingler's Jour., **207**, 405–7.

Constanz der Lichtspectren.

>Jahresber. d. Chemie (1869), 174.

Sur le spectre anormal de la lumière.

>Klercker (de). Comptes Rendus, **89**, 784; Phil. Mag., (5) **8**, 571–2; Beiblätter, **4**, 273–4.

Lichtspectren.

>Lecoq de Boisbaudran (F.). Ber. chem. Ges., **3**, 140, 503, 572.

Zur Theorie des Lichtes.

>Lommel (E.). Ann. Phys. u. Chem., n. F. **16**, 427–41.

Emploi du spectroscope pour distinguer une lumière plus faible dans une plus forte.

>Seguin. Comptes Rendus, **68**, 1822.

Chastaing's neue Theorie der chemischen Wirkung des Lichtes.
> Vogel (H. W.). Ber. chem. Ges., **10**, 1638–44; Beiblätter, **1**, 681 (Abs.).

Les observations spectroscopiques à la lumière monochromatique.
> Zenger (Ch. V.). Comptes Rendus, **94**, 155; Amer. Jour. Sci., (3) **23**, 322.

LIGHTNING.

(Look under Electricity.)

LIMITS.

Limites des couleurs dans le spectre.
> Listing. Ann. Chim. et Phys., (4) **13**, 460.

Limites des couleurs dans le spectre.
> Thalén (Rob.). Ann. Chim. et Phys., (4) **18**, 218.

LINES OF THE SPECTRUM.

Welchen Stoffen die Fraunhofer'schen Linien angehören.
> Angström (A. J.). Ann. Phys. u. Chem., **117**, 296–302.

Die Fraunhofer'schen Ringe, die Quetelet'schen Streifen und verwandte Erscheinungen.
> Exner (K.). Sitzungsber. d. Wiener Akad., **76** II, 522.

Bestimmung des Brechungs-und Farbenzerstreuungs-Vermögens verschiedener Glasarten.
> Fraunhofer (Jos.). Denkschr. d. k. Akad. d. Wiss. zu München, Band **V** (1814–15), 193–226, mit drey Kupfertafeln, München, 1817, 4°.

Note on the theoretical explanation of Fraunhofer's lines.
> Hartshorne (H.). Jour. Franklin Inst., **75**, 88–43; **105**, 38; Les Mondes, **45**, 517–22; Beiblätter, **2**, 561.   '

Die Zusammensetzung des Spectrums.
> Jahresber. d. Chemie, **1**, 197; **5**, 126, 131; **8**, 123.

Ueber die Fraunhofer'schen Linien.
> Jahresber. d. Chemie, **3**, 154; **4**, 152; **5**, 124; **6**, 167; **7**, 137.

Anwendung der Fraunhofer'schen Linien als chemisches Reagens.
> Jahresber. d. Chemie, **5**, 125.

Künstliches Spectrum einer Fraunhofer'schen Linie.
> Jahresber. d. Chemie (1868), 124.

Newton, Wollaston, and Fraunhofer's lines.
> Johnson (A.). Nature, **26**, 572; Beiblätter, **7**, 65–6 (Abs.).

On certain remarkable groups in the lower spectrum.
> Langley (S. P.). Proc. Amer. Acad., **14**, 92.

Erklärung der Linien und Streifen in den Lichtspectren.
> Lecoq de Boisbaudran (F.). Ber. chem. Ges., **2**, 614.

Mutual attraction of spectral lines.
> Peirce (C. S.). Nature, **21**, 108; Beiblätter, **4**, 278 (Abs.).

On spectral lines of low temperature.
> Salisbury (The Marquis of). Phil. Mag., (4) **45**, 241–5; Jour. Chem. Soc., (2) **11**, 711 (Abs.); Amer. Jour. Sci., (3) **6**, 141–2.

The relation between spectral lines and atomic weights.

Vogel (E.).   Pharmaceutical Jour. Trans., (3) 6, 464–5.

Darstellung eines Spectrums mit einer Fraunhofer'schen Linie.

Wüllner (A.).   Ann. Phys. u. Chem., 135, 174.

LIQUIDS.

Pouvoirs absorbants des corps pour la chaleur; solutions dans l'eau, etc.

Aymonnet. Comptes Rendus, **83**, 971.

Ueber eine einfache Methode zur approximativen Bestimmung der Brechungsexponenten flüssiger Körper.

Bodynski (J.). Carl's Repertorium, **18**, 502–4; Beiblätter, **6**, 932 (Abs.).

Molecular-Refraction flüssiger organischer Verbindungen von hohem Dispersifvermögen.

Brühl (J. W.). Ann. Phys. u. Chem., **235**, 1–106; Ber. chem. Ges., **19**, 2746 (Abs.); Jour. Chem. Soc., **52**, 191 (Abs.).

Spectroscopische Untersuchung der Constanten von Lösungen.

Burger (H.). Ber. chem. Ges., **11**, 1876.

Methoder til at maale Brydningsforholdet for farvede Vaedsker (Ueber die Messung des Brechungsverhältnisses gefärbter Flüssigkeiten).

Christiansen (C.). Oversigt kgl. Danske Vidensk. Selsk. Forh. (1882), 217–50; Ann. Phys. u. Chem., n. F. **19**, 257–67; Nature, **28**, 308 (Abs.).

Nouvelle méthode de détermination des indices de réfraction des liquides.

Croullebois (M.). Ann. Chim. et Phys., (4) **22**, 189–50.

Recherches sur le pouvoir réfringent des liquides.

Damien (B. C.). Ann. de l'École normale, (2) **10**, 233–304; Beiblätter, **5**, 579–84 (Abs.); Jour. de Phys., **10**, 394–401, 431–34 (Abs.).

On the specific refraction and dispersion of light by liquids.

Gladstone (J. H.). Rept. British Assoc. (1881), 591; Nature, **24**, 468 (Abs.); Beiblätter, **6**, 21 (Abs.).

Ueber Regenbogen, gebildet durch Flüssigkeiten von verschiedenen Brechungsexponenten.

Hammerl (H.). Sitzungsber. d. Wiener Akad., **86** II, 206–15; Beiblätter, **7**, 383–5 (Abs.).

Preliminary notice of experiments concerning the chemical constitution of saline solutions.

Hartley (W. N.). Proc. Royal Soc., **22**, 241–3; Chem. News, **29**, 148.

On the action of heat on the absorption spectra and chemical constitution of saline solutions.

> Hartley (W. N.). Proc. Royal Soc., **23**, 872–3; Phil. Mag., (5) **1**, 244–5; Ber. chem. Ges., **8**, 765 (Abs.).

Application des franges de Talbot à la détermination des indices de réfraction des liquides.

> Hurion. Comptes Rendus, **92**, 452–3.

Spectren gefärbter Lösungen.

> Jahresber. d. Chemie, **15**, 84.

Ueber die Constitution von Lösungen.

> Krüss (G.). Ber. chem. Ges., **10**, 1248–9; Jour. Chem. Soc., **42**, 1018 (Abs.); Nature, **26**, 568; Beiblätter, **6**, 677 (Abs.); Amer. Jour. Sci., (3) **24**, 141 (Abs.).

Ueber das Absorptionsspectrum der flüssigen Untersalpetersäure.

> Kundt (A.). Ann. Phys. u. Chem., (2) **7**, 64 (Abs.); Jour. Chem. Soc., (2) **9**, 185 (Abs.).

Ueber den Einfluss des Lösungsmittels auf die Absorptionsspectra gelöster absorbirender Mittel.

> Kundt (A.). Sitzungsber. d. Münchener Akad. (1877), 284–62; Ann. Phys. u. Chem., n. F. **4**, 34–54.

Recherches sur l'illumination des liquides, etc.

> Lallemand. Comptes Rendus, **69**, 182.

Ueber die Molecularrefraction flüssiger organischer Verbindungen.

> Landolt (H.). Sitzungsber. d. Wiener Akad. (1882), 62–91; Ann. Phys. u. Chem., **213**, 75–112; Beiblätter, **7**, 843; Ber. chem. Ges., **15**, 1031–40; Jour. Chem. Soc., **42**, 909 (Abs.).

Absorption des Lichtes durch gefärbte Flüssigkeiten.

> Melde (F.). Ann. Phys. u. Chem., **124**, 91; **126**, 264.

Observations on the colour of fluorescent solutions.

> Morton (H.). Amer. Jour. Sci., (3) **2**, 198–9, 355–7; Jour. Chem. Soc., (2) **9**, 992 (Abs.); **10**, 27 (Abs.); Chem. News, **24**, 77.

Ueber die Aenderung des Volumens und des Brechungsexponenten von Flüssigkeiten durch hydrostatischen Druck.

> Quincke (G.). Ann. Phys. u. Chem., n. F. **19**, 401–85; Sitzungsber. d. Berliner Akad. (1883), 409 (Abs.); Nature, **28**, 808 (Abs.).

Ueber eine neue Flüssigkeit von hohem specifischen Gewicht, hohem
Brechungsexponenten und grosser Dispersion.

> Rohrbach (C.). Ann. Phys. u. Chem., n. F. 1, 169–74; Amer. Jour.
> Sci., (3) 26, 406 (Abs.); Jour. Chem. Soc., 46, 145 (Abs.).

On the absorption bands in the visible spectrum produced by certain colourless liquids.

> Russell (W. J.) and Lapraik (W.). Jour. Chem. Soc., 39, 168–73;
> Amer. Jour. Sci., (3) 21, 500 (Abs.); Nature, 22, 868–70; Beiblätter,
> 5, 44–5.

Ueber die Absorption des Lichtes durch Flüssigkeiten.

> Schönn (J. L.). Ann. Phys. u. Chem., n. F. 6, 267–70.

Untersuchungen über die Abhängigkeit der Molecularrefraction flüssiger
Verbindungen von ihrer chemischen Constitution.

> Schröder (H.). Ber. chem. Ges., 15, 994–8; Jour. Chem. Soc., 42,
> 910 (Abs.).

Fernere Untersuchungen über die Abhängigkeit der Molecularrefraction
flüssiger Verbindungen von ihrer chemischen Zusammensetzung.

> Schröder (H.). Sitzungsber. d. Münchener Akad. (1882), 57–104;
> Ann. Phys. u. Chem., n. F. 15, 636–75; 18, 148–75; Jour. Chem.
> Soc., 42, 1153 (Abs.); 44, 588 (Abs.).

Sur les spectres d'absorption ultra-violets des différents liquides.

> Soret (J. L.). Arch. de Genève, (2) 60, 298–300; Beiblätter, 2, 30
> (Abs.).

Zur Spectralanalyse gefärbter Flüssigkeiten, Gläser und Dämpfe.

> Stein (W.). Jour. prackt. Chemie, 10, 868–84; Jour. Chem. Soc., (2)
> 13, 412 (Abs.).

Méthode nouvelle pour déterminer l'indice de réfraction des liquides.

> Terquem et Trannin. Comptes Rendus, 78, 1843–5; Dingler's Jour.,
> 212, 552–4; Jour. de Phys., 4, 282–8; Ann. Phys. u. Chem., 157,
> 302–9.

Ueber eine Methode zur Untersuchung der Absorption des Lichtes durch
gefärbte Lösungen.

> Tumlirz (O.). Wiener Anzeigen (1882), 165 (Abs.); Beiblätter, 7,
> 895 (Abs.); Chem. News, 49, 201 (Abs.).

Absorption spectra of certain organic liquids.

> Wolff (C. H.). Chem. News, 47, 178.

LITHIUM.

Ueber quantitative Bestimmung des Lithiums mit dem Spectral-Apparat.
>> Ballmann (H.). Zeitschr. analyt. Chemie, 14, 297–301; Jour. Chem.
>> Soc., 1876, 2, 550 (Abs.).

On the presence of lithium in meteorites.
>> Bunsen. Phil. Mag., (4) 23, 474.

Existence de la lithine et de l'acide borique dans les eaux de la mer Morte.
>> Dieulafait. Comptes Rendus, 94, 1352–54; Jour. Chem. Soc., 42, 1087
>> (Abs.); Ann. Chim. et Phys., (5) 25, 145–67.

La lithine, la strontiane et l'acide borique dans les eaux minérales de
Contrexeville et Schinznach (Suisse).
>> Dieulafait. Comptes Rendus, 95, 999–1001; Jour. Chem. Soc., 44,
>> 801 (Abs.).

Les salpêtres naturels du Chili et du Pérou au point de vue du rubidium,
du cæsium, du lithium et de l'acide borique.
>> Dieulafait. Comptes Rendus, 98, 1545–8; Chem. News, 50, 45 (Abs.).

On the blue band in the lithium spectrum.
>> Franckland. Phil. Mag., (4) 22, 472.

Recherches photométriques sur le lithium.
>> Gouy. Comptes Rendus, 83, 269; 85, 70.

Transparence des flammes colorées pour leur propres radiations; lithium,
etc.
>> Gouy. Comptes Rendus, 86, 1078.

Spectrum des Lithiums in der Wasserstofflamme.
>> Jahresber. d. Chemie, 15, 80.

Funkenspectrum von kohlensäuren Lithium.
>> Jahresber. d. Chemie (1873), 152.

Sels de lithine en solution.
>> Lecoq de Boisbaudran (F.). Spectres Lumineux, Paris, 1874, p. 56,
>> planche VI.

Spectre du lithium.
>> Lecoq de Boisbaudran. Comptes Rendus, 77, 1152; Bull. Soc. chim.
>> Paris, n. s. 21, 125.

On the spectra of magnesium and lithium.

> Liveing (G. D.) and Dewar (J.). Proc. Royal Soc., **30**, 93_9; Bei-
> blätter, **4**, 866 (Abs.).

Note on the order of reversibility of the lithium lines.

> Liveing (G. D.) and Dewar (J.). Proc. Royal Soc., **35**, 76; Chem.
> News, **47**, 188.

Sur les spectres des vapeurs aux températures élévées, lithium.

> Lockyer (J. N.). Comptes Rendus, **78**, 1790; Nature, **30**, 78; Chem.
> News, **30**, 98.

Sur l'origine de l'arsénic et de la lithine dans les eaux sulfatées calciques.

> Schlagdenhauffen. Jour. de Pharm., (5) **6**, 457–68; Jour. Chem. Soc.,
> **44**, 802 (Abs.).

On the flame of lithia.

> Talbot (H. Fox). Phil. Mag., (3) **4**, 11.

De la présence de la lithine dans le sol de la Limagne et des eaux minérales de l'Auvergne. Dosage de cet alcali au moyen du spectroscope.

> Truchot (P.). Comptes Rendus, **78**, 1022_4; Ber. chem. Ges., **7**, 653
> (Abs.).

The blue band in the lithium spectrum.

> Tyndall and Franckland. Phil. Mag., (4) **22**, 151, 472.

LONGITUDINAL RAYS.

Note sur les raies longitudinales observées dans le spectre prismatique par M. Zantedeschi.

> Babinet. Comptes Rendus, **35**, 418. (Look below.)

Raies longitudinales du spectre.

> Porro. Comptes Rendus, **35**, 479.

Sur les lignes longitudinales du spectre.

> Wartmann (E.). Arch. des Sciences phys. et nat., **7**, 33; **10**, 302; Phil. Mag., **32**, 499.

Sur les causes des lignes longitudinales du spectre.

> Zantedeschi (F.). Archives des Sciences phys. et nat., **12**, 43; Corresp. scient. di Roma, No. **9**, 69.

LUMINOUS SPECTRA.

Observations sur le rayonnement des corps lumineux.

> Baudrimont. Comptes Rendus, **33**, 496.

Divers effets lumineux qui résultent de l'action de la lumière sur les corps.

> Becquerel (E.). Comptes Rendus, **45**, 817.

Constitution du spectre lumineux.

> Lecoq de Boisbaudran (F.). Comptes Rendus, **69**, 445, 606, 657, 694; **73**, 658.

Recherches d'analyse spectrale.

> Volpicelli. Comptes Rendus, **57**, 571.

Sur les causes des effets lumineux, etc.

> Volpicelli. Comptes Rendus, **69**, 730.

MAGNESIUM.

Lead and magnesium spark spectrum, magnesium spark spectrum, magnesium arc spectrum.

> Capron (J. R.)., Photographed Spectra, London, 1877, p. 84, 85, 86.

Détermination des longueurs d'onde des radiations très réfrangibles du magnésium, du cadmium, du zinc et de l'aluminium.

> Cornu (A.). Archives de Genève, (8) 2, 119–126; Beiblätter, 4, 84 (Abs.); Jour. de Phys., 10, 425–31.

Renversement des raies spectrales du magnésium.

> Cornu (A.). Comptes Rendus, 73, 832.

Recherches sur le spectre du magnésium en rapport avec la constitution du Soleil.

> Fiévez (C.). Bull. de l'Acad. de Belgique, (2) 50, 91–8; Beiblätter, 4, 789 (Abs.); Ann. Chim. et Phys., (5) 23, 866–72.

Spectre de chlorure de magnésium.

> Gouy. Comptes Rendus, 84, 231.

Spectre continu des sels de magnésie.

> Gouy. Comptes Rendus, 84, 878.

Spectrum des Magnesiumlichtes.

> Jahresber. d. Chemie, 18, 96; 23, 174; 25, 145.

Chlorure de magnésium en solution.

> Lecoq de Boisbaudran (F.). Spectres Lumineux, Paris, 1874, p. 85, planche XII.

Permanganate de potasse en solution.

> Lecoq de Boisbaudran (F.). Spectres Lumineux, Paris, 1874, p. 106, planche XVI.

Ueber eine empfindliche spectralanalytische Reaction auf Thonerde und Magnesia.

> Lepel (F. von). Ber. chem. Ges., 9, 1641.

Ueber den Nachweis der Magnesia mit Hülfe des Spectroskops.

> Lepel (F. von). Ber. chem. Ges., 9, 1845; 10, 159; Bull. Soc. chim. Paris, n. s. 28, 478; Jour. Chem. Soc., 1877, 1, 676; Beiblätter, 1, 240 (Abs.).

Der Alkannafarbstoff, ein neues Reagens auf Magnesiumsalze.

Lepel (F. von). Ber. chem. Ges., **13**, 768-8.

Pflanzenfarbstoffe als Reagentien auf Magnesiumsalze.

Lepel (F. von). Ber. chem. Ges., **13**, 766-8; Jour. Chem. Soc., **40**, 68 (Abs.).

On the spectra of magnesium and lithium.

Liveing (G. D.) and Dewar (J.). Proc. Royal Soc., **30**, 98-9; Beiblätter, **4**, 366 (Abs.).

Investigations on the spectrum of magnesium.

Liveing (G. D.) and Dewar (J.). Proc. Royal Soc., **32**, 189-203; Nature, **24**, 118.

Die dichroïtische Fluorescenz des Magnesiumplatincyanürs.

Lommel (E.). Ann. Phys. u. Chem., n. F. **8**, 684; **9**, 108; **13**, 247.

Osservazioni delle inversioni della coronale 1474 k, e delle b del magnesio fatte nel Osservatorio di Palermo.

Riccò (A.). ' Mem. Spettr. ital., **10**, 148-51.

Spectre du magnésium dans l'arc voltaïque.

Secchi (A.). Comptes Rendus, **77**, 178.

Spectre du magnésium.

Secchi (A.). Comptes Rendus, **82**, 275.

Magnésium dans la chromosphère du Soleil.

Tacchini (P.). Comptes Rendus, **75**, 28, 430; Phil. Mag., (4) **44**, 159-60.

Présence du spectre du magnésium sur le bord entière du Soleil.

Tacchini (P.). Comptes Rendus, **76**, 1577.

Nouvelles observations relatives à la présence du magnésium sur le bord du Soleil, et réponse à quelques points de la théorie émise par M. Faye.

Tacchini (P.). Comptes Rendus, **77**, 606-9.

Nouvelles observations relatives à la présence du magnésium sur le bord du Soleil.

Tacchini (P.). Comptes Rendus, **82**, 1385-7.

Spectre du magnésium sur la surface du Soleil.

Vicaire (E.). Comptes Rendus, **76**, 1540.

Ueber eine empfindliche Spectralreaction auf Magnesium.

> Vogel (H. W.).　Ber. chem. Ges., **9**, 1641; Jour. Chem. Soc., 1877, **1**,
> 742 (Abs.); Beiblätter, **1**, 240 (Abs.); Bull. Soc. chim. Paris, n. s.
> **28**, 475.

Die Purpurin-Thonerde-Magnesia-Reaction.

> Vogel (H. W.).　Ber. chem. Ges., **10**, 157, 878.

MANGANESE.

Sur l'effet du manganèse sur la phosphorescence du calcium carbonate.

Becquerel (E.). Comptes Rendus, **103**, 1098–1101; Jour. Chem. Soc., **52**, 190 (Abs.).

Ueber das Absorptionsspectrum des übermangansauren Kalis, und seine Benutzung bei chemisch-analytischen Arbeiten.

Brücke (E.). Chemisches Centralblatt, (8) **8**, 139–148; Jour. Chem. Soc., **34**, 242 (Abs.).

Manganese arc spectrum.

Capron (J. R.). Photographed Spectra, London, 1877, p. 36.

On the light reflected by potassium permanganate.

Conroy (Sir J.). Proc. Royal Soc., **2**, 340–4; Phil. Mag., (5) **6**, 454–8; Jour. Chem. Soc., **36**, 425 (Abs.).

Spectre de l'azotate de manganèse.

Gouy. Comptes Rendus, **84**, 281; Chem. News, **35**, 107.

Absorptionslinien der Manganlösungen.

Hoppe-Seyler. Jour. prackt. Chemie, **90**, 308.

Spectra of manganese in blowpipe beads.

Horner (Charles). Chem. News, **25**, 189.

Anwendung der dunklen Linien des Spectrums als Reagens auf Mangansäure.

Jahresber. d. Chemie, **5**, 125.

Absorptionsspectrum des Mangansuperchlorids.

Jahresber. d. Chemie (1869), 184.

Chlorure de manganèse en solution, étincelle courte; do., étincelle moyenne; do., dans le gaz.

Lecoq de Boisbaudran (F.). Spectres Lumineux, Paris, 1874, p. 110, 114, 120, planches XVII, XVIII.

Fluorescence des composés de manganèse dans la vide sous l'influence de l'arc voltaïque.

Lecoq de Boisbaudran (F.). Comptes Rendus, **103**, 468–471; Jour. Chem. Soc., **52**, 8 (Abs.); Beiblätter, **11**, 87.

Das Absorption der Mangansäure nicht die Umkehrung einer dürch
Manganchlorür gefärbten Flamme.

Müller (J.).   Ann. Phys. u. Chem., **128**, 835.

Spectrum von Mangan.

Simmler (R. Th.).   Ann. Phys. u. Chem., **115**, 425.

Das von übermangansaurem Kali reflectirte Licht.

Wiedemann (E.).   Ann. Phys. u. Chem., **151**, 625.

MAPS.

Recherches sur les spectres des métalloïdes.

> Angström (A. J.) et Thalén (T. R.). Upsal., E. Berling, 1875, 4°.
> Extrait des Nova Acta Reg. Soc. Sc. Upsal., Ser. III, Vol. IX.
> Avec deux planches.
> (Wave-lengths. Spectra of carburetted hydrogen; of carbonic oxide;
> bioxide of nitrogen; of light at the negative pole; of oxygen; of
> carbon; of hydrogen; some isolated rays of carburetted hydrogen,
> and of carbonic oxide.)

Sur le spectre normal du Soleil, partie ultra-violette.

> Cornu (A.). Paris, Gauthier-Villars, 1881, 4°. Extrait des Annales
> de l'École normale supérieure, (2) 9 (1880). Avec deux planches.
> (Wave-lengths.)

Étude du spectre solaire.

> Fievez (Ch.). Bruxelles, F. Hayez, 1882, 4°.
> (Wave-lengths. Lines 6899 to 4522.)
> Extrait des Annales de l'Observatoire royal de Bruxelles, n. sér., t. IV.

*Étude de la région rouge (A–C.) du spectre solaire.

> Fievez (Ch.). F. Hayez, Bruxelles, 1883, 4°. Extrait des Annales de
> l'Observatoire royal de Bruxelles, n. sér., t. V. Avec deux planches.
> (Wave-lengths. Lines 7500 to 6500.)

Studien auf dem Gebiete der Absorptionsspectralanalyse.

> Hasselberg (B.). St. Pétersbourg, et à Leipzig (L. Voss), 1878, 4°.
> Mit vier Karten. Mém. Acad. imp. des Sci. de St. Pétersbourg, (7)
> 26, No. 4.
> (Wave-lengths. Absorptionspectra of hypernitric acid at different
> densities, and absorptionspectrum of bromine.)

Ueber die Spectra der Cometen, und ihre Beziehung zu denjenigen gewisser Kohlenverbindungen.

> Hasselberg (B.). St. Pétersbourg, 1880, Leipzig (G. Haessel), 4°. Mit
> einem Tafel. Mém. de l'Acad. imp. St. Pétersbourg, (7) 28, No. 2.

Untersuchungen über das zweite Spectrum des Wasserstoffs.

> Hasselberg (B.). St. Pétersbourg, 1882, Leipzig (G. Haessel), 4°. Mém.
> de l'Acad. imp. St. Pétersbourg, (7) 30, No. 7. Mit einem Tafel.
> (Wave-lengths.)

Untersuchungen über das Sonnenspectrum und die Spectren der chemischen Elemente.

> Kirchhoff (G.). Besondere Abdrücke aus den Abhandlungen der Berliner Akademie der Wissenschaften, 1861 und 1862. I. Theil, Dümmler, Berlin, 1864, 4°. II. Theil, Dümmler, Berlin, 1875, 4°. Mit vier Tafeln.
> (He used an arbitrary scale.)

Recherches sur le spectre solaire ultra-violet, et sur la détermination des longueurs d'onde, suivies d'une note sur les formules de dispersion.

> Mascart (E.). Extrait des Annales scientifiques de l'École normale supérieure, t. I (1864), Paris, Gauthier-Villars, 1864, 4°.

Recherches sur la détermination des longueurs d'onde.

> Mascart (E.). Paris, Gauthier-Villars, 1866, 4°. Extrait des Annales de l'École normale supérieure, t. IV. Avec un planche.

[A photographic map of the solar spectrum is being prepared by Prof. Rowland, and some parts of it have been distributed, viz: wave-lengths 0.0003675 to 0.0005796.]

Mémoire sur la détermination des longueurs d'onde des raies métalliques.

> Thalén (Rob.). Upsal., W. Schultz, 1868, 4°. Mit zwei Tafeln. Extrait des Nova Acta Reg. Soc. Sci. Upsal., Ser. III, Vol. VI.
> (Gives the wave-lengths of the bright rays of the metals.)

Le spectre d'absorption de la vapeur d'iode.

> Thalén (Rob.). Upsal., Ed. Berling, 1869, 4°. Avec trois planches.

[Thollon's map of the solar spectrum is in Vol. I of the Annales de l'Observatoire de Nice, which is about to appear. Vol. II will contain a smaller map or sheets of the group B.]

MERCURY.

Mercury spark spectrum.
> Capron (J. R.). Photographed Spectra, London, 1877, p. 87.

Spectre du cinabre, de l'oxide de mercure, de l'iodure de mercure.
> Lallemand (A.). Comptes Rendus, **78**, 1272.

Bichlorure de mercure en solution, étincelle.
> Lecoq de Boisbaudran (F.). Spectres Lumineux, Paris, 1874, p. 169, planche XIV.

On the dispersion of a solution of mercuric iodide.
> Liveing (G. D.). Proc. Philosoph. Soc. Cambridge, **3**, 258–60; Beiblätter, **4**, 610 (Abs.).

Spectrum of mercury at elevated temperatures.
> Lockyer (J. N.). Chem. News, **30**, 98; Nature, **30**, 78; Comptes Rendus, **78**, 178.

Emissionsspectra der Haloïdverbindungen des Quecksilbers.
> Peirce (B. O.). Ann. Phys. u. Chem., n. F. **6**, 597.

Ueber die Spectren des Wasserstoffs, Quecksilbers, und Stickstoffs.
> Vogel (H. W.). Monatsber. d. Berliner Akad. (1879), 586–604; Beiblätter, **4**, 125–80; Amer. Jour. Sci., (3) **19**, 406 (Abs.).

METALS.

Researches on the spectra of the metalloids.

> Angström (A. J.) and Thalén (Rob.). Acta Soc. Upsala, (3) 9;
> Nature, **15**, 401 (Abs.); Beiblätter, **1**, 85–47; Bull. Soc. chim. Paris,
> n. s. **25**, 183.

Spectres d'émission infra-rouges des vapeurs métalliques.

> Becquerel (H.). Comptes Rendus, **97**, 71–4; **99**, 874; Chem. News,
> **48**, 46 (Abs.); Nature, **28**, 287 (Abs.); Beiblätter, **7**, 701 (Abs.);
> Amer. Jour. Sci., (3) **26**, 821 (Abs.); **28**, 459 (Abs.); Ber. chem.
> Ges., **16**, 2487 (Abs.); Jour. Chem. Soc., **46**, 1 (Abs.); Zeitschr. f.
> analyt. Chemie, **23**, 49 (Abs.); Phil. Mag., Oct., 1884.

Procédé pour obtenir en projection les raies des métaux et leur renversement.

> Boudréaux. Jour. de Phys., **3**, 806.

Ueber die electrische Spectra der Metallen.

> Brassack. Zeitschr. f. d. Gesellsch. f. Naturwiss, **9**, 185.

Dissociation of the metalloid elements.

> Brodie (B. C.). Nature, **21**, 491–2.

Discoveries of the new alcaline metals.

> Bunsen (R.). Ber. d. Berliner Akad., 10 Mai, 1860; Chem. News, **3**,
> 132.

Kleinste im Inductionsfunken durch die Spectralanalyse noch erkennbare
Gewichtsmenge verschiedener Metalle; do., im Bunsen'schen Gasflamme; Vergleich beider.

> Cappel (E.). Ann. Phys. u. Chem., **139**, 631.

Some experiments on metallic reflection with the spectroscope.

> Conroy (Sir J.). Proc. Royal Soc., **28**, 244.

On the projection of the spectra of the metals.

> Cooke (J. P.). Amer. Jour. Sci., (2) **40**, 243.

Renversement des raies spectrales des vapeurs métalliques.

> Cornu (A.). Comptes Rendus, **73**, 332; Bull. Soc. chim. Paris, n. s.
> **15**, 5.

On the means of increasing the intensity of metallic spectra.

> Crookes (W.). Chem. News, **5**, 234.

Analyse des spectres colorés par les métaux.

Debray (M. H.). Comptes Rendus, 54, 169.

Sur l'emploi de la lumière Drummond et sur la projection des raies brilliants des flammes colorées par les métaux.

Debray (M. H.). Ann. Chim. et Phys., (8) 65, 331.

Remarques sur les métaux nouveaux de la gadolinite, et de la samarskite; holmium ou philippine, thulium, samarium, décipium.

Delafontaine. Comptes Rendus, 90, 221.

Recherches sur l'influence des éléments électronégatifs sur le spectre des métaux, avec planches des spectres de chloride de cuivre et de bromide de cuivre.

Diacon (E.). Ann. Chim. et Phys., (4) 6, 1.

Sur les spectres des métaux alcalins.

Diacon et Wolf. Mém. de l'Acad. de Montpellier, 1868; Comptes Rendus, 55, 884.

Spectres des métalloïdes des familles du soufre, du chlore et de l'azote.

Ditte. Bull. Soc. chim. Paris, n. s. 16, 229.

On the use of the prism in qualitative analysis. (Gives the absorption spectra of many coloured metallic salts.)

Gladstone (J. H.). Jour. Chem. Soc. (1858), 10, 79.

Recherches sur les spectres des métaux à la base des flammes.

Gouy. Comptes Rendus, 84, 231-4; Phil. Mag., (5) 3, 238-40; Chem. News, 35, 107-8; Beiblätter, 1, 238 (Abs.); Bull. Soc. chim. Paris, n. s. 28, 352.

Das electrische Verhalten der im Wasser oder in Salzlösungen getauchten Metalle bei Bestrahlung durch Sonnen-oder Lampen-Licht.

Hankel (W.). Ann. Phys. u. Chem., n. F. 1, 410.

Investigation by means of photography of the ultra-violet spark spectra emitted by metallic elements and their combinations under varying conditions.

Hartley (W. N.). Chem. News, 48, 195.

Beiträge zur Spectroscopie der Metalloïde.

Hasselberg (B.). Bull. Acad. St. Pétersbourg, 27, 405-17.

Auflösung heller Streifen in Metallspectren.

Jahresber. d. Chemie., 15, 29.

Unterschiede in den Spectren bei Anwendung der Metalle oder der Chlor-metalle.

> Jahresber. d. Chemie, **15**, 31, 82.

Constanz der Metallspectren.

> Jahresber. d. Chemie, **15**, 32.

Electrische Metallspectren.

> Jahresber. d. Chemie, **15**, 83; **16**, 104, 106, 107, 118; **17**, 115; **18**, 90, 91.

Einfluss nichtmetallischer Elemente auf die Spectra der Metalle.

> Jahresber. d. Chemie, **18**, 87.

Umkehrung der hellen Spectrallinien der Metalle, insbesondere des Natriums in dunkle.

> Jahresber. d. Chemie, **18**, 90.

Objectivdarstellung der Metallspectren.

> Jahresber. d. Chemie, **26**, 147.

Spectren der Metalloïden.

> Jahresber. d. Chemie, **26**, 149.

Metallspectra.

> Jahresber. d. Chemie, **28**, 122.

Absorptionspectra von Metalldämpfen.

> Jahresber. d. Chemie, **28**, 124, 125.

Quelques spectres métalliques; plomb, chlorure d'or, thallium, lithium.

> Lecoq de Boisbaudran (F.). Comptes Rendus, **77**, 1152; Bull. Soc. chim. Paris, n. s. **21**, 125-6.

Sur un nouveau ordre des spectres métalliques.

> Lecoq de Boisbaudran (F.). Comptes Rendus, **100**, 1487-40; Jour. Chem. Soc., **48**, 949 (Abs.).

Spectra of metallic compounds.

> Leeds (A. R.). Jour. Franklin Inst., **90**, 194.

Reversal lines of metallic vapours.

> Liveing (G. D.) and Dewar (J.). Proc. Royal Soc., (No. I) **27**, 132-6; (No. II) **27**, 350-4; (No. III) **27**, 494-6; (No. IV) **28**, 852-3; (No. V) **28**, 867-72; (No. VI) **28**, 471-5; (No. VII) **29**, 402-6. Beiblätter, **2**, 261 (Abs.), 490 (Abs.); **3**, 710 (Abs.); **4**, 364 (Abs.).

On the disappearance of some spectral lines and the variations of metallic spectra due to mixed vapours.

> Liveing (G. D.) and Dewar (J.). Proc. Royal Soc., **33**, 428–34; Jour. Chem. Soc., **44**, 2–3 (Abs.); Beiblätter, **6**, 676 (Abs.).

Spectral lines of the metals developed by exploding gases.

> Liveing (G. D.) and Dewar (J.). Phil. Mag., (5) **18**, 161–78.

On the circumstances producing the reversal of the spectral lines of metals.

> Liveing (G. D.) and Dewar (J.). Proc. Philosoph. Soc. Cambridge, **4**, 256–65; Beiblätter, **7**, 530 (Abs.).

Quantitative analysis of certain alloys by means of the spectroscope.

> Lockyer (J. N.) and Roberts (W. C.). Proc. Royal Soc., **21**, 507–8; Phil. Trans., **164**, 495–9; Phil. Mag., (4) **47**, 811 (Abs.); Jour. Chem. Soc., (2) **12**, 495 (Abs.); Ber. chem. Ges., **6**, 1426 (Abs.).

On the absorption spectra of metals volatilized by the oxyhydrogen flame.

> Lockyer (J. N.) and Roberts (W. C.). Proc. Royal Soc., **23**, 344–9; Phil. Mag., (5) **1**, 284–9; Jour. Chem. Soc., 1872, **2**, 156 (Abs.).

On a new method of studying metallic vapours.

> Lockyer (J. N.). Proc. Royal Soc., **22**, 371–8; **29**, 266–72; Beiblätter, **4**, 86 (Abs.).

Notice sur les nouveaux métaux obtenus du gadolinite.

> Mendelejeff. Jour. Soc. phys. chim. russe, **13**, 517–20; Bull. Soc. chim. Paris, **38**, 139–48.

Spectra der Haloïdsalze.

> Mitscherlich (A.). Ann. Phys. u. Chem., **121**, 474.

De l'influence de la température sur les spectres des métalloïdes.

> Monckhoven (D. von). Comptes Rendus, **95**, 520.

Sur le spectre des métaux alcalins dans les tubes de Geissler.

> Salet (G.). Comptes Rendus, **82**, 223–6, 274–5; Nature, **13**, 814; Phil. Mag., (5) **1**, 331–8; Jour. Chem. Soc., 1876, **1**, 863 (Abs.); Ann. Phys. u. Chem., **158**, 329–334.

Sur les spectres des métalloïdes.

> Salet (G.). Ann. Chim. et Phys., (4) **28**, 5–71; Chem. News, **27**, 59, 178 (Abs.).

On the spectra of the metalloids.

> Schuster (A.). Phil. Trans. (1879), **170**, 37–54; Proc. Royal Soc., **27**, 383–8 (Abs.); Beiblätter, **1**, 289; **2**, 432 (Abs.); **3**, 749 (Abs.); Jour. Chem. Soc., **38**, 430 (Abs.); Nature, **15**, 447–8.

Les spectres du fer et de quelques autres métaux dans l'arc voltaïque.

> Secchi (A.). Comptes Rendus, **77**, 178; Chem. News, **28**, 82.

Recherches sur l'absorption des rayons ultra-violets par diverses substances; nouvelle étude des spectres d'absorption des métaux
terreux.

> Soret (J. L.). Arch. de Genève, (8) **4**, 261–92; Beiblätter, **5**, 124
> (Abs.).

Sur la fluorescence des sels des métaux terreux.

> Soret (J. L.). Comptes Rendus, **88**, 1077–8; Jour. Chem. Soc., **36**,
> 862 (Abs.); Beiblätter, **3**, 620 (Abs.).

Mémoire sur la détermination des longueurs d'onde des raies métalliques;
spectres des métaux dessinés d'après leurs longueurs d'onde.

> Thulén (R.). Ann. Chim. et Phys., (4) **18**, 202.

Optische Eigenschaften dünner metallischen Schichten.

> Voigt (W.). Ann. Phys. u. Chem., (2) **25**, 95–114.

Leichte Umkehrung der Natriumlinie.'

> Weinhold (A.). Ann. Phys. u. Chem., **142**, 321.

Ueber die Absorption und Brechung des Lichtes in metallisch undurchsichtigen Körpern.

> Wernicke (W.). Monatsber. d. Berliner Akad. (1874), 728-37; Ann.
> Phys. u. Chem., **155**, 87–95.

Electrische Spectra der Metalle.

> Willigen (S. M. von der). Ann. Phys. u. Chem., **106**, 619.

METEOROLOGICAL.

The spectroscope and weather forecasting.
> Abercromby (R.). Nature, **26**, 572–3.

Rain-band Spectroscopy.
> Bell (L.). Amer. Jour. Sci., (8) **30**, 347.

A plea for the rain-band.
> Capron (J. R.). Observatory (1882), 42–7, 71-/; Beiblätter, **6**, 485 (Abs.).

The spectroscope as an aid to forecasting the weather.
> Cory (F. W.). Quar. Jour. Meteorolog. Soc., **9**, 284–9.

Ueber Regenbogen gebildet durch Flüssigkeiten von verschiedenen Brechungsexponenten.
> Hammerl (H.). Sitzungsber. d. Wiener Akad., **86** II, 206–15; Beiblätter, **7**, 883 (Abs.).

Spectroscopic observation of the red-coloured sky at sunset, 1884, Jan. 9, 5 h. 20 min.
> Konkoly (N. von). Monthly Notices Astronom. Soc., **44**, 250–1.

Observations, à propos d'une note récente de M. Reye sur les analogies qui existent entre les taches solaires et les tourbillons de notre atmosphère.
> Marié-Davy. Comptes Rendus, **77**, 1227–9.

The green Sun.
> Manley (W. R.). Nature, **28**, 611–12.

Observations on the rain-band from June, 1882, to Jan., 1883.
> Mill (H. R.). Proc. Royal Soc. Edinburgh, **12**, 47–56.

Note sur les cyclones terrestres et les cyclones solaires.
> Parville (H. de). Comptes Rendus, **77**, 1280–3.

The solar spectrum in a hail-storm.
> Romanes (C. H.). Nature, **25**, 507; Beiblätter, **6**, 486 (Abs.).

The spectroscope and the weather.
> Smith (C. Mitchie). Nature, **12**, 866.

The green Sun.
> Smith (C. Mitchie). Nature, **29**, 28.

The remarkable sunsets.

> Smith (C. Mitchie). Nature, **29**, 381–2.

Spectroscopic prevision of rain with a high barometer.

> Smith (C. Piazzi). Nature, **12**, 281–2, 252–3; Ann. Phys. u. Chem.,
> **157**, 175 (Abs.).

The warm rain-band in the daylight spectrum.

> Smyth (C. Piazzi). Nature, **14**, 9.

Three years' experimenting in spectrum analysis.

> Smith (C. Piazzi). Nature, **22**, 193.

Spectroscopic weather discussions.

> Smyth (C. Piazzi). Nature, **26**, 551–4; Beiblätter, **6**, 877 (Abs.).

Rain-band spectroscopy attacked again.

> Smyth (C. Piazzi). Nature, **29**, 525; Zeitschr. d. oesterreicher Ges. f.
> Meteorol., **14**, 151–2.

Precédé pour déterminer la direction et la force du vent; suppression des
girouettes; application aux cyclones.

> Tarry (H.). Comptes Rendus, **77**, 1117–20.

The use of the spectroscope in meteorological observations.

> Upton (Winslow). U. S. Signal Service Notes (1882), No. 4; Mem.
> Spettr. ital., **13**, 113–18.

MICROSCOPIC SPECTRA.

Prismatic examination of microscopic objects.

> Huggins (William). Trans. Roy. Microscopical Soc. (1865): Quar.
> Jour. Microscopical Sci., July, 1865.

Anwendung der Spectralanalyse auf mikroscopische Untersuchungen.

> Jahresber. d. Chemie (1867), 105.

MINERAL WATERS.

La lithine, la strontiane et l'acide borique dans les eaux minérales de Contrexeville et Schinznach (Suisse).

> Dieulafait. Comptes Rendus, 95, 999-1001; Jour. Chem. Soc., 44, 301 (Abs.).

Existence de l'acide borique en quantité notable dans les lacs salés de la période moderne et dans les eaux salines naturélles, qu'elles soient ou non en relation avec des produits éruptifs.

> Dieulafait. Ann. Chim. et Phys., (5) 25, 145-87.

Untersuchung einiger Mineralwässer und Soole mittelst Spectralanalyse.

> Redtenbacher (Jos.). Sitzungsber. d. Wiener Akad., 44 II, 137, 151, 153-4.

Sur l'origine de la lithine et de l'arsénic dans les eaux sulfatées calciques.

> Schlagdenhauffen. Jour. de Pharm., (5) 6, 457-63; Jour. Chem. Soc., 44, 302 (Abs.).

Spectral-reactionen bündnerischen Gesteine und Mineralwässer.

> Simmler (R. Th.). Ann. Phys. u. Chem., 115, 434-48.

De la présence de la lithine dans le sol de la Limagne et dans les eaux minérales d'Auvergne. Dosage de cet alcali au moyen du spectroscope.

> Truchot (P.). Comptes Rendus, 78, 1022-4; Ber. chem. Ges., 7, 658.

MINIUM.

Spectre du minium.

> Lallemand (A.). Comptes Rendus, 78, 1272.

MOLYBDENUM.

Molybdenum arc spectrum.

Capron (J. R.). Photographed Spectra, London, 1877? p. 87.

MOSANDRUM.

Le mosandrum, un nouvel élément.

Smith (J. Lawrence). Comptes Rendus, **87**, 148–51; note par M. Delafontaine, Comptes Rendus, **87**, 600–2, and Jour. Chem. Soc., **36**, 117 (Abs.).

MULTIPLE SPECTRA.

Multiple Spectra.

Lockyer (J. N.). Nature, **22**, 4–7, 309–12, 562–5; Beiblätter, **5**, 118–22 (Abs.).

NICKEL.

Nickel arc spectrum; nickel spark spectrum; bismuth and nickel spark spectrum.

Capron (J. R.). Photographed Spectra, London, 1877, p. 20, 38.

Salpetersaure Nickellösung als Absorptionspräparat.

Emsmann (H.). Ann. Phys. u. Chem., Ergänzungsband, 1874, 6, 884; Phil. Mag., (4) 46, 829; Jour. Chem. Soc., (2) 12, 118.

Spectrum von Nickel.

Jahresber. d. Chemie, (1872) 145, (1873) 154.

Chlorure de nickel en solution, étincelle.

Lecoq de Boisbaudran (F.). Spectres Lumineux, Paris, 1874, p. 188, planche XIX.

Ueber die Erkennung des Kobalts neben Eisen und Nickel.

Vogel (H. W.). Ber. chem. Ges., 12, 2318–16; Beiblätter, 4, 278 (Abs.); 5, 118 (Abs.).

NIOBIUM.

Niobium arc spectrum.

Capron (J. R.). Photographed Spectra, London, 1877, p. 88.

NITROGEN.

Spectrum von Stickoxyd, und von Stickstoff.

Angström (A. J.). Ann. Phys. u. Chem., **94**, 156–7.

Spectre de l'acide azotique fumant.

Becquerel (H.). Comptes Rendus, **85**, 1227.

Spectre de l'azote.

Becquerel (H.). Comptes Rendus, **90**, 1407.

Spectre du protoxyde de l'azote.

Becquerel (H.). Comptes Rendus, **90**, 1407.

Absorption spectrum of nitrogen peroxide.

Bell (L.). Amer. Chem. Jour., **7**, 32–4; Jour. Chem. Soc., **48**, 949 (Abs.).

Observations of the lines of the solar spectrum, and on those produced by the Earth's atmosphere and by the action of nitrous acid gas.

Brewster (Sir D.). Phil. Mag., (8) **8**, 384.

Carattere spettroscopico della soluzione ammoniacale di carminio, di cocciniglia e di altre sostanze.

Campani (G.). Gazz. chim. ital., **1**, 471–2; Jour. Chem. Soc., (2) **9**, 1096 (Abs.); Ber. chem. Ges., **5**, 287.

Nitrogen spectra.

Capron (J. R.). Photographed Spectra, London, 1877, p. 55.

Sur le spectre d'absorption de l'acide pernitrique.

Chappuis (J.). Comptes Rendus, **94**, 946–8; Jour. Chem. Soc., **42**, 1017 (Abs.); Beiblätter, **6**, 483 (Abs.); Amer. Jour. Sci., (8) **24**, 58 (Abs.); Jour. de Phys., (2) **3**, 48.

Spectre des bandes de l'azote, son origine.

Deslandres (H.). Comptes Rendus, **101** (1885), 1256–60; Jour. Chem. Soc., **50**, 189 (Abs.).

Spectre de l'azote.

Deslandres (H.). Comptes Rendus, **103**, 375–9; Jour. Chem. Soc., **50**, 957 (Abs.); Beiblätter, **11**, 36 (Abs.).

Spectrum von Ammoniak und von Schwefelammon.

Dibbits (H. C.). Ann. Phys. u. Chem., **122**, 518, 534.

Les lacs salpêtres naturels du Chili et du Pérou.

> Dieulafait. Comptes Rendus, **98**, 1545-8; Chem. News, **50**, 45 (Abs.).

Spectres appartenant aux familles de l'azote et du chlore.

> Ditte (A.). Comptes Rendus, **73**, 738; Bull. Soc. chim. Paris, n. s. **16**, 229.

Salpetersaure Nickellösung.

> Emsmann (H.). Ann. Phys. u. Chem., Ergänzungsband, **6** (1873), 334; Jahresber. d. Chemie (1873), 154.

Recherches sur l'intensité relative des raies spectrales de l'hydrogène et de l'azote en rapport avec la constitution des nébuleuses.

> Fiévez (C.). Bull. Acad. Belgique, (2) **49**, 107-113; Phil. Mag., (5) **9**, 309-12; Beiblätter, **4**, 461 (Abs.); Ann. Chim. et Phys., (5) **20**, 179-85; Jour. Chem. Soc., **40**, 69-70.

Action of nitrates on the blood.

> Gamge (A.). Phil. Trans. (1868), 589; Jour. prackt. Chemie, **105**, 287; Ber. chem. Ges., **9**, 833.

Sur les raies d'absorption produites dans le spectre par les solutions des acides hypoazotiques.

> Gernez (D.). Comptes Rendus, **74**, 465-8; Jour. Chem. Soc., (2) **10**, 280 (Abs.); Ber. chem. Ges., **5**, 218; Bull. Soc. chim. Paris, n. s. **17**, 257.

Note sur le prétendu spectre d'absorption spécial de l'acide azoteux.

> Gernez (D.). Bull. Soc. Philom., (7) **5**, 42.

The refraction equivalents of nitrogen, etc., in organic compounds.

> Gladstone (J. H.). Proc. Royal Soc., **31**, 827-330; Ber. chem. Ges., **14**, 1553 (Abs.).

Spectres de l'azotate de cuivre, de l'azotate de manganèse, de l'azotate de plomb.

> Gouy. Comptes Rendus, **84**, 281; Chem. News, **35**, 107.

Spectre de l'azotate d'argent.

> Gouy. Comptes Rendus, **84**, 281.

Azotate.

> Gouy. Comptes Rendus, **85**, 70.

Zur Spectroscopie des Stickstoffs.

> Hasselberg (B.). Mém. de l'Acad. de St. Pétersbourg, (7) **32**, 50 pp. sep.; Beiblätter, **9**, 578 (Abs.).

Ueber die Spectralerscheinungen des Phosphorwasserstoffs und des Ammoniaks.

> Hofmann (K. B.). Ann. Phys. u. Chem., 147, 92–101; Jour. Chem. Soc., (2) 11, 840 (Abs.).

Spectrum des Stickstoffs.

> Jahresber. d. Chemie, 16 (1868), 110; 25 (1872), 142. 144, 145.

Absorptionsspectrum des Dampfs der salpetrigen-und untersalpeter-Säure.

> Jahresber. d. Chemie, 22 (1869), 183.

Spectroscopische Untersuchung der Absorptionsspectren der flüssigen Untersalpetersäure.

> Jahresber. d. Chemie, 23 (1870), 172; 25 (1872), 137.

Absorptionsspectrum des Didymnitrats.

> Jahresber. d. Chemie, 23 (1870), 321.

Absorptionsspectrum der Ammoniakflamme.

> Jahresber. d. Chemie, 25 (1872), 142, 143.

Ueber das Absorptionsspectrum der flüssigen Untersalpetersäure.

> Kundt (A.). Ann. Phys. u. Chem., 141, 157–9; Zeitschr. f. analyt. Chem., (2) 7, 64 (Abs.); Jour. Chem. Soc., (2) 9, 185 (Abs.).

Azotate d'argent en solution, étincelle.

> Lecoq de Boisbaudran (F.). Spectres Lumineux, Paris, 1874, p. 167, planche XXV.

Constitution des spectres lumineux.

> Lecoq de Boisbaudran (F.). Comptes Rendus, 70, 144, 974, 1090.

Spectre du nitrate de didyme.

> Lecoq de Boisbaudran (F.) et Smith (Lawrence). Comptes Rendus, 88, 1167.

Spectre du nitrate de décipium.

> Lecoq de Boisbaudran (F.). Comptes Rendus, 89, 212.

Spectre du nitrate de samarium.

> Lecoq de Boisbaudran (F.). Comptes Rendus, 89, 212.

Spectre de l'ammoniaque par renversement du courant induit.

> Lecoq de Boisbaudran (F.). Comptes Rendus, 101, 42–5.

Spectres des vapeurs aux températures élévées, nitrogène.

> Lockyer (J. N.). Comptes Rendus, 78, 1790; Chem. News, 30, 98.

Sur les spectres de l'acide azoteaux et du peroxyde d'azote.

> Luck (E.). Bull. Soc. chim. Paris, n. s. 13, 498.

Absorption bands of nitrous acid gas.

> Miller (W. Hallows). Phil. Mag., (3) **2**, 881.

Benützung des Ammoniaks zur Spectralanalyse.

> Mitscherlich. Jour. prackt. Chemie, **86**, 14.

Die Spectren der salpetrigen und der untersalpetrigen Säure.

> Moser (J.). Ann. Phys. u. Chem., n. F. **2**, 189–40.

Spectrum von Stickgas, und von Stickoxydul.

> Plücker. Ann..Phys. u. Chem., **105**, 76, 81.

Spectra am negativen Pol im Stickstoff und Wasserstoffröhren; Modification beider Röhren nach langem Gebrauch.

> Reitlinger (E.). Ann. Phys. u. Chem., **141**, 185.

Spectrum einer Lösung von salpetersauren Didymoxyd.

> Rood (O. N.). Ann. Phys. u. Chem., **117**, 350.

Sur le spectre de l'azote et sur celui des métaux alcalins dans les tubes de Geissler.

> Salet (G.). Comptes Rendus, **82**, 223–6, 274–5; Nature, **13**, 314; Phil. Mag., (5) **1**, 331–3; Jour. Chem. Soc., 1876, **1**, 863–4 (Abs.); Ann. Phys. u. Chem., **158**, 829–34.

Spectrum des electrischen Glimmlichts in atmosphärischer Luft; Stickstoff gibt je nach der Temperatur drei Spectra.

> Schimkow (A.). Ann. Phys. u. Chem., **129**, 513–16.

Ueber die Absorption des Lichts durch Ammoniak, etc.

> Schönn (J. L.). Ann. Phys. u. Chem., Ergänzungsband, **8** (1878), 670–5; Jour. Chem. Soc., **34**, 693 (Abs.).

On the spectrum of nitrogen.

> Schuster (A.). Proc. Royal Soc., **20**, 484–7; Phil. Mag., (4) **44**, 537–41; Ann. Phys. u. Chem., **147**, 106–12; Amer. Jour. Sci., (3) **5**, 131 (Abs.); Jour. Chem. Soc., (2) **11**, 840 (Abs.).

Bestimmung der Salpetersäure auf spectralanalytischem Wege.

> Settegast (H.). Zeitschr. f. analyt. Chemie, **20**, 116–117.

Spectres d'absorption ultra-violets des éthers azotiques et azoteux.

> Soret (J. L.) et Rilliet (Alb. A.). Comptes Rendus, **89**, 747.

Spectrum of nitrogen.

> Stearn (C. H.). Nature, **7**, 463.

Spectrum von Stickstoff.

> Vogel (H. C.). Ann. Phys. u. Chem., **146**, 578.

Ueber allmähliche Ueberführung des Bandenspectrums des Stickstoffs in
ein Linienspectrum.

> Vogel (H. C.). Sitzungsber. d. Münchener Akad. (1879), 171–207;
> Ann. Phys. u. Chem., n. F. 8, 590–628.

On the changes produced in the position of the fixed lines in the spectrum
of hyponitric acid by changes in density.

> Weiss (A.). Phil. Mag., (4) 22, 80.

Ueberinstimmung der Absorptionsspectra von Untersalpetersäure mit den
Spectren dessen Dampfes.

> Wüllner (A.). Ann. Phys. u. Chem., 120, 159.

Die beiden Stickstoffspectra nicht durch Unterschiede der Temperatur,
sondern der Entladungsart erklärbar.

> Wüllner (A.). Ann. Phys. u. Chem., 135, 526.

Spectra des Stickstoffs unter hohem Druck.

> Wüllner (A.). Ann. Phys. u. Chem., 137, 356.

Das Spectrum des Stickstoffs ist vielfach; Antwort auf Angström.

> Wüllner (A.). Ann. Phys. n. Chem., 144, 520.

NOMENCLATURE.

Spectroscopic Nomenclature.

> Herschel (J.).   Nature, 5, 499–500; 6, 488–4.

Spectroscopic Nomenclature.

> Young (C. A.).   Nature, 6, 101.

OPTICS.

(With special reference to the spectroscope.)

Optische Untersuchungen.

> Angström (A. J.).   Ann. Phys. u. Chem., **94**, 141 ; Phil. Mag., (4) **9**, 327.

Zwei optische Beobachtungsmethoden.

> Christiansen (C.).   Ann. Phys. u. Chem., **141**, 470.

Optische Untersuchungen einiger Reihen isomorpher Substanzen.

> Christiansen (C.) und Topsoë (Haldor).   Ann. Phys. u. Chem., Ergänzungsband, **6** (1874), 499.

Die optischen Eigenschaften von fein vertheilten Körpern.

> Christiansen (C.).   Ann. Phys. u. Chem., n. F. **23**, 298.

Ueber einen optischen Versuch.

> Ditscheiner (L.).   Ann. Phys. u. Chem., **129**, 840.

Optical Notes.

> Gibbs (Wolcott).   Proc. Amer. Acad., vol. **10** ; Ann. Phys. u. Chem., **156**, 120.

Optische Controversen.

> Ketteler (E.).   Ann. Phys. u. Chem., n. F. **18**, 387–421, 631–63.

Elementare Behandlung einiger optischen Probleme.

> Lommel (E.).   Ann. Phys. u. Chem., **156**, 578–90.

Die Newton'schen Staubringe.

> Lommel (E.).   Ann. Phys. u. Chem., n. F. **8**, 194.

Zur Theorie des Lichtes.

> Lommel (E.).   Ann. Phys. u. Chem., n. F. **16**, 427.

Optische Experimental-Untersuchungen.   Ueber das Verhalten des polarisirten Lichtes bei der Beugung.

> Quincke (G.).   Ann. Phys. u. Chem., **149**, 273–324.

Investigations in optics, with special reference to the spectroscope.

> Rayleigh (Lord).   Phil. Mag., (5) **8**, 261–274, 403–11, 477–86 ; **9**, 40–55 ; Beiblätter, **4**, 860.

OSMIUM.

On the spectrum of osmium.

Fraser (W.). Chem. News, 8, 34.

Spectrum des Osmiums.

Jahresber. d. Chemie, **16** (1868), 112.

OXYGEN.

The acceleration of oxidation caused by the least refrangible end of the spectrum.

> Abney (W. de W.). Proc. Royal Soc., **27**, 291, 451.

Spectres des gaz simples; l'oxygène.

> Angström (A. J.). Comptes Rendus, **73**, 369.

Spectrum von Sauerstoff.

> Angström (A. J.). Ann. Phys. u. Chem., **94**, 155.

Sauerstoff hat nur ein Spectrum; die vielfachen rühren bei Bemengungen her.

> Angström (A. J.). Ann. Phys. u. Chem., **144**, 302, 304.

Recherches expérimentales sur la polarization rotatoire magnétique dans les gaz; oxygène.

> Becquerel (H.). Comptes Rendus, **90**, 1407.

Ueber das Verhalten von Blut und Ozon zu einander.

> Binz (C.). Medicinalisches Centralblatt, **20**, 721-5; Chem. Central-blatt (1882), 810-11; Jour. Chem. Soc., **44**, 486-7 (Abs.).

Oxygen spectra.

> Capron (J. R.). Photographed Spectra, London, 1877, p. 65-7.

Spectre d'absorption de l'ozone.

> Chappuis (J.). Comptes Rendus, **91**, 985; **94**, 858-60; Chem. News, **45**, 163 (Abs.); Jour. Chem. Soc., **42**, 1017 (Abs.); Beiblätter, 6. 482 (Abs.); Amer. Jour. Sci., (3) **24**, 56 (Abs.).

Étude spectroscopique sur l'ozone.

> Chappuis (J.). Ann. de l'École normale, (2) **11**, 187-87; Beiblätter, **7**, 458 (Abs.).

Étude sur la part de la lumière dans les actions chimiques et en particu-lier dans les oxydations.

> Chastaing (P.). Ann. Chim. et Phys., (5) **11**, 145-223; Jour. Chem. Soc., 1877, **2**, 818 (Abs.); Beiblätter, **1**, 517-20 (Abs.).

On the coïncidence of the bright lines of the oxygen spectrum with bright lines in the solar spectrum.

> Draper (H.). Monthly Notices Astronom. Soc., **39**, 440-7; Amer. Jour. Sci., (3) **18**, 262-76; Beiblätter, **4**, 275 (Abs.); Comptes Ren-dus, **88**, 1332 (Abs.).

Dark lines of oxygen in the spectrum of the Sun.

> Draper (J. C.). Amer. Jour. Sci., (3) **16**, 256; (3) **17**, 448; Nature, **18**, 654; note by Barker (G. F.), Amer. Jour. Sci., (3) **17**, 162–6; Nature, **19**, 852–3; Beiblätter, **3**, 188 (Abs.).

Sur la production des groupes telluriques fondamentaux A et B du spectre solaire par une couche absorbante d'oxygène.

> Egoroff (N.). Comptes Rendus, **97**, 555; Amer. Jour. Sci., (3) **26**, 477.

Spectre d'absorption de l'oxygène.

> Egoroff (N.). Comptes Rendus, **101**, 1143–45; Jour. Chem. Soc., **50**, 189 (Abs.).

Sauerstoffausscheidung von Pflanzenzellen im Mikrospectrum.

> Engelmann (T. W.). Pflüger's Archiv. f. Physiologie, **27**, 485–90; Chem. News, **47**, 11 (Abs.); Beiblätter, **7**, 877 (Abs.).

On the combustion of hydrogen and carbonic oxide in oxygen under great pressure.

> Franckland. Proc. Royal Soc., **16**, 419.

The refraction equivalents of oxygen, etc., in organic compounds.

> Gladstone (J. H.). Proc. Royal Soc., **31**, 327–30; Ber. chem. Ges., **14**, 1553 (Abs.).

The absorption spectrum of ozone.

> Hartley (W. N.). Jour. Chem. Soc., **39**, 57–60; Ber. chem. Ges., **14**, 672 (Abs.); Beiblätter, **5**, 505 (Abs.).

On the absorption of solar rays by atmospheric ozone.

> Hartley (W. N.). Jour. Chem. Soc., **39**, 111–28; Ber. chem. Ges., **14**, 1340 (Abs.); Beiblätter, **5**, 505 (Abs.).

Einfacher Versuch zur Demonstration der Sauerstoffausscheidung durch Pflanzen im Sonnenlichte.

> Hoppe-Seyler (F.). Zeitschr. f. physiol. Chemie, **2**, 425–6; Ber. chem. Ges., **12**, 701 (Abs.); Jour. Chem. Soc., **36**, 819 (Abs.).

Sur les spectres d'absorption de l'oxygène.

> Janssen (J.). Comptes Rendus, **102**, 1352–3; Jour. Chem. Soc., **50**, 749 (Abs.); Beiblätter, **11**, 93.

Spectre de l'oxyde de cuivre.

> Lallemand (A.). Comptes Rendus, **78**, 1272.

Sur les spectres de l'acide azoteux et du peroxyde de l'azote.

> Luck (E.). Bull. Soc. chim. Paris, n. s. **13**, 498.

Oxygen in the Sun.

>Meldola (R.).   Nature, **17**, 161-2 ; Beiblätter, **2**, 91.

Das Sauerstoffspectrum und die electrischen Lichterscheinungen verdünnter Gaze in Röhren mit Flüssigkeitselectroden.

>Paalzow (A.).   Ann. Phys. u. Chem., n. F. **7**, 130.

Ueber das Sauerstoffspectrum.

>Paalzow (A.) und Vogel (H. W.).   Ann. Phys. u. Chem., n. F. **13**, 836–8.

Spectrum von Sauerstoff.

>Plücker.   Ann. Phys. u. Chem., **104**, 126 ; **105**, 78.

Spectrum of Oxygen.

>Schuster (A.).   Phil. Trans., **170** (1879), 87–54 ; Proc. Royal Soc., **27**, 388–8 (Abs.); Beiblätter, **2**, 492 (Abs.); **3**, 749 (Abs.); Jour. Chem. Soc., **38**, 430.

Spectre d'acide oxalique.

>Senarmont (H. de).   Ann. Chim. et Phys., (3) **41**, 386.

Constitution of the lines forming the low temperature spectrum of Oxygen.

>Smyth (C. Piazzi).   Trans. Roy. Soc. Edinburgh, **30**, 419–25 ; Phil. Mag., (5) **13**, 380–87 ; Nature, **25**, 408 (Abs.); Jour. de Phys., (2) **2**, 289 (Abs.).

Spectrum von Sauerstoff.

>Vogel (H. C.).   Ann. Phys. u. Chem., **146**, 576.

Photographische Beobachtungen des Sauerstoffspectrums.

>Vogel (H. C.).   Ber. chem. Ges., **12**, 832 ; Amer. Chem. Jour., **1**, 71.

Drei Spectra bei Sauerstoff.

>Wüllner (A.).   Ann. Phys. u. Chem., **135**, 515.

Spectra des Wasserstoffs.

>Wüllner (A.).   Ann. Phys. u. Chem., **137**, 350 ; n. F. **8**, 253.

PALLADIUM.

Palladium arc spectrum; palladium spark spectrum.
> Capron (J. R.). Photographed Spectra, London, 1877, p. 89.

Chlorure de palladium en solution, étincelle.
> Lecoq de Boisbaudran (F.). Spectres Lumineux, Paris, 1874, p. 184, planche XXVII.

PARAGENIC SPECTRA.

Sur la paragénie.
> Babinet. Cosmos, 25, 898.

On paragenic spectra.
> Brewster (Sir D.). Phil. Mag., January, 1866.

PHILIPPIUM.

On philippium.
> Brown (W. G.). Chem. News, 38, 267-8; Jour. Chem. Soc., 36, 204 (Abs.).

Sur un nouveau métal, le philippium.
> Delafontaine. Comptes Rendus, 87, 559-61; Amer. Jour. Sci., (8) 17, 61 (Abs.); Jour. Chem. Soc., 36, 116-17 (Abs.); Beiblätter, 3, 197 (Abs.).

PHOSPHORESCENCE.

On the violet phosphorescence in calcium sulphide.

> Abney (W. de W.). Proc. Physical Soc., **5**, 85–8; Nature, **35**, 355 (Abs.); Phil. Mag., (5) **13**, 212–14; Jour. Chem. Soc., **42**, 677 (Abs.); Beiblätter, **6**, 888 (Abs.); Jour. de Phys., (2) **2**, 287–8.

Propriétés de la lumière des pyrophores, examen spectroscopique.

> Aubert et Dubois. Comptes Reudus, **99**, 477.

Pouvoir phosphorescent de la lumière électrique.

> Becquerel (E.). Comptes Rendus, **8**, 217.

Réfringibilité des rayons qui excitent la phosphorescence dans les corps.

> Becquerel (E.). Comptes Rendus, **69**, 994.

Analyse de la lumière émise par les composés d'uranium phosphorescents.

> Becquerel (E.). Ann. Chim. et Phys., (4) **27**, 539–79; Comptes Rendus, **75**, 296–808; Jour. Chem. Soc., (2) **11**, 25 (Abs.); Amer. Jour. Sci., (3) **4**, 486 (Abs.).

Sur l'observation de la partie infra-rouge du spectre solaire, au moyen des effets de phosphorescence.

> Becquerel (E.). Comptes Rendus, **96**, 1215; Ann. Chim. et Phys., (5) **10**, 5–18; Jour. de Phys., **6**, 187.

Les spectres des corps phosphorescents.

> Becquerel (E.). La Lumière, tome I, 207.

Étude spectrale des corps rendus phosphorescents par l'action de la lumière ou par les décharges électriques.

> Becquerel (E.). Comptes Rendus, **101**, 205–210.

Effets du manganèse sur la phosphorescence du calcium carbonate.

> Becquerel (É.). Comptes Rendus, **103**, 1098.

Phosphorescence de l'alumine.

> Becquerel (E.). Comptes Rendus, **103**, 1224; Amer. Jour. Sci., (3) **33**, 808 (Abs.); Jour. Chem. Soc., **52**, 409 (Abs.); Chem. News, **55**, 99 (Abs.).

Étude des radiations infra-rouges au moyen des phénomènes de phosphorescence.

> Becquerel (H.). Comptes Rendus, **96**, 1215; Ann. Chim. et Phys., (5) **30**, 5–68; Beiblätter, **8**, 120 (Abs.).

Maxima et minima d'extinction de la phosphorescence sous l'influence des radiations infra-rouges.

> Becquerel (H.). Comptes Rendus, **96**, 1858.

Résultats de ses recherches sur les effets de phosphorescence.

> Becquerel (H.). Bull. Soc. franç. de Physique (1883), 24–5.

Sur les variations des spectres d'absorption et des spectres d'émission par phosphorescence d'un même corps.

> Becquerel (H.). Comptes Rendus, **102**, 106–10.

Sur de nouveaux procédés pour étudier la radiation solaire, tant directe que diffuse, dans ses rapports avec la phosphorescence.

> Biot. Comptes Rendus, **8**, 259, 315.

Spectrum of the light emitted by the glow-worm.

> Conroy (Sir J.). Nature, **26**, 319; Beiblätter, **6**, 880 (Abs.).

De la lumière verte et phosphorescente du choc moléculaire.

> Crookes (W.). Comptes Rendus, **88**, 283–4.

Discontinuous phosphorescent spectra in high vacua.

> Crookes (W.). Proc. Royal Soc., **32**, 206–18; Chem. News, **43**, 237–9; Nature, **24**, 89; Comptes Rendus, **92**, 1281–3; Beiblätter, **5**, 511₋18; Ann. Chim. et Phys., (5) **23**, 555.

Les vibrations de la matière et les ondes de l'ether dans la phosphorescence et la fluorescence.

> Fuvé. Comptes Rendus, **86**, 289–94.

Wirkung der verschiedenen Theile des Spectrums auf phosphorescirende Substanzen.

> Jahresber. d. Chemie, **1** (1847), 164.

Spectren des Lichts phosphorescirender Thiere.

> Jahresber. d. Chemie, **17** (1864), 115.

Spectrum des Phosphorenzlichts von Chlorophan, Phosphorit und Flusspath.

> Kindt. Ann. Phys. u. Chem., **131**, 160; Phil. Mag., Dec., 1867.

Phosphorescence de l'alumine.

> Lecoq de Boisbaudran (F.). Comptes Rendus, **103**, 1224–7; Jour. Chem. Soc., **52**, 191 (Abs.).

Sichtbare Darstellung des Brennpunctes der ultrarothen Strahlen durch Phosphorescenz.

> Lommel (E.). Ann. Phys. u. Chem., (2) **26**, 157–9; Phil. Mag., (5) **20**, 547.

Beobachtungen über Phosphorescenz.

> Lommel (E.). Ann. Phys. u. Chem., (2) **30**, 473-87; Jour. Chem. .
> Soc., **52**, 410 (Abs.).
> (Gives the phosphorescent spectra of 16 substances prepared by Dr.
> Schuchardt and with Balmain's paint.)

Lumière phosphorescent des cucuyos.

> Pasteur. Comptes Rendus, **59**, 509; Ann. Phys. u. Chem., **124**, 192;
> Jour. prackt. Chemie, **93**, 381.

Ueber die Phosphorescenz der organischen und organisirten Körper.

> Radziszewski (B.). Ann. Chem. u. Pharm., **203**, 305-36; Beiblätter,
> **4**, 620 (Abs.).

Spectrum of the light of the glow-worm.

> Spiller (J.). Nature, **26**, 343; Beiblätter, **6**, 880.

On the causes of a light border frequently noticed in photographs just
outside the outline of a dark body seen against the sky; with
some introductory remarks on phosphorescence.

> Stokes (G. G.). Proc. Royal Soc., **34**, 63-68; Nature, **26**, 142-3; Bei-
> blätter, **6**, 682 (Abs.).

Sur les causes déterminantes de la phosphorescence du sulfure de calcium.

> Verneuil (A.). Comptes Rendus, **103**, 501-4; Beiblätter, **11**, 253.

Un composé de calcium sulphide ayant une phosphorescence violette.

> Verneuil (A.). Comptes Rendus, **103**, 600-3; Jour. Chem. Soc., **52**,
> 2 (Abs.).

PHOSPHORUS.

Coloration de la flamme et de ses composés, spectre du phosphore.
>Christofle (P.) et Beilstein (F.). Comptes Rendus, **56**, 899; Ann.
>Chim. et Phys., (4) **3**, 281.

Spectre du phosphate.
>Gouy. Comptes Rendus, **85**, 70.

Ueber phosphorhaltigen Stahl.
>Greiner (A.). Dingler's Jour., **217**, 33–41; Jour. Chem. Soc., 1876,
>**1**, 454–7 (Abs.).

Ueber die Spectralerscheinungen des Phosphorwasserstoffs und des Ammoniaks.
>Hofmann (K. B.). Ann. Phys. u. Chem., **147**, 92–101; Jour. Chem.
>Soc., (2) **11**, 840 (Abs.).

Spectra of phosphoric acid blowpipe beads.
>Horner (C.). Chem. News, **29**, 66.

Spectrum des Phosphors.
>Jahresber. d. Chemie, **16** (1863), 111; **17** (1864), 109; **23** (1870), 173.

Absorptionsspectrum des Phosphorwasserstoffs.
>Jahresber. d. Chemie, **25** (1872), 142.

Spectrum des Phosphorescenzlichts von Phosphorit.
>Kindt. Ann. Phys. u. Chem., **131**, 160.

Sur la diffusion lumineuse du phosphore de cuivre obtenu sans précipitation.
>Lallemand (A.). Comptes Rendus, **79**, 693.

Phosphate d'erbine, émission.
>Lecoq de Boisbaudran (F.). Spectres Lumineux, Paris, 1874, p. 92,
>97, planche XIV.

Sur les spectres des vapeurs aux températures élévées; phosphore.
>Lockyer (J. N.). Comptes Rendus, **78**, 178, 1790; Nature, **30**, 98.

Expériences spectrales tendant à démontrer la nature composé du phosphore.
>Lockyer (J. N.). Comptes Rendus, **89**, 514–15; Beiblätter, **4**, 182
>(Abs.).

Spectrum des Phosphors, etc.

Mulder.   Jour. prackt. Chemie, **91**, 111.

Recherche du soufre et du phosphore par le spectroscope.

Salet (G.).   Bull. Soc. chim. Paris, n. s. **13**, 289.

Spectres du phosphore et des composés de silicium.

Salet (G.).   Comptes Rendus, **73**, 1056-59.

Sur les spectres du phosphore et du soufre.

Seguin (J. M.).   Comptes Rendus, **53**, 1272 ; Phil. Mag., (4) **23**, 416

PLATINUM.

Platinum arc spectrum.

Capron (J. R.). Photographed Spectra, London, 1877, p. 89.

Spectre de chlorure de platine.

Gouy (J. R.). Comptes Rendus, **84**, 281 ; Chem. News, **35**, 107.

Distribution of heat in the spectra of various scources of radiation ; platinum.

Jacques (W. W.). Proc. Amer. Acad., **14**, 156.

Die optische Eigenshaften der Platincyanüre.

König (W.). Ann. Phys. u. Chem., n. F. **19**, 491.

Spectre du noir de platine.

Lallemand (A.). Comptes Rendus, **78**, 1272.

Chlorure de platine en solution, étincelle.

Lecoq de Boisbaudran (F.). Spectres Lumineux, Paris, 1874, p. 181, planche XXVII.

Spectre du platine incandescent.

Masson (A.). Comptes Rendus, **32**, 127.

On the character and intensity of the rays emitted by glowing platinum.

Nichols (E. L.). Amer. Jour. Sci., (3) **18**, 446–68.

Radiation du platine incandescent, spectre du platine.

Violle (J.). Comptes Rendus, **88**, 171.

Intensités lumineuses des radiations émises par le platine incandescent.

Violle (J.). Comptes Rendus, **92**, 866–8, 1204–6 ; Beiblätter, **5**, 508 (Abs.).

POLARIZED LIGHT.

Die Phasenveränderung des parallel zur Einfallsebene polarisirten Lichts durch Reflexion.

> Glan (P.). Ann. Phys. u. Chem., **156**, 243.

Polarizationswinkel des Fuchsins.

> Glan (P.). Ann. Phys. u. Chem., n. F. **7**, 321.

Absorption und Emission des polarisirten Lichtes.

> Kirchhoff (G.). Ann. Phys. u. Chem., **109**, 29%.

Sur l'illumination des corps transparents par la lumière polarisée.

> Lallemand (A.). Comptes Rendus, **69**, 917.

Sur la polarization rotatoire du quartz.

> Soret (J. L.). Arch. de Genève, (3) **8**, 5–59, 97–182, 201–28; Jour. de Phys., (2) **2**, 331–6 (Abs.).

Elliptische Polarization des Lichtes und ihre Beziehung zu den Oberflächenfarben der Körper.

> Wiedemann (E.). Ann. Phys. u. Chem., **151**, 1.

Ueber die elliptische Polarization des von durchsichtigen Körpern reflectirten Lichtes.

> Wernicke (W.). Ann. Phys. u. Chem., (2) **30** (1887), 452–69.

POTASSIUM.

Absorptionsspectrum des übermangansauren Kalis und seine Benützung bei chemisch analytischen Arbeiten.

> Brücke (E.). Sitzungsber. d. Wiener Akad., **74** III, 428; Chem. Centralblatt, (3) **9**, 189–43; Jour. Chem. Soc., **34**, 242 (Abs.).

On the light reflected by potassium permanganate.

> Conroy (Sir J.). Proc. Physical Soc., **2**, 340–44; Phil. Mag., (5) **6**, 454–8; Jour. Chem. Soc., **36**, 425 (Abs.).

Transparence des flammes colorées pour leurs propres radiations; la double raie du potassium.

> Gouy. Comptes Rendus, **86**, 1078.

Spectrum des Kaliums.

> Jahresber. d. Chemie, **16** (1868), 112.

Linien von Kalium.

> Kirchhoff (G.). Ann. Phys. u. Chem., **110**, 173.

Permanganate de Potasse en solution, absorption.

> Lecoq de Boisbaudran (F.). Spectres Lumineux, Paris, 1874, p. 108, planche XVI.

Sulfate de potasse fondu, étincelle; chlorure de potassium dans le gaz.

> Lecoq de Boisbaudran (F.). Spectres Lumineux, Paris, 1874, p. 48, planche V.

On the spectra of sodium and potassium.

> Liveing (G. D.) and Dewar (J.). Proc. Royal Soc., **29**, 398–402; Beiblätter, **4**, 868 (Abs.).

Sur le chromocyanure de potassium.

> Moissan (H.). Comptes Rendus, **93**, 1079–81; Chem. News, **45**, 22 (Abs.); Ber. chem. Ges., **15**, 248 (Abs.).

Absorption spectra of sodium and potassium at low temperatures.

> Roscoe (H. E.) and Schuster (A.). Proc. Royal Soc., **22**, 362.

Modifications of the spectrum of potassium which are effected by the presence of phosphoric acid.

> Thudichum (J. L. W.). Proc. Royal Soc., **30**, 278–86.

Ueber das von übermangansaurem Kali reflectirten Licht.

> Wiedemann (E.). Ber. d. k. sächs. Ges. d. Wiss. zu Leipzig, **25**, 367–70; Ann. Phys. u. Chem., **151**, 625–28; Phil. Mag., (4) **48**, 231–33; Jour. Chem. Soc., (2) **13**, 120 (Abs.).

PRESSURE.

De l'influence de la pression sur les raies du spectre.

> Cailletet (L.). Bull. Soc. chim. Paris, n. s. **18**, 218; Ber. chem. Ges., **5**, 482; Comptes Rendus, **74**, 1282.

Gasspectren bei steigendem Druck.

> Jahresber. d Chemie, **22** (1869), 178.

Einfluss des Drucks auf das Spectrum.

> Jahresber. d. Chemie, **25** (1872), 142.

Effect of pressure on the character of the spectra of gases.

> Stearn (C. H.) and Lee (G. H.). Proc. Royal Soc., **21**, 282.

RADIATION.

Réflexions à l'occasion d'une experience de M. Dumas relative à la formation d'un acide nouveau sous l'influence de la radiation solaire.

Biot. Comptes Rendus, 8, 622.

Sur les radiations chimiques de la lumière.

Biot. Comptes Rendus, 12, 170.

Radiant Matter Spectroscopy; the Bakerian lecture.

Crookes (W.). Proc. Royal Soc., 35, 262; Chem. News, 47, 261; 49, 159, 169, 181, 194, 205; 51, 301.

Détermination du pouvoir éclairant des radiations simples.

Crova (A.) et Lagarde. Comptes Rendus, 93, 959; Jour. de Phys., (2) 1, 162-9.

De la loi d'absorption des radiations de toute espèce à travers les corps, et de son emploi dans l'analyse spectrale quantitative.

Gòvi (G.). Comptes Rendus, 85, 1046-9, 1100-3; Phil. Mag., (5) 5, 78-80; Jour. Chem. Soc., 34, 190 (Abs.); Beiblätter, 2, 342 (Abs.).

On the relation between the radiating and absorbing powers of different bodies for light and heat.

Kirchhoff (G.). Phil. Mag., (4) 20, 1.

Ueber Ausstrahlung und Absorption.

Lecher (E.). Sitzungsber. d. Wiener Akad., 85 II, 441-90; Ann. Phys. u. Chem., n. F. 17, 477-518.

The dynamical theory of radiation.

Schuster (A.). Phil. Mag., (5) 12, 261-6; Beiblätter, 5, 793.

RED END OF THE SPECTRUM.

Photography of the red end of the spectrum.
> Abney (W. de W.). Nature, 13, 432; Chem. News, 40, 311.

Work in the infra-red of the spectrum.
> Abney (W. de W.). Nature, 27, 15.

Atmospheric absorption in the infra-red of the solar spectrum.
> Abney (W. de W.) and Festing (Lieut. Col.). Nature, 28, 45.

Wave-lengths of A, a and other prominent lines in the red and infra red of the visible spectrum.
> Abney (W. de W.). Chem. News, 48, 283.

Sur l'observation de la partie infra-rouge du spectre solaire au moyen des effets de la phosphorescence.
> Becquerel (E.). Comptes Rendus, 83, 249.

Étude de la région infra-rouge du spectre.
> Becquerel (H.). Comptes Rendus, 96, 121.

Étude des radiations infra-rouges, au moyen des phénomènes de phosphorescence.
> Becquerel (H.). Comptes Rendus, 96, 1215; Nature, 29, 227; Amer. Jour. Sci., (3) 26, 321; Ann. Chim. et Phys., (5) 30, 5.

Maxima et minima d'extinction de la phosphorescence sous l'influence des radiations infra-rouges.
> Becquerel (H.). Comptes Rendus, 96, 1858.

Sichtbare Darstellung der ultrarothen Strahlen.
> Lommel (E.). Ann. Phys. u. Chem., (2) 26 (1885), 157.

Eine Wellenlängenmessung im ultrarothen Sonnenspectrum.
> Pringsheim (E.). Ann. Phys. u. Chem., n. F. 18, 32.

Visible representation of the ultra-red rays.
> Tyndall. Phil. Mag., (5) 20 (1885), 547; Amer. Jour. Sci., (3) 31, 150.

REFRACTION.

Ueber die Bestimmung des specifischen Brechungsvermögens fester Kor-
per in ihren Lösungen.

> Bedson (P. P.) and Williams (W. C.). Ber. chem. Ges., **14**, 2540-56;
> Jour. Chem. Soc., **42**, 351 (Abs.); Beiblätter, **6**, 91-3 (Abs.); Jour.
> de Phys., (2) **1**, 377 (Abs.).

Réfrangibilité des rayons qui excitent la phosphorescence dans les corps.

> Becquerel (Ed.). Comptes Rendus, **69**, 994.

Spectrum der Brechbaren Strahlen.

> Crookes (W.). Cosmos, **8**, 90; Ann. Phys. u. Chem., **97**, 621.

Sur la double réfraction circulaire et la production normale des trois systèmes de franges des rayons circulaires.

> Croullebois. Comptes Rendus, **92**, 520.

Sur la variation des indices de réfraction dans les mélanges de sels iso-
morphes.

> Dufet (H). Comptes Rendus, **86**, 881-4; Jour. Chem. Soc., **34**, 681-2.

Variation des indices de réfraction du quartz sous l'influence de la température.

> Dufet (H.). Comptes Rendus, **98**, 1265; Jour. de Phys., **10**, 513-19;
> Bull. Soc. minéral., **4**, 191-6; **6**, 76-80, 287.

Die brechbarsten oder unsichtbaren Lichtstrahlen im Beugungsspectrum
und ihre Wellenlänge.

> Eisenlohr (W.). Ann. Phys. u. Chem., **98**, 358.

Beugungsspectrum auf fluorescirenden Substanzen.

> Eisenlohr (W.). Ann. Phys. u. Chem., **99**, 168.

Ueber die Aenderung der Brechungsexponenten isomorpher Mischungen,
mit deren chemischer Zusammensetzung.

> Fock (A.). Zeitschr. Krystallogr. u. Mineralog., **4**, 583-608; Bei-
> blätter, **4**, 662-4 (Abs.).

Experimentaluntersuchungen über die Intensität des gebeugten Lichtes.

> Fröhlich (J.). Ann. Phys. u. Chem., n. F. **15**, 575-613; Jour. de
> Phys., (2) **1**, 559 (Abs.).

Recherches sur le réfraction de la lumière.

> Gouy. Ann. Chim. et Phys., (6) **8** (1886), 145-92; Beiblätter, **11**
> (1887), 95 (Abs.).

Das Auge empfindet alle Strahlen die brechbarer sind als die Rothen.

Helmholtz (H.).　Ann. Phys. u. Chem., **94**, 205.

The refractive index and specific inductive capacity of transparent insulating media.

Hopkinson (J.).　Proc. Royal Soc., **5**, 88–40.

Aenderung des Moleculargewichtes und Molecularrefractionsvermögen.

Janowsky (J. V.).　Sitzungsber. d. Wiener Akad., **81** II, 589–58; **82** II, 147–58.

Sur la relation du pouvoir réfringent et la composition des composés organiques.

Kanonnikoff (J.).　Ber. chem. Ges., **16**, 3047–51 (Abs.); Jour. Soc. phys. chim. russe, **15**, 484–79; Bull. Soc. chim. Paris, **41**, 818 (Abs.); Beiblätter, **8**, 875 (Abs.).

Sur les relations entre la composition et le pouvoir réfringent des composés chimiques.　Second mémoire.

Kanonnikoff (J.).　Jour. Soc. phys. chim. russe, **16**, 119–31; Ber. chem. Ges., **17**, Referate, 157–9 (Abs.); Nature, **30**, 84 (Abs.); Beiblätter, **8**, 493–6 (Abs.); Bull. Soc. chim. Paris, **41**, 549 (Abs.); Jour. Chem. Soc., **46**, 1–2 (Abs.).

Experimentaluntersuchung über den Zusammenhang zwischen Refraction und Absorption des Lichtes.

Ketteler (E.).　Ann. Phys. u. Chem., n. F. **12**, 481–519.

Constanz des Refractionsvermögens.

Ketteler (E.).　Ann. Phys. u. Chem., (2) **30** (1887), 285–99.

Ueber Prismenbeobachtungen mit streifend einfallendem Licht, und über eine Abänderung der Wollaston'schen Bestimmungsmethode für Lichtbrechungsverhältnisse.

Kohlrausch (F.).　Ann. Phys. u. Chem., n. F. **16**, 603.

Abhängigkeit des Brechungsquotienten der Luft von der Temperatur.

Lang (V. von).　Ann. Phys. u. Chem., **153**, 450.

Theorie der Doppelbrechung.

Lommel (E.).　Ann. Phys. u. Chem., n. F. **4**, 55.
(Look below, under Voigt.)

Sur la réfraction des gaz.

Mascart.　Comptes Rendus, **78**, 417; Ann. Phys. u. Chem., **153**, 153.

Wellenlänge und Brechungsexponent der äussersten dunklen Wärme-strahlen des Sonnenspectrums.

> Müller (J.). Ann. Phys. u. Chem., **115**, 543; Berichtigung dazu, **116**, 644.

Bei zunehmender Verdünnung der Gaze erlöschen zuerst die minder brechbaren Strahlen.

> Plücker. Ann. Phys. u. Chem., **116**, 27.

Report of the committee, consisting of Dr. J. H. Gladstone, Dr. W. R. E. Hodgkinson, Mr. Carleton Williams, and Dr. P. P. Bedson (Sec-retary), appointed for the purpose of investigating the Method of Determining the Specific Refraction of Solids from their solutions.

> Report of the British Association, 1881, 155.

Indices de réfraction ordinaire et extraordinaire du quartz pour les ray-ons de différentes longueurs d'onde jusqu'à l'extrème ultra-violet.

> Sarasin (E.). Archives de Genève, (2) **61**, 109–19; Comptes Rendus, **85**, 1230–2 (Abs.); Beiblätter, **2**, 77–8 (Abs.).

Indices de réfraction de spath d'Islande.

> Sarasin (E.). Arch. de Genève, (3) **8**, 392–4; Jour. de Phys., (2) **2**, 369–71.

Indices de réfraction ordinaire et extraordinaire du spath d'Islande pour les rayons de diverses longueurs d'onde jusqu'à l'extrème ultra-violet.

> Sarasin (E.). Comptes Rendus, **95**, 680.

Indices de réfraction du spath-fluor pour les rayons de différentes long-ueurs d'onde.

> Sarasin (E.). Comptes Rendus, **97**, 850.

Untersuchungen über die Abhängigkeit der Molecularrefraction von der chemischen Constitution der Verbindungen.

> Schroder (H.). Ber. chem. Ges., **14**, 2518–16; Jour. Chem. Soc., **42**, 851 (Abs.).

Indices de réfraction des aluns cristallisés.

> Soret (Ch.). Comptes Rendus, **99**, 867.

On a method of destroying the effects of slight errors of adjustment in ex-periments of changes of refrangibility due to relative motions in the line of sight.

> Stone (E. J.). Proc. Royal Soc., **31**, 881.

Indices de réfraction des liquides.

> Terquem et Trannin. Jour. de Phys., **4**, 222; Ann. Phys. u. Chem., **157**, 302.

Brechungsvermögen und Verbrennungswärme.

> Thomsen (J.). Ber. chem. Ges., **15**, 66–69; Jour. Chem. Soc., **42**, 567 (Abs.); Beiblätter, **6**, 877 (Abs.).

Bemerkungen zu Hrn. Lommel's Theorie der Doppelbrechung.

> Voigt (W.). Ann. Phys. u. Chem., n. F. **17**, 468.

Methode zur Bestimmung des Brechungsexponenten von Flüssigkeiten und Glasplatten.

> Wiedemann (E.). Ann. Phys. u. Chem., **158**, 875.

RHABDOPHANE.

Analysis of rhabdophane, a new British mineral.

> Hartley (W. N.). Jour. Chem. Soc., **41**, 210–20; Chem. News, **45**, 40 (Abs.).

Analysis of rhabdophane, a new British mineral.

> Liveing (G. D.) and Dewar (J.). Jour. Chem. Soc., **41**, 210–220; Chem. News, **45**, 40 (Abs.).

RHODIUM.

Rhodium arc spectrum.

> Capron (J. R.). Photographed Spectra, London, 1877, p. 40.

RUBIDIUM.

Observations on cæsium and rubidium.
> Allen (O. D.).   Amer. Jour. Sci., Nov., 1862; Phil. Mag., (4) 25, 189.

Les salpêtres naturels du Chili et du Pérou au point de vue du rubidium.
> Dieulafait.   Comptes Rendus, 98, 1545–8; Chem. News, 50, 45 (Abs.).

Spectre du rubidium.
> Gouy.   Comptes Rendus, 86, 1078.

Beschreibung der Metallen Cæsium und Rubidium.
> Kirchhoff und Bunsen.   Ann. Phys. u. Chem., 113, 337; Phil. Mag.,
> (4) 22, 498; 24, 46.

Chlorure de rubidium dans le gaz.
> Lecoq de Boisbaudran (F.).   Spectres Lumineux, Paris, 1874, p. 46,
> planche IV.

RUTHENIUM.

Ruthenium arc spectrum.
> Capron (J. R.).   Photographed Spectra, London, 1877, p. 40.

Professor Young and the presence of ruthenium in the chromosphere.
> Roscoe (H. E.).   Nature, 9, 5.

SALT.

Blue flame from common salt.

     Gladstone (J. H.). Nature, **19**, 582.

Sur les caractères des flammes chargées de poussières salines.

     Gouy. Comptes Rendus, **85**, 489.

Preliminary notice of experiments concerning the chemical constitution of saline solutions.

     Hartley (W. N.). Proc. Royal Soc., **22**, 241–8; Chem. News, **29**, 148.

On the action of heat on the absorption spectra and chemical constitution of saline solutions.   '

     Hartley (W. N.). Proc. Royal Soc., **23**, 872–3; Ber. chem. Ges., **8**, 765 (Abs.); Phil. Mag., (5) **1**, 244–5.

Ausschluss des Kochsalzes.

     Jahresber. d. Chemie, **16** (1863), 114.

Absorptionsspectren von Salzlösungen.

     Jahresber. d. Chemie, **27** (1874), 96.

On the optical properties of rock salt.

     Langley (S. P.). Amer. Jour. Sci., **26** (1885), 477; Jour. de Phys., (2) **5**, 188 (Abs.).

Blue flame from common salt.

     Smith (A. P.). Nature, **19**, 483; **20**, 5; Chem. News, **39**, 141; Jour. Chem. Soc., **36**, 497 (Abs.).

Propriétés modulaires des pouvoirs réfringents dans les solutions salines.

     Valson (C. A.). Comptes Rendus, **76**, 224–6; Jour. Chem. Soc., (2) **11**, 460 (Abs.).

SAMARIUM.

Om Samarium.

> Clève (P. T.). Ofversigt. k. Vetensk. Akad. Förhandl., **40**, No. 7,
> 17–26; Beiblätter, **8**, 264 (Abs.); Jour. Chem. Soc., **43**, 362–70;
> Chem. News, **48**, 74–6; Ber. chem. Ges., **16**, 2493 (Abs.); Comptes
> Rendus, **97**, 94.

Mutual extinction of the spectra of yttrium and samarium.

> Crookes (W.). Comptes Rendus, **100**, 1495–7; Jour. Chem. Soc., **48**,
> 1025 (Abs.).

Remarques sur les métaux nouveaux de la gadolinite et de la samarskite;
holmium ou philippium, thulium, Samarium, décipium.

> Delafontaine. Comptes Rendus, **90**, 221.

Recherches sur le samarium, radical d'une terre nouvelle extraite de la
samarskite.

> Lecoq de Boisbaudran (F.). Comptes Rendus, **89**, 212–14; Ber. chem.
> Ges., **12**, 2160 (Abs.); Beiblätter, **3**, 872 (Abs.).

Om de lysande spectra hos Didym och Samarium.

> Thalén (R.). Ofversigt. k. Vetensk. Akad. Förhandl., **40**, No. 7, 3–16;
> Jour. de Phys., (2) **2**, 446–9; Ber. chem. Ges., **16**, 2760 (Abs.); Bei-
> blätter, **7**, 893–5 (Abs.).

SAMARSKITE.

New elements in gadolinite and samarskite.

    Crookes (W.).   Proc. Royal Soc., **40**, 502–9; Jour. Chem. Soc., **52**, 884 (Abs.).

Remarques sur la samarskite.

    Delafontaine.   Comptes Rendus, **90**, 221.

Nouvelles raies spectrales observées dans des substances extraites de la samarskite.

    Lecoq de Boisbaudran (F.).   Comptes Rendus, **88**, 322.

Sur les terres de la samarskite.

    Marignac (C.).   Comptes Rendus, **90**, 899–908.

Sur les spectres d'absorption du didyme et de quelques autres substances extraites de la samarskite.

    Soret (J. L.).   Comptes Rendus, **88**, 422–4.

SCANDIUM.

Scandium ne donne pas de spectre.
>Clève (P. T.). Comptes Rendus, **89**, 420.

Sur le scandium, élément nouveau.
>Nilson (L. F.). Comptes Rendus, **88**, 645–8; Amer. Jour. Sci., (8)
>**17**, 478 (Abs.); Beiblätter, **3**, 859 (Abs.).

On Scandium, en ny jordmetall. (Ueber Scandium, ein neues Erdmetall.)
>Nilson (L. F.). Oefversigt af k. Vetensk. Akad. Förhand., **36** III,
>45–51; Ber. chem. Ges., **12**, 554–7; Jour. Chem. Soc., **36**, 601 (Abs.);
>Beiblätter, **4**, 42 (Abs.).

Sur quelques sels caractéristiques du scandium, et sur leurs spectres.
>Nilson (L. F.). Comptes Rendus, **91**, 118.

Raies brilliantes spectrales du métal scandium.
>Thalén (R.). Comptes Rendus, **91**, 45–8; Jour. Chem. Soc., **38**, 685
>(Abs.).

Spektralundersökningar rörande Skandium, Ytterbium, Erbium och Thulium.
>Thalén (R.). Oefversigt af k. Vetensk. Akad. Förhand., **38**, No. 6,
>18–21; Jour. de Phys., (2) **2**, 85–40; Chem. News, **47**, 217 (Abs.);
>Jour. Chem. Soc., **44**, 954 (Abs.).

Spectraluntersuchungen über Scandium.
>Thalén (R.). Oefversigt k. Vetensk. Akad. Förhand. (Stockholm),
>1881, No. 6; Beiblätter, **11**, 249.

SECONDARY SPECTRUM.

Secondary Spectrum.
>Rood (O. N.). Amer. Jour. Sci., (8) **6**, 172.

SELENIUM.

Effect of light upon selenium.

> Adams (W. G.). Proc. Royal Soc., **23**, 585; Ann. Phys. u. Chem., **159**, 625.

Nouvelle note sur la propriété spécifique du sélénium à l'égard des radiations thermiques.

> Assche (F. van). Comptes Rendus, **97**, 945.

Selenium and tellurium spark spectrum; selenium and iron spark spectrum; selenium and aluminium spark spectrum; iron meteoric arc spectrum.

> Capron (J. R.). Photographed Spectra, London, 1877, p. 32, 38, 40.

Spectre du sélénium.

> Ditte. Comptes Rendus, **73**, 628.

Spectre d'absorption du vapeur de l'acide sélénieux.

> Gernez (D.). Comptes Rendus, **74**, 803; Bull. Soc. chim. Paris, n. s. **18**, 172.

Absorptionsspectrum des Bromselens und des Chlorselens.

> Jahresber. d. Chemie, **17** (1864), 109; **25** (1872), 189, 140.

Spectrum des Selens.

> . Mulder. Jour. prackt. Chemie, **91**, 111.

Spectrum von Selenwasserstoff.

> Plücker. Ann. Phys. u. Chem., **113**, 276, 278.

Spectres du sélénium et du tellure.

> Salet (G.). Comptes Rendus, **73**, 742, 743.

Ueber die Refraction und Dispersion des Selens.

> Sirks (J. L.). Ann. Phys. u. Chem., **143**, 429–39; Ann. Chim. et Phys., (4) **26**, 286 (Abs.).

SILICIUM.

Silicic fluoride spectrum; silicic quartz spectrum.

Capron (J. R.). Photographed Spectra, London, 1877, p. 75, 76.

Spectre du fluorure de silicium dans les tubes de Geissler.

Chautard (J.). Comptes Rendus, **82**, 278.

Das Aufleuchten, die Phosphorescenz und Fluorescenz des Flussspaths.

Hagenbach (E.). Naturforscherversammlung in München, 1877; Ber. chem. Ges., **10**, 2282 (Abs.).

Line spectra of boron and silicon.

Hartley (W. N.). Proc. Royal Soc , **35**, 301-4; Chem. News, **48**, 1-2; Jour. Chem. Soc., **46**, 242 (Abs.); Beiblätter, **8**, 120.

Spectrum des Phosphorescenzlichts von Flussspath.

Kindt. Ann. Phys. u. Chem., **131**, 160.

Ueber eine empfindliche spectralanalytische Reaction auf Thonerde.

Lepel (F. von). Ber. chem. Ges., **9**, 1641.

Spectres des composés de silicium.

Salet. Comptes Rendus, **73**, 1056-9.

Indices de réfraction du spath fluor.

Sarasin (E.). Arch. de Genève, (3) **10**, 308-4.

Spectre du fluorure de silicium.

Séguin (J. M.). Comptes Rendus, **54**, 993.

Spectre du silicium.

Troost et Hautefeuille. Comptes Rendus, **73**, 620; Bull. Soc. chim. Paris, n. s. **16**, 229.

Spectre du silicium sur la surface du Soleil.

Vicaire (E.). Comptes Rendus, **76**, 1540.

Absorptionsspectrum des Granats und Rubins; Erkennung von Thonerde neben Eisensalzen.

Vogel (H. W.). Ber. chem. Ges., **10**, 873-5; Jour. Chem. Soc., 1877, **2**, 269 (Abs.); Beiblätter, **1**, 242 (Abs.).

Ueber eine empfindliche spectralanalytische Reaction auf Thonerde.

Vogel (H. W.). Ber. chem. Ges., **9**, 1641.

Spectra des Fluorsiliciums und des Siliciumwasserstoffs.

Wesendonck (K.). Ann. Phys. u. Chem., n. F. **21**, 427-37; Jour. Chem. Soc., **46**, 649 (Abs.).

SILVER.

Effect of the spectrum on silver chloride.

> Abney (W. de W.). Rept. British Assoc., 1881, 594; Chem. News, 44 (1881), 184.

Effect of the spectrum on the haloid salts of silver and on mixtures of the same.

> Abney (W. de W.). Proc. Royal Soc., 33, 164-86; Jour. Chem. Soc., 42, 565 (Abs.); Chem. News, 44 (1881), 297.

Comparative effect of different parts of the spectrum on silver salts.

> Abney (W. de W.). Proc. Royal Soc., 40, 251-2; Jour. Chem. Soc., 50, 749 (Abs.); see preceding reference.

Action des rayons différemment réfrangibles sur l'iodure et le bromure d'argent; influence des matières colorantes.

> Becquerel (E.). Comptes Rendus, 79, 185-90; Jour. Chem. Soc., (2) 13, 80 (Abs.).

Silver spark spectrum; silver arc spectrum; silver and copper (alloy) arc spectrum.

> Capron (J. R.). Photographed Spectra, London, 1877, p. 42, 43.

Sur l'indice de réfraction du chlorure d'argent naturel.

> Cloiseaux (Des). Bull. Soc. minéral. de France, 5, 25.

Renversement des raies spectrales de l'argent.

> Cornu (A.). Comptes Rendus, 73, 882.

De l'action des différentes lumières colorées sur une couche de bromure d'argent impregnée de diverses matières colorantes organiques.

> Cros (Ch.). Comptes Rendus, 88, 879-81; Jour. Chem. Soc., 36, 504 (Abs.).

Les salpêtres naturels du Chili et du Pérou.

> Dieulafait. Comptes Rendus, 98, 1545-8; Chem. News, 50, 45 (Abs.).

Wellenlänge der auf Iodsilber chemisch wirkenden Strahlen.

> Eisenlohr (W.). Ann. Phys. u. Chem., 99, 162.

Salpetersaure Nickellösung als Absorptionspräparat.

> Emsmann (H.). Ann. Phys. u. Chem., Erganzungsband, 6 (1874), 334-5; Phil. Mag., (4) 46, 829-30; Jour. Chem. Soc., (2) 12, 113.

Spectre de l'azotate de l'argent.

> Gouy. Comptes Rendus, **84**, 231 ; Chem News, **35**, 107.

Spectroscopische Untersuchung der Absorptionsspectren der flüssigen Untersalpetersäure.

> Jahresber. d. Chemie, **23** (1870), 172.

Ueber das Absorptionsspectrum der flüssigen Untersalpetersäure.

> Kundt (A.). Ann. Phys. u. Chem., **141**, 157-9; Zeitsch. analyt. Chemie, (2) **7**, 64 (Abs.); Jour. Chem. Soc., (2) **9**, 185 (Abs.).

On the action of the less refrangible rays of light on silver iodide and silver bromide.

> Lea (M. Carey). Amer. Jour. Sci., (3) **9**, 269-78; Jour. Chem. Soc., 1876, **1**, 28 (Abs.).

Note on the sensitiveness of silver bromide to the green rays as modified by the presence of other substances.

> Lea (M. Carey). Amer. Jour. Sci., (8) **11**, 459-64.

On the sensitiveness to light of various salts of silver.

> Lea (M. Carey). Amer. Jour. Sci., (8) **13**, 869-71; Jour. Chem Soc., 1877, **2**, 690 (Abs.); Beiblätter, **1**, 405 (Abs.).

On the theory of the action of certain organic substances in increasing the sensitiveness of silver haloids.

> Lea (M. Carey). Amer. Jour. Sci., (8) **14**, 96-9; Beiblätter, **1**, 563 (Abs.).

Azotate de l'argent en solution, étincelle.

> Lecoq de Boisbaudran (F.). Spectres Lumineux, Paris, 1874, p. 167, planche XXV.

Ueber die Lichtempfindlichkeit der Silberhaloïdsalze und den Zusammenhang von optischer und chemischer Licht.

> Schultz-Selback (C.). Ann. Phys. u. Chem., **143**, 161-71; Ber. chem. Ges., **4**, 210 (Abs.); Jour. Chem. Soc., (2) **9**, 302 (Abs.); Phil. Mag., (4) **41**, 549 (Abs.); Ann. Chim. et Phys., (4) **26**, 280 (Abs.).

Chemische und mechanische Veränderung der Silberhaloïdsalze durch das Licht.

> Schultz-Selback (C.). Ann. Phys. u. Chem., **143**, 439-49; Ber. chem. Ges., **4**, 343-5; Phil. Mag., (4) **41**, 550-2.

Bestimmung der Salpetersäure und Phosphorsäure auf spectralanalytischem Wege.

> Settegast (H.). Zeitschr. analyt. Chemie, **20**, 116-17.

Azione dei raggi solari sui composti aloidi d'argento.

> Tommasi (D.). Rend. del R. Ist. Lomb., **11**, 652-8; Beiblätter, **3** 621-2 (Abs.).

Sur la radiation de l'argent au moment de sa solidification.

> Violle (J.). Comptes Rendus, **96**, 1083–5; Chem. News, **47**, 213 (Abs.); Beiblatter, **7**, 457 (Abs.).

Ueber die Lichtempfindlichkeit des Bromsilbers für die sogenannten chemisch unwirksamen Farben.

> Vogel (H. W.). Ber. chem. Ges., **6**, 1302-6; Ann. Phys. u. Chem. **150**, 458–9; Jour. Chem. Soc., (2) **12**, 217 (Abs.); Amer. Jour. Sci. (3) **7**, 140–1; Phil. Mag., (4) **47**, 273–77; Bull. Soc. chim. Paris, n s. **21**, 233.

Ueber die chemische Wirkung des Lichtes auf reines und gefärbtes Bromsilber.

> Vogel (H. W.). Ber. chem. Ges., **8**, 1635–6; Jour. Chem. Soc., 1876, **1**, 510 (Abs.); Amer. Jour. Sci., (3) **11**, 215–16 (Abs.).

Neue Beobachtungen über die Lichtempfindlichkeit des Bromsilbers.

> Vogel (H. W.). Ber. chem. Ges., **9**, 667–70; Jour. Chem. Soc., 1876, **2**, 265 (Abs.).

Ueber die Empfindlichkeit trockner Bromsilberplatten gegen das Sonnenspectrum.

> Vogel (H. W.). Ber. chem. Ges., **14**, 1024-8; Jour. Chem. Soc., **40**, 773 (Abs.); Beiblätter, **5**, 521 (Abs.).

Ueber die verschiedenen Modificationen des Bromsilbers und Chlorsilbers.

> Vogel (H. W.). Ber. chem. Ges., **16**, 1170–9; Beiblätter, **7**, 536 (Abs.).

Ueber die chemische Wirkung des Sonnenspectrums auf Silberhaloïdsalze.

> Vogel (H. W.). Ann. Phys. u. Chem., **153**, 218–50; Jour. Chem. Soc., (2) **13**, 326 (Abs.).

Ueber die Brechung und Dispersion des Lichtes in Iod-, Brom-und Chlor-Silber.

> Wernicke (W.). Ann. Phys. u. Chem., **142**, 560–73; Jour. Chem. Soc., (2) **9**, 653–4 (Abs.); Ann. Chim. et Phys., (4) **26**, 287 (Abs.).

SODIUM.

Spectrum of sodium.

> Abney (W. de W.). Chem. News, **44**, 8.

Note on the spectrum of sodium.

> Abney (W. de W.). Proc. Royal Soc., **32**, 448.

Reversal of the sodium lines.

> Ackroyd (W.). Chem. News, **36**, 164–5.

Lumière jaune de la flamme de sodium.

> Becquerel (H.). Comptes Rendus, **90**, 1407.

Spectronatromètre.

> Champion (P.), Pellet (H.) et Grenier (M.). Comptes Rendus, **76**, 707–11; Jour. Chem. Soc., (2) **11**, 984–5 (Abs.).
> (Look below, under Janssen.)

Spectre de la soude dans les tubes de Geissler.

> Chautard (J.). Comptes Rendus, **82**, 278.

Renversement des raies spectrales du sodium.

> Cornu (A.). Comptes Rendus, **73**, 882; Jour. de Phys., **1**, 206.

Ueber die Opacität der gelben Natronflamme für Licht von ihrer eignen Farbe.

> Crookes (W.). Ann. Phys. u. Chem., **112**, 844.

Indices de réfraction des dissolutions aqueuses d'acide acétique et d'hyposulfite de soude.

> Damien. Comptes Rendus, **91**, 823–5; Beiblätter, **5**, 41.

Das Verhältniss der Intensitäten der beiden Natriumlinien.

> Dietrich (W.). Ann. Phys. u. Chem., n. F. **12**, 519.

Spectre de sodium.

> Fizeau (H.). Comptes Rendus, **54**, 498; Ann. Phys. u. Chem., **116**, 492.

Recherches photométriques sur le sodium.

> Gouy. Comptes Rendus, **83**, 269; **85**, 70; **86**, 878, 1078.

Ueber ein einfaches Verfahren die Umkehrung der farbigen Linien der Flammenspectra, insbesondere der Natriumlinie, subjectiv dazustellen.

> Günther (O.). Ann. Phys. u. Chem., n. F. **2**, 477.

22 T

Sur l'emploi de la lumière monochromatique, produite par les sels de soude, pour apprécier les changements de couleur de la teinture de tournesol, dans les essais alkalimétriques.

> Henry (L. d'). Comptes Rendus, **76**, 222-4; Ann. Chem. u. Pharm., **169**, 272; Dingler's Jour., **207**, 405-7.

Soda flames in coal fires.

> Herschel (J.). Nature, **27**, 78, 108.

Spectrum des Natriums.

> Jahresber. d. Chemie, **15** (1862), 29, 30.

Umkehrung der hellen Spectrallinien der Metalle, insbesondere des Natriums, in dunkle.

> Jahresber. d. Chemie, **18** (1865), 90.

Note sur l'analyse spectrale quantitative, à propos de la communication précédente de M. M. Champion, Pellet et Grenier.

> Janssen (J.). Comptes Rendus, **76**, 711-13; Jour. Chem. Soc., (2) **11**, 1258 (Abs.).

Chemische Analyse durch Spectralbeobachtungen; Linien von Natrium.

> Kirchhoff (G.) und Bunsen (R.). Ann. Phys. u. Chem., **110**, 161-87.

Ueber anomale Dispersion im glühenden Natriumdamp.

> Kundt (A.). Ann. Phys. u. Chem., n. F. **10**, 321-5; Phil. Mag., (5) **10**, 58-7.

Sulfate de soude fondu, étincelle; sels de soude dans le gaz; sels de soude et de lithine dans le gaz.

> Lecoq de Boisbaudran (F.). Spectres Lumineux, Paris, 1874, p. 54, 55, planche V, VI.

Reversal of the lines of the metallic vapours, sodium.

> Liveing and Dewar. Nature, **24**, 206; **26**, 466.

On the spectra of sodium and potassium.

> Liveing (G. D.) and Dewar (J.). Proc. Royal Soc., **29**, 398-402; Beiblätter, **4**, 368 (Abs.).

Note on some phenomena attending the reversal of lines.

> Lockyer (J. N.). Proc. Royal Soc., **28**, 428-32; Beiblätter, **3**, 608 (Abs.).

Note on the spectrum of sodium.

> Lockyer (J. N.). Proc. Royal Soc., **29**, 140; Chem. News, **39**, 243.

Spectrum of sodium at elevated temperatures.

> Lockyer (J. N.). Chem. News, **30**, 98.

Sur les raies de la vapeur de sodium.
> Lockyer (J. N.). Comptes Rendus, 88, 1124.

Die Natriumline gehört dem Metall an.
> Mitscherlich (A.). Ann. Phys. u. Chem., 116, 505.

Absorption spectra of sodium and potassium at low temperatures.
> Roscoe (H. E.) and Schuster (A.). Proc. Royal Soc., 22, 362.

Indice du quartz pour les raies du sodium.
> Sarasin (Éd.). Comptes Rendus, 85, 1280.

Et spectres du fer et quelques autres métaux dans l'arc voltaïque; sodium.
> Secchi (A.). Comptes Rendus, 77, 173; Chem. News, 28, 82.

Spectre du sodium.
> Secchi (A.). Comptes Rendus, 82, 275.

Propriétés optiques de sous carbonate de soude et de hyposulfite de soude.
> Senarmont (H. de). Ann. Chim. et Phys., (3) 41, 336.

Sur le déplacement des raies du sodium, observé dans le spectre de la grande comète de 1882.
> Thollon et Gouy. Comptes Rendus, 96, 371.

Leichte Umkehrung der Natriumlinie.
> Weinhold (A.). Ann. Phys. u. Chem., 142, 321; Phil. Mag., (4) 41, 404.
> (See Soret. Arch. de Genève, (2) 41, 64–5.)

Sur la dispersion du chromate de soude à 4 H₂ O.
> Wyrouboff (G.). Bull. Soc. minéral. de France, 5, 160–1.

Re-reversal of sodium lines.
> Young (C. A.). Nature, 21, 274–5; Beiblätter, 4, 370.

STRONTIUM.

Ueber den Einfluss der Temperatur auf die Brechungsexponenten der naturlichen Sulfate des Baryum, Strontium und Blei.

> Arzruni (A.). Zeitschr. Krystallogr. u. Mineral., **1**, 165–192; Jahrb. f. Mineral., 1877, 526 (Abs.); Jour. Chem. Soc., **34**, 189 (Abs.).

Strontium spark spectrum.

> Capron (J. R.). Photographed Spectra, London, 1877, p. 44.

La strontiane dans les eaux minérales de Contrexeville et Schinznach (Suisse).

> Dieulafait. Comptes Rendus, **95**, 999–1001; Jour. Chem. Soc., **44**, 801 (Abs.).

Recherches photométriques sur le strontium.

> Gouy. Comptes Rendus, **83**, 269.

Spectre de chlorure de strontium.

> Gouy. Comptes Rendus, **84**, 231.

Recherches photométriques; spectre du strontium.

> Gouy. Comptes Rendus, **85**, 70.

Sur les caractères des flammes chargées du chlorure de strontium.

> Gouy. Comptes Rendus, **85**, 489.

Spectre continu du strontium.

> Gouy. Comptes Rendus, **86**, 878, 1078.

Spectrum von Strontium.

> Jahresber. d. Chemie, **23** (1870), 174.

Chlorure de strontium en solution, étincelle; dans le gaz; dans le gaz chargé de H Cl.

> Lecoq de Boisbaudran (F.). Spectres Lumineux, Paris, 1874, p. 69, planche IX; p. 72 et 75, planche X.

Linien von Strontium.

> Kirchhoff (G.) und Bunsen (R.). Ann. Phys. u. Chem., **110**, 174.

SULPHUR.

On the violet phosphorescence in calcium sulphide.

> Abney (W. de W.). Proc. Physical Soc., **5**, 85–8; Nature, **35**, 355
> (Abs.); Phil. Mag., (5) **13**, 212–14; Jour. Chem. Soc., **42**, 677
> (Abs.); Beiblätter, **6**, 388 (Abs.); Jour. de Phys., (2) **2**, 287 (Abs.).

Spectres des gaz simples; soufre.

> Angström (A. J.). Comptes Rendus, **73**, 369; Ann. Phys. u. Chem.,
> **94**, 159.

Spectre du sulfure de carbone.

> Becquerel (H.). Comptes Rendus, **85**, 1227.

Sulphur spectrum, sulphuric acid spectrum, sulphur quartz spectrum.

> Capron (J. R.). Photographed Spectra, London, 1877, p. 68, 74, 75.

Spectrum von Schwefel.

> Dibbits (H. C.). Ann. Phys. u. Chem., **122**, 527–34.

Spectre du soufre.

> Ditte (A.). Comptes Rendus, **73**, 622–4; Bull. Soc. chim. Paris, n. s.
> **16**, 229.

Spectres d'absorption des vapeurs de soufre.

> Gernez (D.). Comptes Rendus, **74**, 803; Bull. Soc. chim. Paris, n. s.
> **17**, 259.

Spectre de sulfate de thallium.

> Gouy. Comptes Rendus, **84**, 831.

Sulfate acide.

> ¶ Gouy. Comptes Rendus, **85**, 70.

Spectrum of murexide.

> Hartley (W. N.). Jour. Chem. Soc., **51** (1887), 199–200.

Spectrum des Schwefels.

> Jahresber. d. Chemie, **16** (1863), 110; **17** (1864), 109; **22** (1869), 181;
> **23** (1870), 173; **25** (1872), 139, 141; **28** (1875), 122.

Spectre du sulfure de plomb.

> Lallemand (A.). Comptes Rendus, **78**, 1272.

Sur la diffusion lumineuse du sulfure de cuivre obtenu sans précipitation.

> Lallemand (A.). Comptes Rendus, **79**, 693.

Die Absorptionsstreifen in Prismen von Schwefelkohlenstoff.

     Lamansky (S.).   Ann. Phys. u. Chem., **146**, 213, 215.

Sur les spectres des vapeurs aux températures élévées ; spectre du soufre.

     Lockyer (J. N.).   Comptes Rendus, **78**, 1790; Nature, **30**, 78; Chemical News, **30**, 98.

Spectrum des Schwefels, Schwefelkohlenstoffs, Schwefelwasserstoffs und Selens.

     Mulder.   Jour. prackt. Chemie, **91**, 111.

Sulla refrazione atomica dello zolfo.

     Nasini (R.).   Gazz. chim. ital., **13**, 296–311; Jour. Chem. Soc., **46**, 149–51 (Abs.); Ber. chem. Ges., **15**, 2878–92; Beiblätter, **7**, 281 (Abs.).

Dampf des wasserfreien Schwefelsäure.

     Plücker.   Ann. Phys. u. Chem., **113**, 276, 278.

Spectrum des Muroxids.

     Reynolds.   Jour. prackt. Chemie, **105**, 359.

De la flamme du soufre, et des diverses lumières utilisables en photographie.

     Riche (A.) et Brady (C.).   Comptes Rendus, **80**, 238–41; Ber. chem. Ges., **8**, 182 (Abs.).

Recherche du soufre par le spectroscope.

     Salet (G.).   Comptes Rendus, **68**, 404; Bull. Soc. chim. Paris, n. s. **11**, 302; Ann. Phys. u. Chem., **137**, 171.

Spectre du soufre.

     Salet (G.).   Comptes Rendus, **73**, 559.

Recherche du soufre et du phosphore par le spectroscope.

     Salet (G.).   Bull. Soc. chim. Paris, n. s. **13**, 289.

Sur la réaction spectroscopique du soufre et sur la flamme de l'hydrogène.

     Salet (G.).   Bull. Soc. chim. Paris, n. s. **14**, 182.

Sur le spectre d'absorption de la vapeur du soufre.

     Salet (G.).   Comptes Rendus, **74**, 865–6; Jour. Chem. Soc., (2) **10**, 882 (Abs.); Ber. chem. Ges., **5**, 828 (Abs.).

Sur les spectres du phosphore et du soufre.

     Séguin (J. M.).   Comptes Rendus, **53**, 1272.

Propriétés optiques d'hyposulfite de soude.

     Sénarmont (H. de).   Ann. Phys. u. Chem., (8) **41**, 336.

TELLURIUM.

Tellurium spark spectrum.

Capron (J. R.).   Photographed Spectra, London, 1877, p. 20, 40, 45.

Spectre du tèllure.

Ditte (A.).   Comptes Rendus, **73**, 622–24.

Sur les spectres d'absorption de tellure, de protochlorure et de protobromure de tellure.

Gernez (D.).   Comptes Rendus, **74**, 1190–2; Jour. Chem. Soc., (2)
10, 665 (Abs.); Phil. Mag., (4) **43**, 478–5; Amer. Jour. Sci., (3) **4**,
59 (Abs.); Bull. Soc. chim. Paris, n. s. **18**, 172.

Spectrum des Tellurs.

Jahresber. d. Chemie, **25** (1872), 140.

Spectre du tellure.

Salet (G.).   Comptes Rendus, **73**, 744.

TERBIUM.

Absorptionsspectrum von Terbiumlösungen.

Delafontaine.   Jour. prackt. Chemie, **94**, 808.

Vergleich der Absorptionsspectra von Didym, Erbium und Terbium.

Delafontaine.   Ann. Phys. u. Chem., **124**, 685; Chem. News, **11**, 253;
Ann. Chim. et Phys., **135**, 194.

Sur un spectre électrique particulier aux terres rares du groupe terbique.

Lecoq de Boisbaudran (F.).   Comptes Rendus, **102**, 153–55; Jour.
Chem. Soc., **50**, 298 (Abs.).

THALLIUM.

Thallium and indium spark spectrum.

    Capron (J. R.).   Photographed Spectra, London, 1877, p. 45, 47.

Renversement des raies spectrales du thallium.

    Cornu (A.).   Comptes Rendus, **73**, 832.

Discovery of thallium.

    Crookes (W.).   Chem. News, **3**, 193.

Thallium and its compounds.

    Crookes (W.).   Jour. Chem. Soc., **17**, 112.

Recherches photométriques sur le thallium.

    Gouy.   Comptes Rendus, **83**, 269.

Spectre de sulfate de thallium.

    Gouy.   Comptes Rendus, **84**, 231.

Spectrum des Thalliums und der Thalliumsalzen.

    Jahresber. d. Chemie, **16** (1863), 112; **26** (1873), 152, 158.

Sur le thallium, nouveau métal dont l'analyse spectrale a fait connaîtr l'existence.

    Lamy (A.).   Comptes Rendus, **54**, 1255; Ann. Chim. et Phys., (3) **67** 885; Ann. Phys. u. Chem., **116**, 495.

Moyen de constater une empoisonnement par le thallium.

    Lamy (A.).   Comptes Rendus, **57**, 442.

Sels de thallium dans le gaz.

    Lecoq de Boisbaudran (F.).   Spectres Lumineux, Paris, 1874, p. 141, planche XXI.

Spectre de thallium.

    Lecoq de Boisbaudran (F.).   Comptes Rendus, **77**, 1152; Bull. Soc chim. de Paris, n. s. **21**, 125.

Note on the spectrum of thallium.

    Miller (W. A.).   Proc. Royal Soc., **12**, 407.

Sur la raie spectrale du thallium.

    Nicklés.   Comptes Rendus, **58**, 182; Ann. Phys. u. Chem., **121**, 836.

Spectre du thallium dans l'arc voltaïque.

    Secchi (A.).   Comptes Rendus, **77**, 178.

THULIUM.

• **Spectre de thulium.**

Clève (P. T.). Comptes Rendus, **89**, 478; **91**, 828.

Remarques sur le thulium.

Delafontaine. Comptes Rendus, **90**, 221.

Examen spectral du thulium.

Thalén (R.). Comptes Rendus, **91**, 376–8; Jour. Chem. Soc., **40**, 849–50 (Abs.); Beiblätter, **4**, 789 (Abs.).

Spectralundersökningar rörande Skandium, Ytterbium, Erbium och Thulium.

Thalén (R.). Oefversigt af k. Vetensk. Acad. Förhand., **38**, No. 6, 18–21; Jour. de Phys., (2) **2**, 35–40; Chem. News, **47**, 217 (Abs.); Jour. Chem. Soc., **44**, 954 (Abs.).

TIN.

Tin arc spectrum; tin and zinc spark spectrum; tin chloride spectrum.

Capron (J. R.). Photographed Spectra, London, 1877, p. 49, 76.

Bichlorure d'étain en solution, étincelle.

Lecoq de Boisbaudran (F.), Paris, 1874, p. 148, planche XXII.

Spectres d'étain et ses composés.

Salet (G.). Comptes Rendus, **73**, 862–3; Jour. Chem. Soc., (2) **9**, 1147–9 (Abs.).

TITANIUM.

Spectre du bichlorûre de titanium.

> Becquerel (H.). Comptes Rendus, **85**, 1227.

Titanium spark spectrum; titanium, aluminium, and palladium spark spectrum; titanium arc spectrum.

> Capron (J. R.). Photographed Spectra, London, 1877, p. 47.

Spectre du titanium.

> Troost et Hautefeuille. Comptes Rendus, **73**, 620; Bull. Soc. chim. Paris, n. s. **16**, 229.

Coïncidence of the spectrum lines of iron, calcium, and titanium.

> Williams (W. Matthieu). Nature, **8**, 46.

URANIUM.

Analyse de la lumière émise par les composés d'uranium phosphorescents.

> Becquerel (E.). Comptes Rendus, **75**, 296–308; Jour. Chem. Soc., (2) **11**, 25 (Abs.); Amer. Jour. Sci., (3) **4**, 486 (Abs.).

Relation entre l'absorption et la phosphorescence des composés d'uranium.

> Becquerel (H.). Comptes Rendus, **101**, 1252–6; Jour. Chem. Soc., **50**, 189 (Abs.).

Uranium arc spectrum.

> Capron (J. R.). Photographed Spectra, London, 1877, p. 50.

Anwendung der dunklen Linien des Spectrums als Reagens auf Uransäure.

> Jahresber. d. Chemie, **5** (1862), 125.

Absorptionsspectren der Uransalzen.

> Jahresber. d. Chemie, **26** (1873), 158.

Investigation of the fluorescent and absorption spectra of the uranium salts.

> Morton (H.) and Bolton (H. C.). Chem. News, **28**, 47–50, 113–16, 164–7, 233–4, 244–6, 257–9, 268–70; **29**, 17–19; Jour. Chem. Soc., (2) **12**, 12–13 (Abs.), 642 (Abs.).

On some remarkable spectra of compounds of zirconia and of the oxides of uranium.

> Sorby (H. C.). Proc. Royal Soc., **18**, 197; Ber. chem. Ges., **3**, 146.

Spectra der Uranlösungen.

> Thudichum. Jour. prackt. Chemie, **106**, 415.

Absorption spectrum of uranine.

> Wiley (H. W.). Amer. Chem. Jour., **1**, 211.

Untersuchungen über das Uran.

> Zimmermann (C.). Ann. Phys. u. Chem., **213**, 285–329; Chem. News, **46**, 172 (Abs.); Zeitschr. analyt. Chemie, **23**, 220 (Abs.).

VANADIUM.

Vanadium arc spectrum.

> Capron (J..). Photographed Spectra, London, 1877, p. 50.

VIOLET AND ULTRA-VIOLET.

Sur l'absorption des rayons ultra-violets par quelques milieux.
> Chardonnet (E. de). Comptes Rendus, **93**, 406.

Vision des radiations ultra-violettes.
> Chardonnet (E. de). Comptes Rendus, **96**, 509–71; Jour. de Phys.,
> **12**, 219.

Sur l'absorption atmosphérique des radiations ultra-violettes.
> Cornu (A.). Jour. de Phys., **10**, 5–16.

Erklärung der ultra-violetten Strahlen des Spectrums.
> Eisenlohr (W.). Ann. Phys. u. Chem., **93**, 628.

Note upon certain photographs of the ultra-violet spectra of elementary
bodies.
> Hartley (W. N.). Jour. Chem. Soc., **41**, 84–90; Chem. News, **43**, 289
> (Abs.); Beiblätter, **5**, 659 (Abs.); **6**, 789 (Abs.).

Investigation by means of photography of the ultra violet spark spectra
emitted by metallic elements and their combinations under varying conditions.
> Hartley (W. N.). Chem. News, **48**, 195; note on the above by
> Wiedemann (E.), Chem. News, **49**, 117; Jour. Chem. Soc., **46**, 801
> (Abs.); Beiblätter, **8**, 581 (Abs.).

Visibility of the ultra-violet rays of the spectrum.
> Herschel (A. S.). Nature, **16**, 22–8.

On the ultra-violet spectra of the elements.
> Liveing (G. D.) and Dewar (J.). Phil. Trans., **174**, 187–222; Proc.
> Royal Soc., **34**, 122 (Abs.); Beiblätter, **6**, 934 (Abs.); **7**, 598, 849–56
> (Abs.); Jour. Chem. Soc., **44**, 262 (Abs.); Proc. Royal Institution,
> **10**, 245–52.

Notes on the absorption of ultra-violet rays by various substances.
> Liveing (G. D.) and Dewar (J.). Proc. Royal Soc., **35**, 71.

Détermination des longueurs d'onde des rayons lumineux et des rayons
ultra-violets.
> Mascart. Comptes Rendus, **58**, 1111.

Visibilité des rayons ultra-violets.
> Mascart. Comptes Rendus, **68**, 402; Ann. Phys. u. Chem., **137**, 163.

Spectres ultra-violets.

> Mascart. Comptes Rendus, **69**, 887.

Sur les moyens propres à la réproduction photographique des spectres ultra-violets des gaz.

> Monckhoven (van). Bull. Acad. Belgique, (2) **43**, 187–92; Beiblätter, **1**, 286 (Abs.).

Fluorescence and the violet end of a projected spectrum.

> Morton (Henry). Chem. News, **27**, 88.

Photographie des durch ein Quarzprisma erhaltenen ultra-violetten Theils des Spectrums.

> Müller (J.). Ann. Phys. u. Chem., **109**, 151.

A comparison of the maps of the ultra-violet spectrum.

> Pickering (E. C.). Amer. Jour. Sci., (8) **32**, 223–6; Beiblätter, **11** (1887), 145 (Abs.).

On the lower limit of the prismatic spectrum, with especial reference to some observations of Sir J. Herschel.

> Rayleigh (Lord). Phil. Mag., (5) **4**, 348–53; Beiblätter, **1**, 682 (Abs.).

Report on the ultra-violet spark spectra emitted by metallic elements.

> Report of the British Association, 1882, p. 148, presented by Prof. Hartley; Nature, **26**, 458.

Nicht alle Quarzprismen verlängern das Spectrum am ultravioletten Ende.

> Salm-Horst (Der Fürst zu). Ann. Phys. u. Chem., **109**, 158.

Experimente über die Sichtbarkeit ultra-violetter Strahlen.

> Sauer (L.). Ann. Phys. u. Chem., **155**, 602.

Ueber ultra-violette Strahlen.

> Schönn (J. L.). Ann. Phys. u. Chem., n. F. **9**, 488–92; **10**, 143–8.

Der ultra-violette Theil des Spectrums lässt sich unmittelbar sichtbar machen.

> Seculic (M.). Ann. Phys. u. Chem., **146**, 157.

Recherches sur l'absorption des rayons ultra-violets par diverses substances.

> Soret (J.). Comptes Rendus, **86**, 708, 1062–4; Arch. de Genève, (2) **63**, 89–112; (8) **4**, 261–92, 377–81; **10**, 429–94; Beiblätter, **2**, 410 (Abs.); **3**, 196 (Abs.); **5**, 124 (Abs.); Jahresber. d. Chemie (1873), 154.

Sur la transparence des milieux de l'œil pour les rayons ultra-violets.

> Soret (J. L.).   Comptes Rendus, **88**, 1012.

Spectres d'absorption ultra-violets des éthers azotiques et azoteux.

> Soret (J. L.) et Rilliet (Alb. A.).   Comptes Rendus, **89**, 747.

Sur la visibilité des rayons ultra-violets.

> Soret (J. L.).   Comptes Rendus, **97**, 814.

Sur l'absorption des rayons ultra-violets par les milieux de l'œil et par quelques autres substances.

> Soret (J. L.).   Comptes Rendus, **97**, 572, 642.

The Change of Refrangibility of Light.   (Gives a drawing of the fixed lines in the solar spectrum in the extreme violet and in the invisible region beyond.)

> Stokes (G. G.).   Phil. Trans. for 1852, part II, 463.

Visibilité des rayons ultra-violets, à l'aide du parallelipipède de dispersion.

> Zenger (Ch. V.).   Comptes Rendus, **98**, 1017.

VOLCANOES.

Observations on Mt. Etna.

> Langley (S. P.).   Amer. Jour. Sci., (8) **20**, 88-4 ; Beiblätter, **4**, 790 (Abs.).

Recherches spectroscopiques sur les fumerolles de l'éruption du Vesuve en avril 1872.

> Palmieri (L.).   Comptes Rendus, **76**, 1427-8.

WATER SPECTRA.

Colour of the Mediterranean and other waters.

Aitken (J.). Proc. Royal Soc. Edinburgh, **11**, 472–88; Jour. Chem. Soc., **42**, 1017 (Abs.); Beiblätter, **6**, 879 (Abs.).

Note on the absorption of sea-water.

Aitken (J.). Proc. Royal Soc. Edinburgh, **11**, 637; Beiblätter, **7**, 872 (Abs.).

Évaporation de l'eau sous l'influence de la radiation solaire ayant traversé des verres colorés.

Baudrimont (A.). Comptes Rendus, **89**, 41–3.

Spectre de l'eau.

Becquerel (H.). Comptes Rendus, **85**, 1227.

The spectroscope in water analysis.

Church (A. H.). Chem. News, **22**, 322.

Indices de réfraction de l'eau en surfusion.

Damien (B. C.). Jour. de Phys., **10**, 198–202.

Untersuchungen einiger Wässer.

Dibbits. Jour. prackt. Chemie, **92**, 38, 50.

Spectre lumineux de l'eau.

Huggins (W.). Comptes Rendus, **90**, 1455.

Spectres d'absorption de la vapeur d'eau.

Janssen (J.). Comptes Rendus, **56**, 538; **60**, 218; **63**, 289; **78**, 995; **95**, 885; Phil. Mag., (4) **32**, 815; Ann. Chim. et Phys., (4) **24**, 215–17; Jour. Chem. Soc., (2) **10**, 280 (Abs.); Jahresber. d. Chemie (1866), 76.

Spectre de la vapeur d'eau.

Lecoq de Boisbaudran (F.). Comptes Rendus, **74**, 1050.

Spectrum of water.

Liveing (G. D.) and Dewar (J.). Proc. Royal Soc., **30**, 580; **33**, 274–6; Jour. Chem. Soc., **44**, 140 (Abs.); Beiblätter, **6**, 481 (Abs.).

Sur la réfraction de l'eau comprimée.

Mascart. Comptes Rendus, **78**, 801–5; Amer. Jour. Sci., (3) **7**, 598; Ann. Phys. u. Chem., **153**, 154–8.

Studî spettrali sub colore delle acque, nota seconda.

>   Riccò (A.).   Mem. Spettr. ital., **8**, 1–10.

Ueber die Absorption des Lichts durch Wasser, etc.

>   Schönn (J. L.).   Ann. Phys. u. Chem., Ergänzungsband, 1878, **8**, 670–5; Jour. Chem. Soc., **34**, 693 (Abs.).

Observations relatives à une communication de M. Crocé-Spinelli sur les bandes de la vapeur d'eau dans le spectre solaire.

>   Secchi (A.).   Comptes Rendus, **78**, 1080.

Sur la couleur de l'eau.

>   Soret (J. L.).   Arch. de Genève, (3) **11**, 276–96; Beiblätter, **8**, 505 (Abs.); Jour. de Phys., **13**, 427.

Spectre d'absorption de l'eau.

>   Soret (J. L.) et Sarasin (Ed.).   Comptes Rendus, **98**, 624; Amer. Jour. Sci., (3) **27**, 485.

Ueber die Absorption des Seewassers.

>   Vogel (H. W.).   Beiblätter, **7**, 582.

WAVE-LENGTHS.

Wave-lengths of A, a and lines in the infra-red of the visible spectrum.

> Abney (W. de W.). Nature, **29**, 190; Chem. News, **48**, 283; Comptes Rendus, **97**, 1206.

Corrections to the computed lengths of waves of light, published in the Philosophical Transactions of the year 1868.

> Airy (G. B.). Phil. Trans., 1872, **142**, 89–109; Proc. Royal Soc., **20**, 21–2 (Abs.).

Wellenlänge Messungen.

> Angström (A. J.). Ann. Phys. u. Chem., **123**, 489; Jahresber. d. Chemie (1865), 85.

La détermination des longueurs d'onde des rayons de la partie infra-rouge du spectre au moyen des effets de phosphorescence.

> Becquerel (E.). Comptes Rendus, **77**, 302; Jahresber. d. Chemie (1873), 160.

Phosphorographie de la région infra-rouge du spectre solaire; longueur d'onde des principales raies.

> Becquerel (H.). Comptes Rendus, **96**, 121.

On the absolute wave-length of light.

> Bell (Louis). Phil. Mag., (5) **23** (1887), 265–82; Amer. Jour. Sci., (3) **33**, 167–82.

Photometrische Untersuchungen.

> Bohn (C.). Ann. Phys. u. Chem., Ergänzungsband, **6** (1874), 386.

Détermination des longueurs d'onde des radiations très réfrangibles.

> Cornu (A). Jour. de Phys., **10**, 425.

Étude spectrométrique de quelques scources lumineuses.

> Crova (A.). Comptes Rendus, **87**, 322.

Comparaison photométrique des scources lumineuses des teintes différentes.

> Crova (A.). Comptes Rendus, **93**, 512; Ann. Chim. et Phys., (6) **6**, 528–45.

Détermination des longueurs d'onde des rayons calorifiques à basse température dans le spectre.

> Desaines (P.) et Curie (P.). Comptes Rendus, **90**, 1506.

Wellenlänge der Fraunhofer Linien.

> Ditscheiner (L.). Ber. d. Wiener Akad., Bd. II, Abth. **1**, 296; Amer. Jour. Sci., (3) **3**, 297–9.

23 T

Die brechbarsten oder unsichtbaren Lichtstrahlen im Beugungspectrum und ihre Wellenlänge.

> Eisenlohr (W.). Ann. Phys. u. Chem., 98, 353 ; 99, 159–62.

Eine Wellenmessung im Spectrum jenseits des Violetts.

> Esselbach (E.). Ann. Phys. u. Chem., 98, 513.

Les vibrations de la matière et les ondes de l'éther dans les combinations photochimiques.

> Favé. Comptes Rendus, 86, 560–5.

On the normal solar spectrum. (Gives the wave-lengths of the principal lines of the solar spectrum.)

> Gibbs (Wolcott). Amer. Jour. Sci., 93, 1.

On the measurement of wave-lengths by means of indices of refraction.

> Gibbs (Wolcott). Amer. Jour. Sci., March, 1869; Phil. Mag., (4) 50, 177. [See also Rep'ts British Association for 1881 and 1884.]

Recherches photométriques sur les flammes colorées.

> Gouy. Comptes Rendus, 83, 269–272; 85, 70, 439; 86, 878, 1078; Ann. Chim. et Phys., (5) 18, 5–101.

Measurements of the wave-lengths of lines of high refrangibility in the spectra of elementary substances.

> Hartley (W. N.) and Adeney (W. E.). Phil. Trans., 175, 63–137; Proc. Royal Soc., 35, 148 (Abs.); Chem. News, 47, 193 (Abs.); Beiblätter, 7, 599 (Abs.).

Zur Reduction der Kirchhoff'schen Spectralbeobachtungen auf Wellenlängen.

> Hasselberg (B.). Bull. Acad. St. Pétersbourg, 25, 131–46; Beiblätter, 3, 79.

Note sur l'analyse spectrale.

> Janssen (J.). Comptes Rendus, 76, 711–13; Jour. Chem. Soc., (2) 11, 1258 (Abs.).

Photometrische Untersuchungen.

> Ketteler (E.) und Pulfrich (C.). Ann. Phys. u. Chem., n. F. 15, 337–378; Amer. Jour. Sci., (3) 23, 486 (Abs.); Monatsber. d. Berliner Acad. (1864), 682.

Ueber die Empfindlichkeit des normalen Auges für Wellenlängenunterschiede des Lichtes.

> König (A.) und Dieterici (C.). Ann. Phys. u. Chem , n. F. 22, 579–89; Jour. de Phys., (2) 4, 323 (Abs.).

Mesure de l'intensité photométrique des raies spectrales.

Lagarde (H.). Comptes Rendus, **95**, 1850.

Recherches photométriques sur le spectre de l'hydrogène.

Lagarde (H.). Ann. Chim. et Phys., (6) **4**, 248-369, planche.

Wave-lengths in the invisible spectrum.

Langley (S. P.). Trans. National Acad. Sci. (1883); Amer. Jour. Sci., (3)·**27**, 169; (3) **30**, 480; Ann. Chim. et Phys., (6) **2**, 145; Ann. Phys. u. Chem., n. F. **22**, 598.

On hitherto unrecognized wave-lengths.

Langley (S. P.). Amer. Jour. Sci., (3) **32**, 83; Phil. Mag., (5) **22** (1886), 149.

Courbe représentant le rapport des longueurs d'ondes aux divisions de mon micromètre.

Lecoq de Boisbaudran (F.). Spectres Lumineux, Paris, 1874, p. 194, planche XXIX.

Comparaison photométrique des diverses parties du même spectre.

Macé de Lépinay (J.). Ann. Chim. et Phys., (5) **24**, 289; **30**, 145; Jour. de Phys., **12**, 64.

Sur une méthode pratique pour la comparaison spectroscopique des scources usuelles diversement colorées.

Macé de Lépinay (J.). Comptes Rendus, **97**, 1428.

Méthode pour mesurer, en longueurs d'onde, de petites épaisseurs.

Macé de Lépinay (J.). Ann. Chim. et Phys., (6) **10**, 68-84; Jour. de Phys., (2) **5**, 405-11.

Détermination de la longueur d'onde de la raie A du spectre.

Mascart. Comptes Rendus, **56**, 188.

Détermination des longueurs d'onde des rayons lumineux et des rayons ultra-violets.

Mascart. Comptes Rendus, **58**, 1111.

Longueurs d'onde de quelques métaux.

Mascart. Ann. de l'École normale, **4** (1866).

Spectralphotometrische Untersuchungen einiger photographischer Sensibilisatoren.

Messerschmidt (J. B.). Ann. Phys. u. Chem., (2) **25**, 655-74; Jour. Chem. Soc., **48**, 1097 (Abs.); Jour. de Phys., (2) **5**, 518.

Sur la détermination des longueurs d'onde calorifiques.

Mouton. Comptes Rendus, **88**, 1078-82; Beiblätter, **3**, 616-18 (Abs.)

Wellenlänge und Brechungsexponent der äusserstern dunklen Wärmestrahlen des Sonnenspectrums.

> Müller (J.). Ann. Phys. u. Chem., 115, 543, Berichtigung dazu, 116,
> 644; Phil. Mag., (4) 26, 259; 30, 76; Jahresber. d. Chemie, 16
> (1868), 191; 18 (1865), 229.

Note on the progress of experiments for comparing a wave-length with a
metre.

> Peirce (O. S.). Amer. Jour. Sci., (3) 18, 51; Beiblätter, 3, 711 (Abs.).

The ghosts in Rutherford's diffraction spectrum.

> Peirce (O. S.). Amer. Jour. Mathematics, 2, 380–47 ; Nature, 20, 99
> (Abs.); Beiblätter, 5, 48–50 (Abs.).

Photometric Researches.

> Pickering (W. H.). Proc. Amer. Acad., 15, 236–50; Beiblätter, 4, 728
> (Abs.).

Photometrische Untersuchungen.

> Pulfrich (C.). Ann. Phys. u. Chem., n. F. 14, 177–218; Amer. Jour.
> Sci., (3) 23, 50 (Abs.); Jour. de Phys., (2) 1, 285 (Abs.).

Tableau de conversion de l'échelle spectrale en longueurs d'onde.

> Salet (G.). Bull. Soc. chim. Paris, n. s. 27, 482.

On the relative wave-lengths of the lines of the solar spectrum.

> Rowland (Henry A.). Phil. Mag., (5) 23 (1887), 257.

Three years' experimenting in mensurational spectroscopy

> Smyth (Piazzi). Nature, 22, 198–5, 222–5.

Mémoire sur la détermination des longueurs d'onde des raies métalliques,
spectres des métaux dessinés d'après leurs longueurs d'onde.
(With a plate giving the lines and wave-lengths of forty-five
metals.)

> Thalén (Rob.). Ann. Chim. et Phys., (4) 18, 202; Nova Acta Reg.
> Soc. Sci. Upsala, (3) 6.

Longueur d'onde des bandes spectrales donnees par les composé du carbone.

> Thollon (L.). Comptes Rendus, 93, 260; Ann. Chim. et Phys., (5)
> 25, 287.

Mesures photométriques dans les différentes régions du spectre.

> Trannin (H.). Jour. de Phys., 5, 297, 849.

Photometrie der Fraunhofer Linien.

> Vierordt (K.). Ann. Phys. u. Chem., n. F. 13, 888–46.

Resultate spectralphotometrischer Untersuchungen.

> Vogel (H. C.). Monatsber. d. Berliner Akad. (1880), 801-11; Beiblätter, 5, 286 (Abs.).

Messung der Wellenlängen des Lichtes mittels Interferenzstreifen im Beugungsstreifen.

> Weinberg (M.). Carl's Repertorium, 19, 148-54; Beiblätter, 7, 299 (Abs.).

Note au sujet d'un mémoire de M. Lagarde.

> Wiedemann (E.). Ann. Chim. et Phys., (6) 7, 143-4.

YELLOW BODIES.

Spectrum gelber Körper.

> Thudichum. Ber. chem. Ges., 2, 68.

YTTERBIUM.

Examen spectrale de l'ytterbine.

Lecoq de Boisbaudran (F.). Comptes Rendus, 88, 1342.

Sur l'ytterbine, nouvelle terre contenue dans la gadolinite.

Marignac (C.). Comptes Rendus, 87, 578–81; Amer. Jour. Sci., (3) 17, 63 (Abs.); Jour. Chem. Soc., 36, 118 (Abs.).

Sur l'ytterbine, terre nouvelle de M. Marignac.

Nilson (L. F.). Comptes Rendus, 88, 642–5; Amer. Jour. Sci., (3) 17, 478 (Abs.); Ber. chem. Ges., 12, 550–3; Jour. Chem. Soc , 36, 601 (Abs.).

Sur quelques caractéristiques de l'ytterbium et sur leurs spectres.

Nilson (L. F.). Comptes Rendus, 91, 56.

Recherches spectrales de l'ytterbium.

Thalén (R.). Jour. de Phys., 12, 85.

Spectres de l'ytterbium et de l'erbium.

Thalén (R.). Comptes Rendus, 91, 326; Beiblätter, 5, 122; Chemical News, 42, 184.

YTTRIUM.

Yttrium arc spectrum.

> Capron (J. R.).   Photographed Spectra, London, 1877, p. 51.

Sur les combinaisons de l'yttrium et de l'erbium.

> Clève (P. T.) et Hoegland (O.).   Bull. Soc. chim. Paris, **18**, 196–201, 289–97; Jour. Chem. Soc., (2) **11**, 186–9. ·

Sur les poids atomiques de l'yttrium.

> Clève (P. T.).   Bull. Soc. chim. Paris, **39**, 120–2; Amer. Jour. Sci., (3) **25**, 381 (Abs.).

On radiant matter spectroscopy.   The detection and wide distribution of yttrium.

> Crookes (W.).   Phil. Trans., **174**, 891–918; Proc. Royal Soc., **35**, 262 (Abs.); Chem. News, **47**, 261 (Abs.); Ber. chem. Ges., **16**, 1689 (Abs.); Jour. Franklin Inst., **86**, 118–128; Beiblätter, **7**, 599 (Abs.); Jour. Chem. Soc., **46**, 241 (Abs.); Chem. News, **49**, 159–60, 169–71, 181–2, 194–6, 205–8; Ann. Chim. et Phys., (6) **3**, 145–87.

Spectre des terres faisant partie du groupe de l'yttria et de la cérite; holmium, philippium, samarium, décipium.

> Soret (J. L.).   Comptes Rendus, **89**, 521–3; **91**, 378; Ber. chem. Ges., **12**, 2267–8; Jour. Chem. Soc., **38**, 7 (Abs.); Chem. News, **40**, 147.

Spectre de l'yttrium.   Avec une planche.

> Thalén (R.).   Jour. de Phys., **4**, 88.

ZINC.

Ueber die optischen Eigenschaften der Zincblende von Santander.   (See under Voigt, below.)

> Calderon (L.).   Zeitschr. Krystallogr. u. Mineralog., 4, 504–17, Beiblätter, 5, 861 (Abs.).

Zinc spectra

> Capron (J. R.).   Photographed Spectra, London, 1877, p. 28, 49, 51, 52.

Déterminations des longueurs d'onde des radiations très réfrangibles du magnésium, du cadmium, du zinc et de l'aluminium.

> Cornu (A.).   Archives de Genève, (8) 2, 119–126; Beiblätter, 4, 34 (Abs.); Jour. de Phys., 10, 425–81; Comptes Rendus, 73, 832.

Spectre du chlorure de zinc.

> Gouy.   Comptes Rendus, 84, 231; Chem. News, 35, 107.

Chlorure de zinc en solution.

> Lecoq de Boisbaudran (F.).   Spectres Lumineux, Paris, 1874, p. 138, planche XX.

Spectrum of zinc at elevated temperatures.

> Lockyer (J. N.).   Chem. News, 30, 98; Proc. Royal Soc., 17, 289; 18, 79; 21, 83; Jahresber. d. Chemie (1872), 145.

Indice du quartz pour les raies du zinc.'

> Sarasin (E.).   Comptes Rendus, 85, 1230.

Ueber den Einfluss einer Krümmung der Prismenflächen auf die Messungen von Brechungsindices, und über die Beobachtungen des Herrn Calderon an der Zincblende.

> Voigt (W.).   Zeitschr. f. Krystallogr. u. Mineral., 5, 118–130; Beiblätter, 5, 861–2 (Abs.).

ZIRCONIUM.

Zirconium arc spectrum; zirconium and palladium spark spectrum; zirconium spark spectrum.

> Capron (J. R.). Photographed Spectra, London, 1877, p. 53.

On zirconia.

> Hannay (J. B.). Jour. Chem. Soc., (2) **11**, 703–10; Ber. chem. Ges., **6**, 571 (Abs.).

Absorption spectra of zircons.

> Linnemann (E.). Monatsber. f. Chemie, **6**, 531–6; Jour. Chem. Soc., **48**, 1178 (Abs.).

On some remarkable spectra of compounds of zirconia and the oxides of uranium.

> Sorby (H. C.). Proc. Royal Soc., **18**, 197; Ber. chem. Ges., **3**, 146.

Spectre du zirconium.

> Troost et Hautefeuille. Comptes Rendus, **73**, 620; Bull. Soc. chim Paris, n. s. **16**, 229.

INDEX OF AUTHORS.

24 T

WOLFF (C. H.). Quantitative Analysis, 51; Absorption, 60; Alkalies, 61; Astronomical in general, 70; Comets, 72, 73, 75; Fixed Stars, 82; Sun-Spots, 128; Fuchsin, 172; Indigo, 176; Cobalt, 196; Copper, 202; Iron, 269; Liquids, 278.

WOLLASTON (Dr.). History, 7; Dark Lines in the Solar Sp., 106; Dark Lines, 206.

WRIGHT (A. W.). Meteors, 83; Aurora, 142; Flame, 239; Iron, 269.

WROTTESLEY (Lord). Books, 10.

WÜLLNER (A.). Analysis, 49; Bromine, 148; Acetylene, 161; Carbonic Acid, 180; Dispersion, 216; Electric, 225; Flame, 239, 240; Fluorescent, 245; Hydrogen, 260; Iodine, 267; Lines of the Spectrum, 275; Nitrogen, 304; Oxygen, 310.

WUNDER (J.). Absorption Sp., 60; Ultra-Marine, 184.

WÜNSCH (C. E.). History, 7.

WURTZ (A.). History, 7.

WYROUBOFF (G.). Dispersion, 216; Sodium, 339.

YOUNG (C. A.). Books, 10; Apparatus, 18; Analysis, 49; Comets, 73, 75, 79; Planets, 88; Solar in general, 99; Bright Lines in the Solar Sp., 102; Corona, 105; Displacement of Solar Sp., 106; Eclipses, 110, 111; Sun-Spots, 128; Inversion, 264; Nomenclature, 305; Sodium, 339.

YOUNG (T.). History, 8.

YUNG (E.). Color, 199.

ZAHN. Apparatus, 33, 38; Quantitative Analysis, 51.

ZANTEDESCHI. History, 8; Apparatus, 32; Solar in general, 99; Longitudinal, 281.

ZENGER (C. V.). Apparatus, 12, 14, 15, 24, 35, 37, 39; Diffraction, 211; Light, 273; Ultra-Violet, 350.

ZENGER (K. W.). Analysis, 49; Photography of Solar Sp., 117.

ZENKER (W.). Apparatus, 33; Solar Protuberances, 122.

ZIMMERMANN (C.). Uranium, 347.

ZÖLLNER (F.). Apparatus, 30, 36, 37; Astronomical in general, 70; Nebulæ, 85; Solar in general, 99; Corona, 105; Dark Lines in the Solar Sp., 106; Solar Protuberances, 122; Solar Rotation, 124; Sun-Spots, 129; Aurora, 142; Dark Lines, 206; Density, 208; Flame, 240; Heat, 254.

ZONA. Comet, 76.

SUPPLEMENT.

As the omission of the authors' names in connection with references to the Jahresberichte der Chemie has been pointed out as a serious defect in the Index, these names are now supplied below.

Jahresber. d. Chemie (1847–'8), 161, analysis, by Draper.

" " (1847–'8), 164, analysis, by Becquerel.

" " (1847–'8), 197, analysis, by Brewster.

" " (1847–'8), 197, analysis, by Airy.

" " (1847–'8), 198, analysis, by Melloni.

" " (1847–'8), 198, analysis, by Brewster.

" " (1847–'8), 221, chlorine and hydrogen, by Favre and Silbermann.

" " (1849), 164, photography of, by Becquerel.

" " (1850), 154, lines in the sp., by Brewster.

" " (1851), 151, longitudinal lines, by Ragona-Scinà.

" " (1851), 134; (1852), 117, interference sp., both by Nobert.

" " (1851), 152, Fraunhofer lines, by Broch.

" " (1851), 152, electric sp., by Masson.

" " (1852), 124, Fraunhofer lines, by Phillips and by Merz.

" " (1852), 125, analysis, by Stokes.

" " (1852), 125, longitudinal lines, by Zantedeschi.

" " (1852), 126, measurements of the sp., by Porro.

" " (1852), 126, 131, analysis, by Helmholtz.

" " (1853), 167, Fraunhofer lines, by Kuhn.

" " (1853), 167, Longitudinal lines, by Salm-Horstmar.

" " (1853), 178, colors, by Grassmann.

" " (1854), 137, Fraunhofer lines, by Heusser.

" " (1854), 197, solar sp. in general, by Becquerel.

" " (1855), 123, analysis, by Helmholtz.

" " (1855), 123, lines of the sp., by Grassmann.

Jahresber. d. Chemie (1862), 33, metallic spectra produced by electric sparks, by W. A. Miller, Stokes, and T. R. Robinson.

"   (1862), 33, spectra of carbon and of fluorine, by Sequin, Attfield, and Swan.

"   (1862), 34, violet coloring given to the flame by various chlorides, by Gladstone.

"   (1862), 34, spectra of colored solutions, by Brewster, Gladstone, and by Rood.

"   (1862), 29, spectrum of sodium, by Wolf et Diacon.

"   (1862), 30, spectrum of lithium in the hydrogen flame, by Wolf et Diacon.

"   (1862), 30, spectra of copper and of lead, by Debray.

"   (1862), 535, spectrum of blood, by F. Hoppe.

"   (1863), 101, photography of the solar spectrum, by Mascart.

"   (1863), 104, 106, 107, photographic effect of electric spectra of metals, by W. A. Miller.

"   (1863), 107, 110, dark lines in the solar spectrum, by Kirchhoff.

"   (1863), 108, note, atmospheric or telluric lines of the solar spectrum, by Jasssen.

"   (1863), 108, note, spectra of the stars, by Secchi.

"   (1863), 109, spectrum of iodine, by A. Wüllner.

"   (1863), 110, accuracy and comparison of spectroscopes, by Bunsen and Kirchhoff, and by J. P. Cooke.

"   (1863), 110, spectra of sulphur and of nitrogen, by Plücker and Hittorf.

"   (1863), 111, spectra of the chlorine metals, by E. Diacon.

"   (1863), 111, spectrum of hydrogen, by Leclancé.

"   (1863), 111, spectra of phosphorus, by Christofle and Beilstein.

"   (1863), 112, use of spectrum analysis in the manufacture of steel, by Roscoe.

"   (1863), 112, spectra of sodium and potassium, by L. M. Rutherfurd.

Jahresber. d. Chemie (1863), 112, spectrum of thallium, by W. A. Miller and by J. P. Gassiot.

" (1863), 112, spectrum of osmium, by W. Fraser.

" (1863), 113, history of spectrum analysis, by G. Kirchhoff and by H. C. Dibbits.

" (1863), 113, spectra of various metals in electricity, by Daniel.

' (1863), 113, spectrum of carbon, by Daniel.

" (1863), 114, apparatus, by Wolcott Gibbs, Littrow, R. Th. Simmler, J. P. Gassiot, H. Osann, B. Valz, and E. Mulder.

" (1864), 108, spectrum analysis of colored solutions, by C. Werner.

" (1864), 108, dark lines of the elements, by R. Bunsen.

" (1864), 109, spectrum of lightning, by L. Grandeau.

" (1864), 109, spectrum of the non-luminous carbon flame, by A. Morren.

" (1864), 109, spectra of phosphorus, sulphur, and selenium, by E. Mulder.

" (1864), 109, spectra of flames, by H. C. Dibbits.

" (1864), 110, spectra of glowing gases and vapours in electricity, by J. Plücker and S. W. Hittorf.

" (1864), 112, spectra of the elements and of their compounds, by A. Mitscherlich.

" (1864), 115, electric spectra of metals, by W. Huggins.

" (1864), 115, spectrum of the light from phosphorescent animals, by Pasteur.

" (1864), 115, note, spectra of the sun, fixed stars, planets, and nebulæ, by Janssen, W. A. Miller, and Huggins.

" (1864), 115, apparatus with 11 sulphide of carbon prisms, by J. P. Gassiot.

" (1864), 115, harmonious results given by the spectroscope, by F. Gottschalk.

" (1865), 85, absorption spectra of colored solutions, by F. Melde.

Jahresber. d. Chemie (1865), 87, influence of non-metallic elements on the spectra of the metals, by E. Diacon.

" (1865), 89, on the flame-spectra of carbon compounds, by A. Morren.

" (1865), 90, change of the bright lines of the metals, especially of sodium into dark lines, by H. G. Madan.

" (1865), 90, 91, electric spectra of metals, by W. Huggins and by Laborde.

" (1865), 91, spectrum analysis by means of electricity, by Brassack.

" (1865), 92, spectrum analysis of electricity, by A. von Waltenhofen.

" (1865), 92, spectra of the sun and of the stars, by Janssen.

" (1865), 94, spectroscopes, by H. Rexroth, J. Browning, J. P. Cooke, L. M. Rutherfurd, W. Huggins, J. P. Gassiot.

" (1865), 96, spectrum of the magnesium light, by A. Schrötter.

" (1866), 76, absorption spectrum of steam, by Janssen.

" (1866), 77, telluric lines of the solar spectrum, by Angström and by Secchi.

" (1866), 78, note, spectra of the stars, by W. Huggins and W. A. Miller.

" (1866), 78, connection of the distance of the spectrum lines with the dimensions of the atoms, by G. Hinrichs.

" (1866), 78, history of spectrum analysis, by Brewster.

" (1866), 78, apparatus, theory of, by L. Ditscheiner; and spectroscopes, by Börsch and A. Forster.

" (1867), 105, apparatus, by J. Müller.

" (1867), 105, application of the spectroscope to microscopical investigations, by H. C. Sorby.

" (1867), 105, production of the spectrum of fluorescent substances, by J. Müller.

" (1867), 105, 106, spectrum of the Bessemer flame, by A. Lielegg and by W. M. Watts.

Jahresber. d. Chemie (1869), 180, spectrum of the aurora, by Angström.

" " (1869), 181, spectrum of sulphur, by G. Salet.

" (1869), 182, spectrum of acetylene, by Berthelot and F. Richard.

" (1869), 182, absorption spectrum of chlorine, by Morren.

" (1869), 183, absorption spectra of steam and of saltpetre, by E. Luck.

" (1869), 184, absorption spectrum of mangansuper-chloride, by E. Luck.

" (1870), 148, spectrum of heat, by Becquerel.

" (1870), 172, spectrum analysis, by A. Kundt.

" (1870), 172, absorption spectra of liquid nitrates, by A. Kundt.

" (1870), 173, spectroscopic examination of sulphur and phosphorus, by Salet.

" (1870), 174, absorption spectrum of iodine vapour, by R. Thalén.

" (1870), 174, spectra of chalk, magnesia, baryta, and strontium, by Huggins.

" (1870), 175, spectrum of fat oils, by J. Müller.

" (1870), 175, influence of temperature on the sensitiveness of spectrum reactions, by E. Cappel.

" (1870), 177, spectra of gases, by A. Secchi.

" (1870), 177, note, spectra of stars, by Leseueur, Hennessey, Secchi, Lockyer, and Young (C. A.).

" (1870), 321, absorption spectrum of nitrates of didymium, by Erk.

" (1870), 930, spectrum analysis in general, by H. C. Sorby.

" (1871), 120, heat spectra of sunlight and limelight, by S. Lamansky.

" (1871), 144–149, spectra of colored bodies, by W. Stein.

" (1871), 150, use of a reflector behind the spectrum apparatus, by H. Fleck.

" (1871), 150, spectrum of calcium, by R. Blochmann.

" (1871), 151, diffraction and dispersion of selenium, by J. L. Sirks.

Jahresber. d. Chemie (1871), 151, diffraction and dispersion in iodide, bromide, and chloride of silver, by W. Wernicke.

" (1871), 153, diffractive power of various liquids, by Croullebois.

" (1871), 153, diffractive power of gases, by Fr. Mohr.

" (1871), 154–160, anomalous dispersion of bodies colored on the surface, by A. Kundt.

" (1871), 160, interference-scale for spectroscopic measurements, by J. Müller and by Sorby.

" (1871), 160, variable spectra, by A. J. Angström.

" (1871), 160–165, spectra of gases, by Angström.

" (1871), 165, spectrum analysis, by G. Salet.

" (1871), 167, spectrum of lightning, by H. Vogel.

" (1871), 168, solar spectrum, by J. Janssen.

" (1871), 169, spectrum of the aurora, by Browning, Zöllner, R. J. Ellery, Lord Lindsay, G. F. Barker, and H. Vogel.

" (1871), 169, comparative investigations of the spectrum, by L. Troost and P. Hautefeuille.

" (1871), 172, absorption by iodine-vapour, by Andrews.

" (1871), 173, inversion of the spectrum lines, by A. Weinhold.

" (1871), 175, illumination, absorption, and fluorescence, by A. Lallemand.

" (1871), 179–189, chemical effects of light, by H. E. Roscoe and T. E. Thorpe.

" (1871), 189, quantitative analysis, by Vierordt.

" (1871), 191, phosphorescence, by A. Forster.

" (1872), 134, ultra-violet rays of the solar spectrum, by Sekulic.

" (1872), 136, absorption spectrum of chlorophyll, by Chautard.

" (1872), 137, absorption spectrum of saltpetre, by D. Gernez.

" (1872), 138, absorption spectrum of chlorine, by Gernez.

" (1872), 139, 141, absorption spectrum of sulphur, by Gernez

Jahresber. d. Chemie (1872), 139, absorption spectra of the chloric acids and of selenium, by D. Gernez.

" (1872), 140, absorption spectra of chloride of selenium, of bromide of selenium, of tellurium, of chloride of tellurium, and of bromide of tellurium, and of alizarine, by D. Gernez.

" (1872), 141, spectrum of iodine and of sulphur, by G. Salet.

" (1872), 141, 143, 144, 145, 146, spectrum of hydrogen, by G. M. Seabroke, Lecoq de Boisbaudran, A. Schuster, L. Cailletet, and E. Villari.

" (1872), 142, spectrum of phosphoretted hydrogen, by K. B. Hofmann.

" (1872), 142, 144, 145, spectrum of nitrogen, by Schuster.

" (1872), 142, spectrum of the flame of ammonia, by K. B. Hofmann.

" (1872), 143, spectrum of ammonia, by A. Schuster.

" (1872), 143, spectra of gases, by Schuster and by Angström.

" (1872), 145, spectra of aluminium, magnesium, zinc, cadmium, cobalt, and nickel, by Lockyer.

" (1872), 145, influence of pressure on the spectrum of the induction spark, by L. Cailletet.

" (1872), 146, spectrum analysis, by C. Horner.

" (1872), 147, solar spectrum, by C. A. Young.

" (1872), 148, spectrum of the aurora, by H. C. Vogel.

" (1872), 148, spectrum of the zodiacal light, by E. Liais.

" (1872), 148, spectrum of lightning, by E. S. Holden.

" (1872), 873, spectrum analysis, by Vierordt.

" (1872), 948, micro-spectroscope, by Timiriasef.

" (1873), 54, use of the spectrum in measuring high temperatures, by J. Dewar and by Gladstone.

" (1873), 146, spectroscopes, by Hartley, Emsmann, Zenger, H. R. Proctor, O. N. Rood, C. A. Young, F. P. Le Roux, Th. Edelmann, R. Hennig and M. M. Champion, Pellet et Grenier.

Jahresber. d. Chemie (1873), 158, absorption spectrum of thallium, by H. Morton.

" (1873), 158, absorption spectrum of uranium salts, by H. Morton and H. C. Bolton.

" (1873), 160, wave-lengths of the spectrum, by E. Becquerel.

" (1873), 160, distribution of chemical effect in the spectrum, by J. W. Draper.

" (1873), 166, albertotype of a photographed diffraction spectrum, by H. Draper.

" (1873), 451, absorption spectrum of anthrapurpurin, by W. H. Perkin.

" (1873), 455, absorption spectrum of chinizarin, by A. Kundt.

" (1874), 96, absorption spectrum of salt solutions, by W. N. Hartley.

" (1874), 152, 153, 154, 155, 156, 157, spectrum analysis, by Lecoq de Boisbaudran, R. Thalén, Ch. Horner, G. Salet, E. Goldstein, J. Chautard, W. de Fonvielle, Th. Hoh, L. Clark, A. J. Angström, S. Lemström, A. Wijkander, A. W. Wright, and E. Hagenbach.

" (1874), 152, apparatus, by S. C. Tisley, J. G. Hofmann, Th. Grubb, F. Kingdon, B. Delachanal and A. Mernset.

" (1874), 958, spectrum analysis of alloys, by J. N. Lockyer and W. C. Roberts.

" (1874), 156–157, fluorescence and absorption, by O. Lubarsch and J. Chautard.

" (1875), 122, metallic spectra, sulphide of carbon spectrum, gas spectra, by Th. Marvin, H. W. Vogel, and A. Wüllner.

" (1875), 122, 123, spectrum of carbon, by W. M. Watts, Piazzi Smyth, and Swan.

" (1875), 123, spectrum of the aurora, by A. S. Herschel and by J. Rand Capron.

" (1875), 123, spectrum of lightning, by L. Clark.

" (1875), 124, 125, absorption spectra of metallic vapours, by J. N. Lockyer and W. Ch. Roberts.

" (1875), 124, absorption spectra, by T. L. Phipson.

Jahresber. d. Chemie (1877), 1034, use of chloride of calcium and of chloride of magnesium in spectroscopy, by A. R. Leeds.

" (1877), 102, distribution of heat in the spectrum of the electric light, by P. Desaines.

" (1877), 182, photographs of ultra-violet gas-spectra, by Van Monckhoven.

" (1877), 182, spectrum of davyum, by S. Kern.

" (1877), 182, spectra of colored flames, by Gouy.

" (1877), 183, spectra of the chemical compounds, by J. Moser.

" (1877), 183, lines of oxygen and nitrogen in the solar spectrum, by H. Draper.

" (1877), 183, spectra of lightning. by J. W. Clark.

" (1877), 184, theory of the dispersion and absorption of light, by E. Ketteler.

" (1877), 184, inversion of the sodium lines, by J. Martenson.

" (1877), 184, absorption spectrum of the garnet and the ruby, by H. W. Vogel.

" (1877), 185, absorption of solutions, by G. Govi.

" (1877), 185, quantitative spectrum analysis, by G. Govi.

" (1877), 195, photography of the infra-red lines of the solar spectrum, by J. W. Draper.

" (1877), 196, dissolution of carbonic acid in plants under the influence of the solar spectrum, by C. Timirjaseff.

" (1877), 1245, photography of the solar spectrum, by H. W. Vogel.

" (1878), 7, comparative spectrum analysis, by N. Lockyer.

" (1878), 67, use of spectrum analysis in determining high temperatures, by A. Crova.

" (1878), 179, apparatus, by Thollon and by A. S. Herschel.

" (1878), 169, conversion of Kirchhoff's scale into wave-lengths, by B. Hasselberg.

" (1878), 169, calculation of the distribution of the spectrum lines, by L. Pfaundler.

Jahresber. d. Chemie (1878), 180, spectroscopic investigation of solutions, by J. Landauer.

" (1878), 181, spectrum of the light of super-manganate of potassium, by J. Conroy.

" (1878), 181, absorption of the ultra-violet rays, by L. Soret.

" (1878), 181, ultra-violet absorption spectra of gadolinite, by J. L. Soret.

" (1878), 182, inversion of the spectrum lines of metallic vapours, by G. D. Liveing and J. Dewar.

" (1878), 185, spectroscopic observations of the sun, by J. N. Lockyer.

" (1878), 185, oxygen in the solar atmosphere, by J. C. Draper.

" (1878), 185, map of the ultra-violet part of the solar spectrum, in continuation of Angström's map, by A. Cornu.

" (1878), 187, photography of the red and infra-red spectrum, by Abney.

" (1878), 188, oxidation hastened by the least refractive end of the spectrum, cause of solarization, by Abney and by Chastaing.

" (1878), 191, flame for spectroscopic observations, by H. Gilm.

" (1879), 10, spectroscopic investigation of the elements, by J. N. Lockyer.

" (1879), 159, nature of spectra, by E. Wiedemann.

" (1879), 160, band and lime spectrum, by A. Wüllner.

" (1871), 163, influence of temperature on the spectra of gases and vapours, by G. Ciamician.

" (1879), 166, limits of the ultra-violet spectrum, by A. Cornu.

" (1879), 161, spectroscopic investigations, by J. N. Lockyer.

" (1879), 1022, quantitative spectrum analysis, by C. H. Wolf.

Jahresber. d. Chemie (1879), 1022, analysis of absorption spectra, by B. Hasselberg.

" (1879), 1023, spectroscopic notes, by H. W. Vogel.

" (1879), 157, character of the rays issuing from glowing platinum, by E. L. Nickols.

" (1880), 201, new method of spectroscopic observation, by J. N. Lockyer.

" (1880), 201, disappearance of lines in the apparatus, by Ch. Fievez.

" (1880), 201, the line H in the spectrum of hydrogen, by J. N. Lockyer.

" (1880), 201, relative intensity of spectrum lines, by J. Rand Capron.

" (1880), 201, harmonic relations in the spectra of gases, by A. Schuster.

" (1880), 202, spectrotelescope, by P. Glan.

" (1880), 203, quantitative spectroscopic researches, by Liveing and Dewar.

" (1880), 205, spectroscopic notes, by C. A. Young.

" (1880), 205, spectroscopic investigations continued, by Ciamician.

" (1880), 206, spectroscopes, by J. E. Reynolds and G. Hüfner.

" (1880), 206, spectrum of the hydrogen flame, by W. Huggins.

" (1880), 206, spectrum of hydrogen and of the carburetted hydrogen flame, by G. D. Liveing and J. Dewar.

" (1880), 206, the helium line D_3 attributed to hydrogen, by E. Spée.

" (1880), 207, absorption spectrum of ozone, by J. Chappuis.

" (1880), 207, spectra of the compounds of carbon with hydrogen and nitrogen, by G. D. Liveing and J. Dewar.

" (1880), 207, fourth note on the spectrum of carbon, by J. N. Lockyer.

" (1880), 207, history of the spectrum of carbon, by G. D. Liveing and J. Dewar.

Jahresber. d. Chemie (1880), 207, spectra of the compounds of carbon with hydrogen and nitrogen, especially the sensitiveness of the spectroscopic reactions of carbo-nitrogen compounds, by G. D. Liveing and J. Dewar.

" (1880), 208, the repeated inversion of the sodium lines, by C. A. Young.

" (1880), 208, method for a constant sodium flame, by Fleck.

" (1880), 208, spectra of magnesium and lithium, by G. D. Liveing and J. Dewar.

" (1880), 209, spectroscopic relations of copper, nickel, cobalt, iron, manganese, and chromium, by Th. Bayley.

" (1880), 209, absorption spectra of the yttrium group, by J. L. Soret.

" (1880), 210, emission spectrum of erbium and ytterbium, by R. Thalén.

" (1880), 211, spectrum of thulium, by R. Thalén.

" (1880), 212, spectrum of scandium, by R. Thalén.

" (1880), 212, displacement of the absorption lines of purpurin in various solutions, by H. Morton.

" (1880), 212, ultra-violet rays, by J. Schönn.

" (1880), 213, limits of the ultra-violet end of the spectrum, by A. Cornu.

" (1880), 213, absorption of the ultra-violet rays by organic bodies, by W. R. Dunstan.

" (1880), 214, the ultra-violet absorption spectra of ytterbium, erbium, holmium, philippium, terbium, samarium, decipium, didymium, and zirconium, by J. L. Soret.

" (1880), 219, photography of the spectra of stars, by Huggins.

" (1880), 219, photographs of the spectrum of bromide of silver, by Abney.

" (1880), 219, photochemistry of silver, by J. M. von Eder.

" (1881), 117, spectroscopic measurement of high temperatures, by A. Crova.

Jahresber. d. Chemie (1881), 117, use of Vierordt's double slit in spectroscopic analysis, by W. Dietrich.

" (1881), 117, spectrophotometer, by A. Crova.

" (1881), 117, phosphorography of the solar spectrum and the ultra-red lines, by J. W. Draper.

" (1881), 118, inversion of spectrum lines, by G. D. Liveing and J. Dewar.

" (1881), 118, disappearance of spectrum lines, by Ch. Fievez.

" (1881), 119, coïncidence of spectrum lines of various elements, by G. D. Liveing and J. Dewar.

" (1881), 119, spectrum of oxygen, by A. Paalzow and H. W. Vogel.

" (1881), 120, spectra of hydrogen and of sulphur, by B. Hasselberg.

" (1881), 120, spectrum of arsenic, by O. W. Huntington.

" (1881), 121, spectra of sodium and calcium, by Abney.

" (1881), 121, relative intensity of the sodium lines D_a and $D\beta$, by W. Dietrich.

" (1881), 121, spectrum of magnesium, by G. D. Liveing and J. Dewar.

" (1881), 122, spectra of magnesium, sodium, copper, baryum, and iron in their harmonic relations, by A. Schuster.

" (1881), 122, spectrum of iron, by J. N. Lockyer.

" (1881), 122, 123, spectra of the carbon compounds, by E. Wesendonck; remarks by A. Wüllner, claiming priority.

" (1881), 123, spectroscopic lines of the arc of Jamin's lamp, by Thollon.

" (1881), 123, spectrum of carbonic acid, by C. Wesendonck.

" (1881), 123, 124, spectrum of acetylene, by A. Wüllner.

" (1881), 125, color of water, by F. Boas.

" (1881), 125, absorption of the solar rays in the atmosphere, by E. Lecher.

Jahresber. d. Chemie (1881), 125, absorption of light in various media, by C. Pulfrich.

"   (1881), 126, molecular structure of carbon compounds and their absortion spectra, by W. N. Hartley.

"   (1881), 127, influence of the molecular arrangement of organic substances on their absorption in the ultra-red part of the spectrum, by Abney and Festing.

"   (1881), 127, the absorption spectrum of ozone, by W. N. Hartley.

"   (1881), 127, absorption spectra of cobalt salts, by W. J. Russell.

"   (1881), 128, absorption bands in the visible spectra of colorless liquids, by W. J. Russell and W. Lapraik.

"   (1881), 128, spectra of terpenes and volatile oils, by W. N. Hartley and A. K. Huntington.

"   (1881), 129, chrysoidine and the allied azo dyestuffs, by J. Landauer.

"   (1881), 129, alkaloid reactions in spectroscopic apparatus, by K. Hock.

"   (1881), 129, absorption of the ultra-violet rays, by De Chardonnet.

"   (1881), 129, passage of rays of small refraction through ebonite, by Abney and Festing.

"   (1881), 130, spectrum of cyanine, by V. von Lang.

"   (1881), 130, 131, 132, discontinuous spectra of phosphorescent bodies, by W. Crookes; E. Becquerel claims priority for a part.

"   (1881), 132, phosphorescence of Balmain's illuminating matter, by E. Dreher.

"   (1881), 133, the light of phosphorescent substances, by E. Obach.

"   (1881), 133, fluorescence, by O. Lubarsch.

"   (1881), 133, comparative effects of light and heat in chemical reactions, by G. Lemoine.

"   (1881), 135, sensitiveness of dry plates of bromide of silver to the solar spectrum, by H. W. Vogel.

Jahresber. d. Chemie (1881), 136, photography in colors, by Ch. Cros and J. Carpenter.

" (1881), 136, effect of the spectrum in radiophony, by E. Mercadier.

" (1881), 137, change from vibrations of light to vibrations of sound, by W. H. Preece.

" (1881), 138, an aragonite prism, by V. von Lang.

" (1881), 139, double refraction in agitated liquids, by A. Kundt and Maxwell.

" (1882), 187, examination of powerful absorbants, by C. Pulfrich.

" (1882), 190, the violet phosphorescence of calcium sulphide, by W. de W. Abney.

" (1882), 285, spectra of the cerite metals, by B. Brauner.

" (1882), 1349, 1350, apparatus, by H. Schulz, Fr. Fuchs, A. Ricco, W. Wernicke, H. Goltzsch, G. G. Stokes, and F. Miller.

" (1882), 183, spectrum of sulphur, chlorine, and sodium in spectroscopic tubes, by B. Hasselberg.

" (1882), 183, spectrum produced in a Geissler tube changed by long use, by B. Hasselberg.

" (1882), 184, comparison of the spectrum of positive light with that of " kathoden " light, by E. Goldstein.

" (1882), 68, absorption spectra of solutions, by G. Krüss.

" (1882), 177, study of the solar spectrum, by Ch. Fievez.

" (1882), 177, distribution of energy in the solar spectrum, observed with his bolometer, by S. P. Langley.

" (1882), 178, distribution of heat in the dark part of the solar spectrum, by P. Desains.

" (1882), 178, spectrum of terbium, by H. E. Roscoe and A. Schuster.

" (1882), 179, spectra of the metalloids, by D. von Monckhoven.

" (1882), 179, ultra-violet spectra of the elements by G. D. Liveing and J. Dewar.

Jahresber. d. Chemie (1882), 180, photographs of the ultra-violet spectra of the elements, by W. N. Hartley.

" (1882), 181, inversion of the metallic lines in too long exposed photographs of spectra, by W. N. Hartley.

" (1882), 181, map of the more refractive part of the spectrum of hydrogen, by G. D. Liveing and J. Dewar.

" (1882), 181, apparatus for the study of glowing vapours, by G. D. Liveing and J. Dewar.

" (1882), 181, displacement of the spectrum lines of hydrogen, by D. von Monckhoven.

" (1882), 182, intensity of the spectrum lines of hydrogen, by H. Lagarde.

" (1882), 183, spectrum of oxygen at low temperatures, by Piazzi Smyth.

" (1882), 184, 185, spectra of carbon and of its compounds, by G. D. Liveing and J. Dewar.

" (1882), 185, spectra of carbon compounds, by K. Wesendonck.·

" (1882), 186, disappearance of spectrum lines and their changes in mixed vapours, by G. D. Liveing and J. Dewar.

" (1882), 186, remarks on Lockyer's theory of dissociation, especially in regard to iron lines in sun-spots, by H. W. Vogel.

" (1882), 187, remarks on Von Lang's examination of powerful absorbants, by C. Pulfrich.

" (1882), 187, absorption spectrum of hypernitric acids, by J. Chappuis.

" (1882), 187, absorption spectrum of ozone, by J. Chappuis.

" (1882), 188, absorption spectrum of the atmosphere, by N. Egoroff.

" (1882), 188, relations of carbon compounds to their absorption spectra, by W. N. Hartley.

" (1882), 189, wave-lengths of various carbon compounds, by Thollon.

" (1882), 189, absorption spectrum of chlorophyll, by W. J. Russell and W. Lapraik.

Jahresber. d. Chemie (1882), 190, absorption curves of liquids, by E. Ketteler and C. Pulfrich.

"   (1882), 190, violet phosphorescence of calcium sulphide, by W. de W. Abney.

"   (1882), 190, origin of phosphorescence, by E. Dreher.

"   (1882), 199, seusitiveness of bromide and chloride of silver to the solar spectrum, by H. W. Vogel.

"   (1882), 201, photography of spectra in connection with new methods of quantitative chemical analysis, by W. N. Hartley.

"   (1883), 1554, duration of the spectroscopic reaction of carbonic acid in the blood, by E. Salfeld.

"   (1883), 1655, apparatus, by H. Schulze, O. Tumlirz, F. Lippich, and W. Ramsay.

"   (1883), 232, a spectrophotometer, by A. Crova.

"   (1883), 240, direct-vision spectroscope, by Ch. V. Zenger.

"   (1883), 1397, energy in the solar spectrum, by C. Timiriaseff.

"   (1883), 240, spectroscopic studies in the ultra-red end, by E. Lommel.

"   (1883), 241, wave-lengths of the extreme warm rays, by E. Pringsheim.

"   (1883), 241, phosphorographic studies in the ultra-red part of the solar spectrum, by H. Becquerel.

"   (1883), 242, on the wave-lengths near the lines A and a in Fievez's map, by W. de W. Abney.

"   (1883), 242, distribution of heat in the solar spectrum, by P. Desains.

"   (1883), 242, selective absorption of the atmosphere and distribution of energy in the solar spectrum, by S. P. Langley.

"   (1883), 243, spectra of sun-spots, by G. D. Liveing and J. Dewar.

"   (1883), 243, spectroscopic observations of sun-spots, by C. A. Young.

"   (1883), 243, emission spectra of metallic vapours, by H. Becquerel.

27 T

r. d. Chemie (1883), 244, ultra-red emission spectra of the metallic vapours, by H. Becquerel.

" (1883), 244, spectra of didymium and samarium,
by R. Thalén.

" (1883), 244, emission spectra of scandium, ytterbium, erbium, and thulium, by Th. Thalén.

" (1883), 245, ultra-violet spectra of the elements, by
W. N. Hartley.

" (1883), 245, method of photographing diffraction
spectra, by W. N. Hartley and W. E. Adeney.

" (1883), 246, ultra-violet emission spectra of the
elements and their compounds photographically
examined, by W. N. Hartley.

" (1883), 246, spectrum of beryllium, by W. N.
Hartley.

" (1883), 246, spectra of boron and silicon, by W. N.
Hartley.

" (1883), 246, 247, absorption spectra of various substances, by G. D. Liveing and J. Dewar.

" (1883), 248, inversion of the spectral lines of the
metals, by G. D. Liveing and J. Dewar.

" (1883), 248, inversion of the hydrogen lines and of
the lithium lines, by G. D. Liveing and J.
Dewar.

" (1883), 248, spectrum of phosphorescent light and
of yttrium, by W. Crookes.

" (1883), 248, spectrum of hydrogen and of acetylene,
by B. Hasselberg.

" (1883), 249, spectrum of hydrogen in the vacuum
tube, by Piazzi Smyth.

" (1883), 249, spectrum of the hydro-carbon flame,
by G. D. Liveing and J. Dewar.

" (1883), 249, absorption and fluorescent spectra of
various bodies, by E. Linhardt.

" (1883), 250, absorption spectrum of sea-water, by
H. W. Vogel and J. Aitken.

" (1883), 250, absorption spectrum of the solution of
iodine in sulphate of carbon, by Abney and
Festing.

Jahresber. d. Chemie (1883), 250, use of selenium in separating the heat rays from the light and the chemical rays, by F. van Assche.

" (1883), 251, absorption of the blood, by J. L. Soret.

" (1883), 251, sight of the ultra-violet rays by man and by vertebrates, by De Chardonnet; remarks by Mascart and by Soret.

" (1883), 252, absorption spectra of organic compounds, by G. Krüss and S. Oeconomides.

" (1883), 253, dissociation of phosphorescence under the influence of the ultra-red rays, by H. Becquerel.

" (1883), 253, phosphorescence of sulphur, by H. Schwarz.

" (1883), 254, phosphorescence of organic bodies, by B. Radzizewski.

" (1883), 254, Stokes's Law of Phosphorescence, maintained by Hagenbach against Lommel and Lubarsch.

" (1883), 254, optical characteristics of the cyanides of platinum, by W. König.

" (1883), 258, sensitiveness of the salts of silver to light, by H. W. Vogel.

" (1883), 258, electro-chemical energy of light, by F. Griveaux.

" (1884), 289, lines peculiar to solar light, by A. Cornu.

" (1884), 294, displacement and inversion of the lines of the spectrum, by Ch. Fievez.

" (1884), 295, cause of the displacement of the lines of the spectrum, by E. Wiedemann and W. N. Hartley.

" (1884), 283, measurement of wave-lengths, by H. Merczyng.

" (1884), 289, 290, wave-lengths and refraction in the invisible part of the spectrum, obtained with the bolometer of his own invention and with a very large Rowland convex grating, by S. P. Langley.

Jahresber. d. Chemie (1884), 291, bands in the ultra-red part of the solar spectrum and the ultra-red spectrum of glowing metallic vapours, by H. Becquerel.

" (1884), 292, spectra of metals, by E. Demarçay.

" (1884), 292, spectroscopic studies of exploding gases, by G. D. Liveing and J. Dewar.

" (1884), 292, spectra of vapours, by J. Parry.

" (1884), 293, phosphorescent spectra, by W. Crookes.

" (1884), 293, spectrum of hydrogen, by B. Hasselberg.

" (1884), 293, spectra of fluoride of silicon and of hydrate of silicon, by K. Wesendonck.

" (1884), 293, influence of temperance on spectroscopic observations, by G. Krüss.

" (1884), 293, changes in the refraction of the H and Mg lines, by Ch. Fievez.

" (1884), 294, displacement and inversion of the spectrum lines, by Ch. Fievez.

" (1884), 295, displacement of the spectrum lines, by E. Wiedemann and W. N. Hartley.

" (1884), 295, spectroscopic studies of dyes, by E. L. Nichols.

" (1884), 296, color of water, by J. L. Soret.

" (1884), 296, absorption spectrum of water, by J. L. Soret and E. Sarasin.

" (1884), 297, absorption spectrum of iodine vapour, by A. Morghen.

" (1884), 297, absorption spectrum of chlorochromic acid, by G. J. Stoney and J. E. Emerson.

" (1884), 297, absorption spectra of æsculine solutions, by K. Wesendonck.

" (1884), 298, absorption spectra of the aromatic series, by J. S. Konic.

" (1884), 298, absorption spectra of the alkaloids, by W. N. Hartley.

" (1884), 298, formula for the dispersion of the ultra-red rays, by A. Wüllner.

" (1884), 1429, influence of the spectrum on the production of carbonic acid gas by plants, by J. Reinke.

Jahresber. d. Chemie (1885), 322, spectroscopic observations of solutions of chloride of cobalt, by W. J. Russell.

"   (1885), 323, absorption spectrum of blue oxalate of potassium, by C. A. Schunk.

"   (1885), 323, absorption spectra in the extreme red, by Abney and Festing.

"   (1885), 323, 324, absorption spectra of various dye-stuffs, by Ch. Girard and Pabst.

"   (1885), 324, absorption spectra of the sub-nitrates, by L. Bell.

"   (1885), 324, absorption spectrum of oxygen, by N. Egoroff.

"   (1885), 324, 325, absorption of atmospheric air and of hydrogen, by J. Janssen.

"   (1885), 325, absorption spectra of the alkaloids, by W. N. Hartley.

"   (1885), 326, absorption spectrum of benzol vapour, by J. S. Konic.

"   (1885), 327, connection between the absorption spectra and the molecular structure of organic compounds, by G. Krüss and Oeconomides.

"   (1885), 328, connection between molecular structure and the absorption of light, by N. von Klobukow.

"   (1885), 329, relations between molecular structure and the absorption of carbon compounds, by W. N. Hartley.

"   (1885), 329, 330, relations between the absorptive power and the emission of phosphorescent rays, by H. Becquerel.

"   (1885), 331, spectroscopy of radiant matter, by W. Crookes.

"   (1885), 332, spectra of samarium and of yttrium, by W. Crookes.

"   (1885), 332, a new kind of metallic spectra and spectra of metallic solutions, by Lecoq de Bois-baudran.

"   (1885), 333, theory of fluorescence, by E. Lommel.

"   (1885), 333, 334, fluorescence, especially of didymium, by E. Lommel.

Jahresber. d. Chemie (1885), 335, fluorescence of naphthalin-red, by K. Wesendonck.

Report of the committee, consisting of Professors Olding, Huntington, and Hartley, appointed to investigate by means of photography the ultra-violet spark spectra emitted by metallic elements and their combinations under varying conditions; drawn up by Professor W. M. Hartley (secretary). Report of the British Association for 1885, pp. 276–284.

Report of the committee, consisting of Professor Sir H. E. Roscoe, Mr. J. N. Lockyer, Professors Dewar, Wolcott Gibbs, Liveing, Schuster, and W. N. Hartley, Captain Abney, and Dr. Marshall Watts (secretary), appointed for the purpose of preparing a new series of wavelength tables of the spectra of the elements and compounds. Report of the British Association for 1885, pp. 288–322, and for 1886, pp. 167–204.

On the spectrum of the Stella Nova visible in the great nebula in Andromeda, by William Huggins. Rept. Brit. Assoc. for 1885, p. 932.

On the solar spectroscopy in the infra-red, by Daniel Draper. Rept. Brit. Assoc. for 1885, p. 935.

On the formation of a pure spectrum by Newton, by G. Griffith. Rept. Brit. Assoc. for 1885, p. 940.

On the absorption spectra of uranium salts, by W. J. Russell and W. Lapraik. Rept. Brit. Assoc. for 1886.

Pritchard's Wedge Photometer, by S. P. Langley, C. A. Young, and E. C. Pickering.